Biochemical Engineering and Bioreactor Design

Prof. (Dr.) C. M. Narayanan

Former Professor of Chemical Engineering,
National Institute of Technology, Durgapur, India.
Email: prof.cmn@gmail.com

Dr. Vikas Narayan

Lead Scientist, Bangalore, India.

Dr. Tripti De

Assistant Professor of Chemical Engineering,
VSB Engineering College (autonomous), Karur, India.

IndiePress

Bangalore

Biochemical Engineering and Bioreactor Design

Printed in India

IndiePress

ISBN: 978-93-6045-601-6

First Printing, 2025

IndiePress

A division of Nasadiya Technologies Private Ltd.

Koramangala, Bengaluru

Karnataka-560029

http://indiepress.in/

Edited by Sakshi Tiwari

Typeset by MAP Systems, Bengaluru

Publishing Consultant: Zoheb Ahmed

Book Cover Design by Keerthipriya PH

Preface

Biochemical engineering is one among the most multidisciplined branches of engineering. It is, in fact, synonymous with Biotechnology or Bioprocess engineering, but is not synonymous with Applied biology or Applied microbiology (a term that is coined by many, these days).

Biochemical engineering or Biotechnology took birth from Chemical engineering and could be called a subdivision of Chemical engineering.

The work of a biochemical engineer or biotechnologist starts where the work of a biologist or microbiologist ends. Isolation and characterization of microbial cultures and different classes of enzymes fall within the ambit of biologists and microbiologists. Design of bioreactors and industrial processes in which these microbial cultures and enzymes are used as biocatalysts, form the responsibility of the biochemical engineer or biotechnologist.

Biochemical reactions are relatively more complex and their kinetic analysis demands a strong mathematical background. Design and analysis of industrial bioreactors based on this kinetic models once again demand use of sophisticated mathematical tools. The mathematical background of a biochemical engineer or biotechnologist must be, therefore, extremely sound, while he needs only a fundamental knowledge of biological sciences.

Biochemical processes and reactions have practically become omnipresent in most process industries of the world. Consequently, they possess very bright future prospects.

The field of biochemical engineering and biotechnology has been adequately and elaborately surveyed in this book, starting from kinetic analysis and development of versatile kinetic models for different microbial fermentation and enzyme-catalyzed processes (Chapter 2), performance analysis and computer aided design of different classes of bioreactors such as conventional stirred tanks and moving bed biofilm reactors, packed bed, fluidized bed, semifluidized bed and inverse fluidized bed bioreactors, sludge bed and sludge blanket (EGSB and UASB) bioreactors and downflow stationary fixed film (DSFF) bioreactors (Chapter 3), analysis and simulation of sterilization systems and processes, both thermal processes and those using fibrous filters, batch and continuous (Chapter 4) to detailed coverage (with condensed flowsheet diagrams) of selected industrial processes, where biotechnology and bioreactors have been successfully employed (Chapter 5). Examples are manufacture of biodiesels, bioplastics, biogas & biomethane, biofertilisers, biodesulfurization of petroleum oils and bioleaching of ores.

Most recent advances in biotechnology and biochemical engineering have been highlighted and prospects for future developments outlined. A good number of numerical examples have been included in Chapters 2, 3 and 4 to illustrate the matter more vividly and explicitly. A rich list of references / bibliography follows each chapter.

It is anticipated that this book shall form an excellent text book for the students of chemical engineering, biotechnology, metallurgical engineering, environmental engineering, energy systems engineering and all related areas. We also hope that this shall constitute an informative reference volume to all academicians, research scientists, industrial consultants and practicing engineers as well.

We are thankful to a large number of fellow-researchers (members of IRG – international research group) from all over the world, mainly from University of Karlsruhe (Germany), Toyama Prefectural University (Japan), Widener University (Philadelphia, USA), Caledonian College of Engineering (Sultanate of Oman) and East China Normal University (Shanghai), for their cooperation and ingenuous assistances.

Our special vote of thanks to the young and not so young students (students of Prof. C. M. Narayanan at National Institute of Technology, Durgapur), who are successful practicing engineers and research scientists today, such as Ms. Shrijita Das (Senior Quality Analyst, First Source Solutions, Hyderabad), Mr. Tamanash Pramanick (Manager, GAIL (India) Limited) and Dr. Subrata Biswas (Assistant Professor of Chemical Engineering, Indian Institute of Technology, Roorkee), who worked hand-in-hand with their Professor towards the successful completion of the manuscript of this book. Special word of gratitude also to Prof. Manju and Mr. Praveen (Government Engineering College, Kozhikode, Kerala) for their assistances during the preparation of the soft copy of the manuscript of this book.

Finally, our compliments to Indie Press, Bangalore for bringing out this book in an appropriate and appealing format.

Suggestions and criticism from our eminent readers are most welcome.

Prof. (Dr.) C. M. Narayanan (prof.cmn@gmail.com)

Dr. Vikas Narayan

Dr. Tripti De

Contents

About the Authors

Prof. (Dr.) C. M. Narayanan is former Professor of Chemical Engineering, National Institute of Technology, Durgapur, India. After 40 (forty) years of service, he retired from NIT in 2016. He received B.Tech in Chemical Engineering from Banaras Hindu University, M.E. from Indian Institute of Science, Bangalore and PhD from Indian Institute of Technology, Kharagpur. He is the author of more than 85 research papers in journals of international repute and has presented more than 45 research papers in international conferences in India and abroad. He has also authored EIGHT well-known books on Chemical Engineering and Bioprocess Engineering, including a Chapter in the Handbook of Food Bioengineering. He has been conferred with ***Bharat Jyoti Award*** and ***Most Eminent Engineering Personality Award*** for his outstanding contribution to chemical and biochemical engineering education and research, particularly in the areas of Computer Aided Design and Software Development and Bioprocess Engineering. He is a member of the editorial board of many prestigious international journals and has visited many countries abroad such as Germany, Switzerland, Japan, USA, Sultanate of Oman, China, Spain, Netherlands, UK and Australia as Visiting scientist and conference delegate. During his tenure at NIT, he had been the faculty advisor to ISTE Students' Chapter for 15 years (since its inception) and has coordinated fourteen (14) All India Students' Engineering Congresses (AISECs) and twelve (12) Intra State Technical Meets (ITMs). Presently practices as a freelance CAD (Software) consultant and Visiting faculty to a number of engineering colleges (institutions). Visit him at: www.profcmn.com

Dr. Vikas Narayan received his B.Tech in Chemical Engineering from Lakshminarayan Institute of Technology, Nagpur, India, M.S. in Bioprocess Engineering from Indian Institute of Technology, Madras and PhD in Chemical Engineering from Denmark Technical University, Denmark. He has served as research engineer in a number of research and consultancy firms of India. With 18+ years of industrial experience, he has been a part of design, debottlenecking and troubleshooting for various industrial projects, with a business impact of 30 MMUSD. He has expertise in reactor design and modelling, which includes first principles-based modelling for a wide range of complex reactor scenarios (catalytic specialty polymer reactors, olefin reactors, bio-film systems, gasifiers), micro-kinetic modeling on complex catalytic systems and hands-on experiments at pilot and lab scale. He also has 3 patents, and 12 research publications. He has also co-authored a book on Advances in Biogas Manufacture, Bioreactor Design and Biogas Utilization (published by Lambert Academic publishing, Mauritius). vikas.chem27@gmail.com

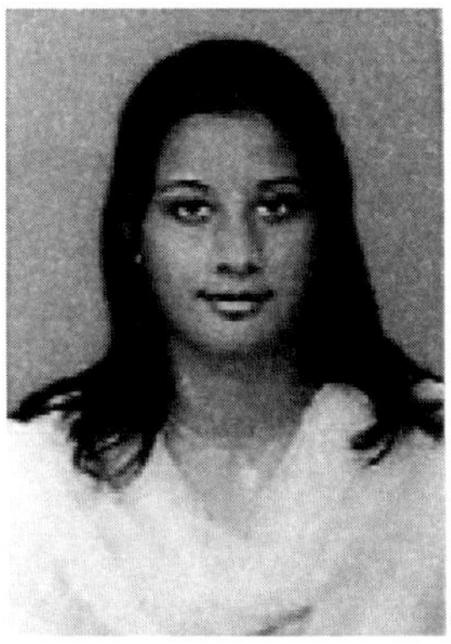

Dr. Tripti De is Assistant Professor of Chemical Engineering, VSB Engineering College (autonomous), Karur, Tamil Nadu, India. She received her B. Tech in Chemical Engineering from West Bengal University of Technology, M. Tech and PhD in Chemical Engineering from National Institute of Technology, Durgapur, India. She has published four research papers in international journals and three in Conference proceedings, has authored one book chapter on Advances in Biodiesel Synthesis and Bioreactor Design (ACS Symposium series, Washington). She is a Life Associate member of Indian Institute of Chemical Engineers and have three patents to her credit. tripti.de2009@gmail.com

Chapter - 1

Introduction

Biochemical engineering or biotechnology is usually termed as a relatively new branch of engineering/technology. However, it remains a fact that the concepts of biochemical engineering and the utilization of biotechnological processes are more than a century old. For example, manufacture of alcohol (ethanol) and alcoholic beverages through microbial fermentation of sucrose/glucose solutions (usually, molasses) has been known for ages. Even in Indian mythology, it has been reported that many saints and demons used to consume wine. So are biological wastewater treatment processes (activated sludge process, generation of biogas through anaerobic digestion of cow dung slurry) which are also almost a century old. Many of the ayurvedic medicines/ herbal medicines, which also possess a very long past history, are being manufactured through microbial fermentation.

In food processing, housewives have been employing many bioconversion processes (knowingly or unknowingly) since ages. Preparation of curd from sterilized milk, idli (the popular breakfast dish of South Indians) from fermented dough are a few examples. Use of yeast (Saccharomyces cerevisiae) as a leavening agent in the making of bread and other bakery products had been practiced even in the ancient times by Greek and Romans. Enzymes have been extensively used in food preparation and processing and even today, food industries are one among the largest practitioners of enzyme technology.

Today, the horizon of biochemical engineering/ biotechnology has widened enormously. Biochemical/ biotechnological processes are being extensively used in many large scale industries such as in petroleum processing (bio-desulfurisation of petroleum oils, biochemical dewaxing), mineral processing (bioleaching of sulfide ores), manufacture of biofuels/ green fuels (biogas, biodiesel, gasohol), manufacture of bioplastics (polylaevolactic acid, poly-β-hydroxybutanoate) and biofertilizers (Phosphate Rich Organic Manure or PROM) and the like. The list is almost unending.

The industrial applications of biotechnology, in fact, received a boost when restriction enzymes were discovered and recombinant DNA technology (also nicknamed as *Gene cloning*) got established itself as a successful means of *synthesizing* microbes in the laboratory. Today, around one hundred varieties of restriction enzymes are available and they possess the distinct characteristic that each of them cuts or cleaves the DNA helix at a specific location/site only (called the recognition site or recognition sequence of that enzyme). With the help of these restriction enzymes and the recombinant DNA (r-DNA) technology, it is possible to genetically modify an existing microbe (called the parent microbe which is mostly the wild microbe that has been isolated from the nature) to form another microbial species with entirely different characteristics (called the daughter microbe or recombinant microbe), The optimum pH and optimum temperature for the growth and multiplication of the recombinant microbe would be distinctly different from those of the parent microbe. Also, it shall release a different enzyme and shall catalyze a different chemical reaction (bioconversion). It, therefore, appears that we have *synthesized* a new microbial species in the laboratory and due to this, the process is being nicknamed as *gene cloning*.

From the millions of microbial families (wild microbes) occurring in nature (microbes exist even in the boiling volcanic lava as well as beneath the icebergs of Antarctica), it had been a herculean task to isolate and culture a specific type of microbe exclusively. The progress in biotechnology and biochemical engineering remained stagnant and the technology more or less got restricted to alcohol making and wastewater treatment for a long period during the past, mainly due to this bottleneck. Today, we could virtually *synthesize* any specific category of microbes in the laboratory and utilise them to catalyse any desired bioconversion process.

It shall be, therefore, not an exaggeration to state that the invention of restriction enzymes and the development of r-DNA technology practically revolutionized the field of biotechnology and biochemical engineering.

Under r-DNA technology, the DNA helix (DNA molecule) is first separated from the cell of a selected microbe (this may be the parent microbe itself that is being genetically modified or any other available microbe) by subjecting them to cell lysis (cell rupture), either by mechanical grinding or by using chemical reagents. From the cell extract so obtained, the proteins and RNA molecules are removed by precipitation and enzyme treatment, thereby leaving behind the cell DNA (or a concentrated solution of cell DNA). We also need a *vector*, which serves as a vehicle to carry the DNA fragment into the cells of parent microbe (host cells) during transformation. Plasmids (circular DNA molecules present in many bacterial cells) are one among the most popularly used vectors. Isolation of plasmid from available bacterial cells is also accomplished by cell lysis and subsequent removal of proteins and RNA molecules from the extract. The isolated cell DNA molecule and the plasmid are now cleaved or fragmented separately using the appropriately chosen restriction enzymes (also called restriction endonuclease enzymes). Once cleaved, the cell DNA helix becomes a straight DNA fragment. This is now joined or fused to the plasmid DNA at its fragmented site with the help of *DNA ligase* enzyme, thereby generating the recombinant DNA molecule. These recombinant DNA molecules are then introduced into the cells of the parent microbe (host cells) by culturing them together. This process is known as *transformation*. We thus obtain a culture of recombinant microbes.

Selected industrial applications of biochemical engineering/ biotechnology are discussed in detail in the last chapter (Chapter-5) of this book.

Before we proceed further, it would be appropriate to eradicate many of the misapprehensions and misinterpretations prevailing among researchers, practicing engineers, students and academicians regarding biotechnology and biochemical engineering [1].

The term biotechnology is being coined by many engineers and scientists without realizing its true jurisdiction. Let us answer a few questions:

1. Are biotechnology and biochemical engineering synonymous?
   Absolutely yes. Biotechnology is nothing but biochemical engineering.

2. Are applied biology and biotechnology more or less synonymous?
   Absolutely not. Biotechnology is a branch of engineering. The work of an engineer starts where that of a biologist ends.
3. Is biotechnology more biology-based or more mathematics-based?
   It is more mathematics-based. A biochemical engineer or biotechnologist needs to possess only a fundamental knowledge of biology. Analysis of biochemical processes (kinetic analysis) and design and analysis of bioreactors (both two phase and three phase systems) involve a large degree of mathematical complexity and accordingly, the biotechnologist must have a strong foundation on mathematical analysis.
4. How are biology, biochemistry and biotechnology inter-related?
   Very simple. The characteristics of a microbial culture (either single culture or complex culture) or those of the selected enzyme are determined and reported by the microbiologist. The biochemist studies the kinetics of the bioconversion process in which the above microbial culture or the enzyme is being used as the biocatalyst. Once the kinetics of the process are known, the biotechnologist (the biochemical engineer) proceeds to design the bioreactor in which the above bioconversion process could be carried out economically on a commercial scale. Apart from the design and performance analysis of the industrial bioreactors, the engineer is also required to select, design and optimize product recovery systems (downstream processing systems). These include design of fractional distillation equipment (binary and multicomponent distillation columns), solvent extraction columns, adsorption equipment, absorption-desorption systems, evaporators (single effect and multi effect evaporators) and condensers, membrane separation systems (ultrafiltration, reverse osmosis, pervaporation/membrane distillation modules) and the like.

From the above discussion, it is clear that the biotechnologist (the biochemical engineer) must be fully proficient in all types of unit operations and unit processes and his responsibilities are very much similar to a practicing chemical engineer in any process industry.

Biochemical processes (bioconversion processes) could be classified into two broad categories:

(a) Microbial processes (also called microbial fermentation processes).
(b) Enzyme - catalyzed processes.

To note that the term *fermentation* is being used to denote any bioconversion process (as a synonym to any bioconversion process and is not restricted to alcohol making).

1.1 Microbial Processes

In microbial fermentation processes, a microbial culture, either a single culture (composed of a single type of microbes) or a complex culture (composed of a variety of microbes), is used as the biocatalyst (which catalyzes the bioconversion of substrate into products). The microbial culture is added (commonly called *seeding*) into the pre-sterilized substrate solution, in which they grow and multiply, release enzyme and catalyze conversion of substrate into product(s). The scheme may be represented as

$$S + [\text{microbes}] \longrightarrow P + [\text{more microbes}] \tag{1.1}$$

Instead of a single substrate, multiple substrates could be involved and the medium may be seeded with a single or complex culture of microbes. Thus,

$$S_1 + S_2 + [\text{microbes}] \longrightarrow P_1 + P_2 + [\text{more microbes}] \tag{1.2}$$

The microbial culture is to be seeded into the bioreactor only at the startup (only at the beginning of the process). This is because as the process continues, microbes continuously grow and multiply in the medium and the strength of the microbial population (number of microbes per unit volume) continuously increases. Since more and more microbial cells are produced during the process and they continue to catalyze bioconversion of additional substrate into product(s), the process of microbial fermentation maybe termed as *autocatalytic* in nature. Due to the same reason, the rate of a microbial process first increases with time, reaches a maximum and thereafter decreases (see figure 1.1).

The autocatalytic nature of the microbial process may be treated as a distinct advantage of this process over enzyme-catalyzed processes. Due to this characteristic, as stated earlier, the microbial culture (biocatalyst) is to be seeded to substrate solution only at the startup of the bioreactor. However, in a microbial fermentation process, the entire substrate is not converted into products. Consumption of substrate occurs for three different purposes:

(a) a part of the substrate is used as food by the microbes for their growth and multiplication,
(b) another part is consumed for cell maintenance (for maintaining the cells healthy and active),
(c) the rest of the substrate is converted into products.

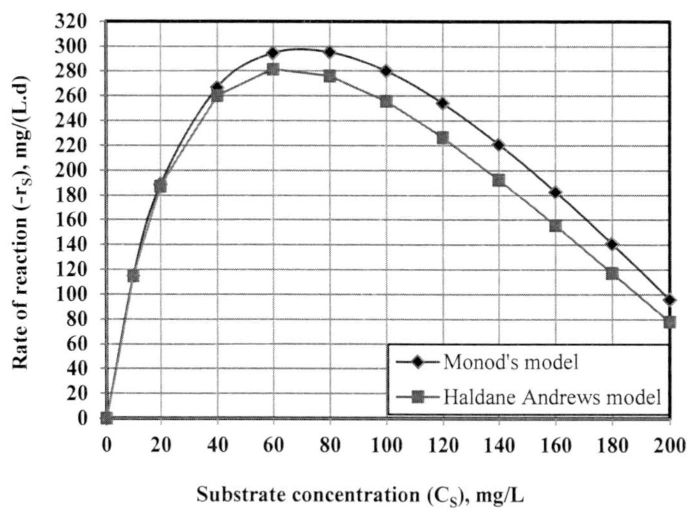

Figure 1.1: Typical rate curves for microbial fermentation processes.

Seeding of microbial culture into the bioreactor is performed in the form of an *inoculum*. For preparing the inoculum (which is usually done in the analytical laboratory), we need the *stock culture* of microbes. This stock culture is prepared in the microbiological laboratory (by the microbiologists), where the isolated microbe or the family of microbes (these may be wild microbes or genetically modified recombinant microbes) is allowed to culture in a very small volume of pre-sterilized, nutrient-rich medium, to form the stock culture. This is preserved in tiny capsules under refrigerated, sterilized atmosphere.

The industrialists purchase the stock culture from the microbiological laboratory and one or two drops of the same (often, half a drop could even serve the purpose) are added to a small volume (10-15 ml) of pre-sterilized, nutrient-rich medium (often, the pretreated substrate itself) taken in a shake flask and the flask is shaken continuously at a low speed for a specific duration of time. The microbes grow and multiply in the medium and thereby form an *inoculum of microbes*. This inoculum of microbes is now seeded to the industrial bioreactor.

The ratio of the volume of inoculum added to the total volume of substrate in the bioreactor is called the *size of inoculum*. This is an important parameter that influences the rate of growth and multiplication of microbes and thus the bioreactor performance. A large size of inoculum helps in minimizing the *lag phase* in microbial growth. When the microbial cells are seeded to the substrate solution, they do not start multiplying immediately. They take some time to acclimatize (to adapt to the new environment) and during this period, the microbes may grow in size, but shall not multiply. This period is called the *lag phase*. During the lag phase, the microbial population remains constant and shall not increase with time (see figure 1.2).

The size of the inoculum used must be, therefore, optimized (the optimum value is determined experimentally by trial in the analytical laboratory). By using the optimum size of inoculum, the lag phase during microbial growth could be reduced to a minimum (or could even be eliminated). Usually, it is around 5-10%.

The lag phase is immediately followed by the exponential growth phase during which the microbial population increases at a fast rate. Microbial death or decay could also occur simultaneously, but the rate of growth and multiplication of microbes shall be distinctly higher than that of cell death or cell decay. The cell mass concentration, (mass of microbial cells per unit volume) thus increases almost exponentially with time. Industrial batch reactors are best operated within this phase. However, the operation of the batch reactor could soon attain a stationary phase within which the rate of growth and multiplication of microbes becomes practically equal to that of cell death or decay. The cell mass concentration (x) then remains more or less constant x_S at and shall not change further with time. It is, however, not advisable to operate the bioreactor within the stationary phase, since the operation could easily slip into the next phase, namely the death phase, when the rate of cell death overtakes that of cell growth and multiplication and the cell mass concentration (x) starts decreasing with time.

Multiple lag phases are not uncommon in microbial fermentation processes. This occurs when the substrate solution is composed of more than one carbon source and the microbes tends to attack/ consume these carbon sources selectively, one at a time. The first lag phase (lag phase- 1) appears when the microbes start attacking carbon source-1 and it is followed by the logarithmic growth phase-1.Thereafter, a second lag phase appears when the microbes proceed to attack carbon source-2 and this is followed by logarithmic phase-2. And so on. This is termed as the *diauxic growth of microbes*.

It is also important to ensure that the microbial population in the bioreactor is not too large (it should not cross the maximum permissible limit). In other words, population control is an important parameter for microbes as well. At excessively large values of cell mass concentration (x), the microbial cells could encounter food deficiency. In such situations, cell lysis (rupture of cells) could occur on large scale and the metabolites liberated from the ruptured

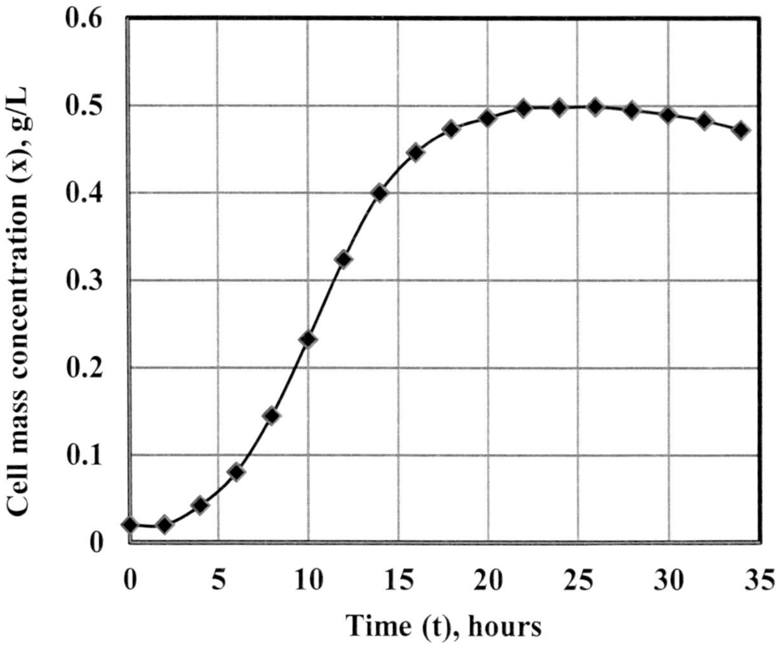

Figure 1.2: Typical growth curve of microbes.

cells (carbohydrates, proteins, amino acids) are used as food by the remaining living cells. In other words, the cells ingest (eat up) each other and a cannibalistic activity develops among them. This tends to lower the effective cell mass concentration (x) in the bioreactor and thereby adversely affect the bioreactor performance. This phenomenon is known as *endogenous decay*. The cell mass concentration (x) in the bioreactor must be, therefore, optimized.

In the case of tank reactors in which suspended growth of microbes occur (microbes grow and multiply while remaining in suspension in the substrate solution), the product solution discharged shall also contain lot of microbial cells in suspension. In other words, microbial cells do leave the bioreactor along with the product solution discharged. It would be, therefore, necessary to separate these microbial cells from the product solution either in a sedimentation tank or by using a membrane filter. The clear product solution that overflows from the sedimentor or that permeates through the membrane filter could be sent to storage. The microbial sludge that is separated is partially recycled back to the bioreactor and the rest is wasted. The recycle of microbial sludge to the bioreactor must be thus accurately controlled (in fact, optimized) such that the cell mass concentration (x) in the bioreactor remains sufficiently large (so as to achieve a sufficiently large rate of bioconversion), but should not be excessively large (to avoid endogenous decay).

Microbial activity could get inhibited by the substrate itself. *Substrate inhibition* occurs in batch bioreactors predominantly at the startup, when the concentration of the substrate is the largest. Unlike chemical reactors in which the higher the reactant concentration, the higher is the reaction rate, in bioreactors, a large substrate concentration could adversely affect microbial growth. Most of the commercially employed microbes can grow and multiply only in dilute aqueous media. For example, for the fermentation of molasses (which is the mother liquor left behind after the crystallization of sugar (sucrose) from sugar cane juice) to produce ethanol, it is first diluted to 10-12% sucrose before feeding to the fermenter. The reason is obvious. Sucrose or cane sugar, by itself, is a preservative and in strong sugar solutions, hardly any microbes can grow and multiply. Similarly, in the manufacture of biogas through anaerobic digestion of any organic matter (say, cow dung), the dung is mixed with water in the ratio 1:1 (by mass) and then fed to the digester/ bioreactor. This, once again is essential to impede substrate inhibition and also to provide the optimum pH (close to 7.0) for the growth and multiplication of methanogenic microbes. The feed solution must be, therefore, adequately diluted prior to feeding to the bioreactor, so that microbial growth shall occur smooth and fast. However, it goes without saying that when the substrate concentration is too low, the rate of bioconversion shall also be significantly low. It is, therefore, essential that the substrate concentration in the feed solution be maintained at an optimum value, at which the threat of substrate inhibition to microbial growth would be at a minimum, but, at the same time, the rate of bioconversion would remain reasonably large.

Any of the products formed during the bioconversion process could also act as a poison to the microbial cells and thereby resist their growth and multiplication and could also cause microbial death. This is termed as *product inhibition*. High concentration of ethanol is injurious to the growth of *Saccharomyces cerevisiae* (yeast cells) and hence, when they are used as the biocatalyst for the fermentation of sucrose or glucose solutions, the ethanol concentration in the bioreactor must have to be maintained at a low value (by continuously discharging the product solution). In the same way, during the manufacture of organic acids (citric acid, lactic acid, acetic acid etc.) from starchy substrates, if the acid formed is allowed to accumulate in the bioreactor, the pH of the reaction mixture could get lowered to much below the optimum pH of the microbes employed and this could hamper their growth and multiplication. In continuous flow reactors therefore, product inhibition to microbial growth can be maintained reasonably low by adjusting the rate of discharge of the product solution. Another means to impede both product inhibition and substrate inhibition to microbial activity is to employ *attached growth* of microbes (discussed subsequently).

Every microbe has an optimum pH and an optimum temperature at which its activity is the highest. *Obligate microbes* are those which are highly sensitive to changes in these operating conditions (pH of the medium, operating temperature), while *facultative microbes* are capable of multiplying in different environments and can thus resist moderate variations in pH and operating temperature. *Methanogenic microbes* (methane forming microbes or *methanogens*) are predominantly obligate in nature, while acid forming bacteria (*acidogens*) are more or less facultative. Based on the optimum temperature for their growth and multiplication, commercially employed microbes may be classified into three broad categories, as given below:

1. Thermophobic microbes, T(opt) = 15-18 C
2. Maesophilic microbes, T(opt) = 30- 45 C
3. Thermophilic microbes, T(opt) = 55-75 C

The above classification encompasses microbes that are presently employed in large scale commercial practices. As stated earlier, it is not impossible to locate microbes under the icebergs or even in the boiling volcanic lava.

Aerobic microbes require a large amount of dissolved oxygen for their growth and multiplication, To be kept in mind that microbes are incapable of absorbing/ utilising gaseous oxygen from the atmosphere, they can only utilise oxygen dissolved in the reaction mixture (substrate solution). In aerobic bioreactors therefore, a large amount of compressed air (pre-sterilized, if necessary) would have to be sparged into the reaction mixture using heavy air compressors, during the entire operation. The solubility of oxygen in water/ aqueous solution is very low and therefore, to maintain an adequate concentration of dissolved oxygen in the reaction mixture, large volumes of atmospheric air would have to be pumped in. This adds to the operating cost of the bioreactor.

A viable solution to this problem is to use *LPO (liquid phase oxygen) technology* [2,3]. In this case, a calculated amount of hydrogen peroxide is added to the reaction mixture, which dissociates in solution and releases nascent oxygen:

$$H_2O_2 \rightarrow H_2O + [O] \tag{1.3}$$

Nascent oxygen so liberated, being enormously reactive, shall meet the entire dissolved oxygen requirement of the process or the microbes. The H2 02 requirement of the process shall also be very low. Usually, a 1.0 - 2.0 mM concentration of H_2O_2 is sufficient to be maintained within the reactor.

By employing LPO technology, the entire cost of operation of air compressors could be dispensed with. The operating cost of the bioreactor thus gets significantly reduced. However, it is very much important to make sure that the concentration of the hydrogen peroxide in the reaction mixture is accurately controlled. Since nascent oxygen is extremely reactive, a slight excess could cause oxidative destruction of microbial cells themselves.

Anaerobic microbes grow and multiply in the absence of oxygen or in presence of very low concentration of dissolved oxygen. Anaerobic bioreactors must be, therefore, kept covered and leakage of atmospheric air/ oxygen into the bioreactor must be prevented.

Anoxic microbes do not need high concentration of dissolved oxygen for their growth and multiplication and hence do not demand external air supply. They utilise the oxygen released during the bioconversion process itself for their growth and multiplication. Denitrifying microbes that reduce dissolved nitrates present in the feed solution into nitrogen gas are good examples in this connection:

$$4NO_3^- + 4H^+ \rightarrow 2N_2 + 2H_2O + 5O_2 \tag{1.4}$$

Oxygen, so released, shall meet the oxygen requirement of the microbes handled.

As stated earlier, in microbial bioreactors, instead of suspended growth of microbes, *attached growth* of microbial cells could also be affected. In tank reactors (agitated vessels), mostly suspended growth of microbes is employed (with the exception of moving bed bioreactors, discussed subsequently). In such cases, the microbial cells leave the reactor along with the product solution and hence, they would have to be separated as sludge and recycled back to the reactor vessel. Further, the substrate solution (reaction mixture) would have to be continuously agitated to ensure adequate contacting between the substrate solution and the microbial cells and also to keep all the cells in suspension. Vigorous agitation using mechanical impellers could cause destruction of microbial cells. These bottlenecks can be completely eliminated by using attached growth bioreactors [3, 4].

Bioreactors employing attached growth of microbes are mostly column reactors and could be classified broadly into two categories:

(1) Biofilm reactors,
(2) Immobilized cell bioreactors.

Most biofilm reactors (with the exception of DSFF bioreactors, discussed subsequently) utilise support particles, such as silica granules, polymer beads or activated carbon particles. These particles are kept soaked in microbial solution for a particular duration of time such that a thin layer called *biofilm* gets formed surrounding each particle (see figure 1.3).

If d_P is the diameter of the support particle and δ the thickness of the surrounding biofilm, then the effective diameter (d_{Pm}) of such a particle-biofilm aggregate shall be

$$d_{Pm} = (d_P + 2\delta) \tag{1.5}$$

Microbes grow and multiply within this biofilm and the thickness of this biofilm does increase. However, when its thickness crosses a particular limit (around 0.3 mm), the film gets detached from the particle surface and falls into the surrounding liquid bulk (substrate solution). This phenomenon is known as *sloughing*. Due to sloughing, the thickness of the biofilm remains more or less constant (at around 0.3 mm). The cell mass concentration in the biofilm shall also remain more or less constant at x_f. Due to the very low volume of the biofilm, the magnitude of x_f shall be substantially large and this helps in attaining significantly high degree of bioconversion in the bioreactor.

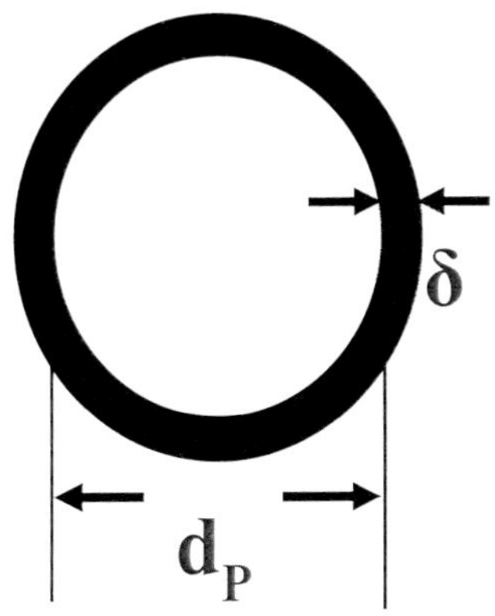

Figure 1.3: Particle – biofilm aggregate.

These particle-biofilm aggregates (of diameter d_{Pm}) could be made to form a packed bed, a fluidized bed or a semifluidized bed inside a tall column. In packed bed biofilm reactors, the feed solution may be admitted from the bottom (upflow operation) or made to trickle down under gravity from the top (trickle bed biofilm reactors). As the solution flows through the bed of particle-biofilm aggregates, the substrate diffuses into the biofilm surrounding each particle, where it is converted into products. Bioconversion thus occurs within the biofilm. The products then diffuse out into the liquid bulk. Packed bed biofilm reactors are restricted to low capacity installations (operation at low feed flow rates).

For high capacity operation (at high feed flow rates), fluidized bed or semifluidized bed biofilm reactors are employed [3,5]. Both of them operate with upflow of feed solution (the feed solution is admitted from the bottom). Design and performance characteristics of these bioreactors are discussed in detail in Chapter-3. It is also possible to use tank reactors that hold the substrate solution in which the particle-biofilm aggregates are kept suspended either by mechanical agitation or bypassing compressed air from below (pneumatic agitation, in case of aerobic processes). These are popularly called *moving bed biofilm reactors*, though a better terminology for them would be *slurry reactors* [6]. In spite of the fact that an additional diffusional resistance does come into play (the substrate has to first diffuse into the biofilm and thereafter the bioconversion occurs in the film), these bioreactors provide higher degree of bioconversion at high capacities (high feed flow rates), as compared to tank reactors that employ suspended growth of microbes. This is due to the large cell mass concentration (x_f) in the biofilm. For example, in a typical packed bed biofilm reactor that receives a feed solution of substrate concentration (C_{S0}) around 50-100 mg/ L, the cell mass concentration (x_f) in the biofilm of particle-biofilm aggregates present in the column could be around 1000 to 4000 mg/L. Further, since neither the substrate nor the product tends to accumulate within the biofilm, substrate inhibition or product inhibition to microbial growth shall be more or less absent. Also, as the microbial cells do not leave the reactor (they remain behind in the biofilm adhering to the particle surface), only the product solution is discharged and the entire cost of separation and recycle of microbial sludge could be avoided.

The so called DSFF (Downflow Stationary Fixed Film) bioreactors [7] employ a vertical stack of pipes or channels (of around 12.5 - 25.4 mm in diameter), the inside surface of each being coated with a thin film of microbial solution (biofilm). The feed solution is admitted from the top and as it flows down each vertical channel, the substrate diffuses into the biofilm, wherein the bioconversion occurs. These biofilm reactors do not use any support particles and due to the downflow mode of operation, their operating cost is also relatively lower.

Sludge bed and sludge blanket bioreactors are relatively newcomers into the family of microbial reactors. An excellent example is the Upflow Anaerobic Sludge Blanket (UASB) bioreactors, widely employed for anaerobic wastewater treatment [3,5,8]. These bioreactors also do not use any support particles, but microbial cells themselves form sludge granules (around 1.0 mm in size) and these granules form a sludge bed at the bottom of the column. As the feed solution flows up through this sludge bed, bioconversion of dissolved substrate occurs and due to the gasses evolved during the process, a *sludge blanket* also gets formed above the sludge bed (which shall be composed of sludge granules as well as gas bubbles). Bioconversion occurs both in the sludge bed as well as in the sludge blanket, though the contribution of the latter is higher. This technology has been successful only with anaerobic microbes and also only with complex culture of microbes. Since microbial sludge granule formation is a slow, evolutionary process, the startup time for these bioreactors is extremely high (often 5-6 months). However, they can operate at high capacities (high feed flow rates), can handle high- strength feedstock (wastewater samples with high BOD) and can affect high degree of bioconversion (90 - 95% BOD destruction).

Design and analysis of UASB bioreactors have also been discussed in detail in Chapter-3.

Immobilization of microbial cells within solid particles has also been successful. Immobilization is conducted in microbial laboratories. Microbial cells maybe immobilized by adsorption on solid surfaces (for example, on the surface of ion exchange resins), or by entrapment within a polymer matrix (within polyacrylamide beads or calcium alginate gel), or by covalent boding to an activated support (made of polyisocyanates or carbodiimide). Once immobilized, cells exhibit high activity and as a result, the immobilized cell bioreactors provide high degree of bioconversion at high capacities, the additional cost of cell immobilization notwithstanding. Separation and recycle of microbial sludge shall not be required. Substrate/ product inhibition to microbial activity shall be more or less absent, nor shall there be any chances of endogenous decay of microbes. Immobilized cell bioreactors can also be designed and constructed as moving bed, packed bed, fluidized bed or semifluidized bed bioreactors (as in the case of biofilm reactors discussed above). All of them are heterogeneous bioreactors.

As stated at the beginning of this chapter, biotechnological processes have found their place in a vast number of industrial practices. A few typical examples are listed in Table 1.1. Details of many of these industrial processes are presented in Chapter-5. Instead of a single culture of microbes (a single category of microbes), complex cultures composed of varieties of microbes are used in many industrial applications such as in aerobic and anaerobic wastewater treatments. In the anaerobic treatment of organic wastes (biological wastes, industrial effluents) to produce biogas (which is principally a mixture of methane and carbon dioxide), cow dung acts as a natural inoculum. The cow dung-water slurry (mixed in 1:1 mass ratio) when confined out of contact with air, shall promote growth of a good number of anaerobic microbes including methanogens and these microbes convert the dissolved organic matter into methane and CO_2 (namely, biogas). This is a property that is exhibited exclusively by cow dung and consequently, when this dung is used as the substrate for biogas production, no separate inoculum is required to be seeded from outside to the bioreactor.

For large scale, commercial production of biogas, alternate substrates could be employed either as single substrate or more preferably, as multiple substrates. Examples are food wastes, agricultural and plant wastes (bagasse, water hyacinth, waste grown algae), cellulosic wastes (wheat/ corn straw, waste paper), poultry litter and piggery wastes etc. [5]. In such cases, a small amount of cow dung is also mixed with the aqueous slurry (substrate solution) and it acts as the inoculum and initiates the growth of the complex culture of microbes. Hydrolytic acidogenic, acetogenic and methanogenic microbes act on the substrate (dissolved organic matter) and convert it to biogas :

$$\text{Organic matter} \xrightarrow[\text{Hydrolytic microbes}]{} \text{glucose/ fructose} + \text{amino acids} \tag{1.6}$$

$$\text{Glucose/fructose} \xrightarrow[\text{acidogens}]{} \text{propanoic/butanoic acids} + \text{ketones, alcohols} + CO_2 + H_2 \tag{1.7}$$

$$\text{Above acids, ketones, alcohols} \xrightarrow[\text{acetogens}]{} CH_3COOH + CO_2 + H_2 \tag{1.8}$$

$$CH_3COOH \xrightarrow[\text{methanogens}]{} CH_4 + CO_2 \tag{1.9}$$

$$CO_2 + 4H_2 \xrightarrow[\text{methanogens}]{} CH_4 + 2H_2O \tag{1.10}$$

A complex culture of aerobic facultative microbes such as genres of *Pseudomonas, Achromobacter*, *Mycobacteria*, *Nitrosomonas* and *Nitrobacter* (the last two being nitrifying bacteria) is used in the *activated sludge process* that is widely used for the aerobic treatment of wastewater (organic wastes). As compared to anaerobic treatment, this process is faster and the organic matter is almost completely decomposed (oxidized), but the gas that is evolved is principally carbon dioxide (no methane is produced). Also, as stated earlier, large volumes of atmospheric air would have to be pumped into the aerobic tank (bioreactor) to meet the oxygen requirement of the process/microbes.

Lactic acid, which can be economically synthesized from molasses and de-proteinized cheese whey (cheese whey is the mother liquor left behind after the separation of fats (casein) from milk) by microbial fermentation (see table-1), is the feedstock for the manufacture of the bioplastic, PLLA (poly laevo lactic acid). It is a biodegradable (bio-compostable) plastic and therefore, disposal of PLLA wastes shall not cause any environmental concern. It strongly resembles polystyrene (and often, polypropylene) in most of its properties and could be used for all industrial/ domestic applications in the place of PS or PP. Manufacture of PLLA bioplastic is discussed in detail in Chapter-5.

Table 1.1 : Typical industrial applications of biotechnology

Sl. No.	Substrate	Microbial Culture	Product(s)	Process
1.	Sucrose/ glucose solutions (eg. molasses)	*Saccharomyces cerevisiae*	Ethanol	Anaerobic
2.	Molasses (sucrose)	*Enterococcus faecalis*	Lactic acid	Anaerobic
3.	De-proteinized cheese whey (lactose)	*Lactobacillus helveticus*	Lactic acid	Anaerobic
4.	Molasses, corn starch	*Aspergillus niger*	Citric acid	Aerobic
5.	n – paraffins (hexanes)	*Pseudomonas sp.*	Adipic acid	Aerobic
6.	De-proteinized cheese whey (lactose)	*Xanthomonas campestris*	Xanthan gum	Aerobic
7.	Soybean meal	*Streptomyces erythreus*	Streptomycin, Erythromycin	Aerobic

8.	Pyrite, chalcopyrite ores	*Thiobacillus ferrooxidant*	Ferrous sulfate or copper sulfate solution (bioleaching or bio- beneficiation)	Aerobic
9.	Rock phosphate ore	*Bacillus megatherium var phosphaticum*	Phosphatic biofertilizer (PROM)	Aerobic
10.	Petroleum oils	*Rhodococcus erythropolis strain*	Desulfurized oil	Aerobic

Source: References (3), (4), (9) and (10).

1.2 Enzyme catalyzed processes

Enzymes are secreted by microorganisms and they are synthesized in microbiological laboratories by culturing the microbes in laboratory flasks/ reactors. All enzymes are proteins (though the reverse is not true) and being proteins, they are water soluble.

Enzymes are very selective biocatalysts (unlike chemical catalysts). As a result, when pure, they catalyze only the desired reaction pathway (hardly any side products/ undesirable products are formed). The product solution shall, therefor, contain essentially the desired product (though may be in dilute solution) and this helps in bringing down the cost of downstream processing (product recovery). On the other hand, enzymes are highly sensitive to contamination. It has been observed that even a 5% contamination could bring down the activity of the enzyme by as much as 50% in certain cases. Enzymes are to be, therefore, used in extremely pure form.

Use of free enzyme (enzyme solution added directly to the bioreactor) is usually discouraged in process industries. This is because the enzyme remains dissolved in the product solution and hence gets discharged from the bioreactor along with the product(s). It, therefore, becomes essential to separate the enzyme from the product solution (say by ultrafiltration) and recycle back to the bioreactor. Recovery and recycle of enzyme (required to be in the very pure form) is fairly expensive and this could thereby bring down the overall economy of the process.

In most industrial practices therefore, enzymes are used in the immobilized form. Immobilization of enzyme could be performed using the same methods as those employed for immobilization of microbial cells (discussed earlier). These can be adsorption on an active solid surface, entrapment within a polymer matrix or by means of covalent bonding to the solid surface. The media used for enzyme immobilization can be polyacrylamide gel, polymethyl methacrylate or polystyrene beads (for entrapment and covalent bonding) and ion exchange resins, silica gel, particles of alumina, clay or activated carbon (for adsorption). Enzymes may also be encapsulated within hollow fibers made of semi-permeable membranes (polyamide, cellulose acetate membranes). Different polymer composites and carbon or silica nanoparticles are also being used for the immobilization of industrial enzymes.

Enzymes are named based on the biochemical reaction they catalyze or based on the substrate on which they act as biocatalyst [11]. Examples are hydrolase, isomerase, dehydrogenase, oxidoreductase enzymes. Cellulase enzyme is used to catalyze cellulose hydrolysis, lipases for the hydrolysis of lipids (fats and oils) to alcohols and carboxylic acids, urease enzyme catalyzes hydrolysis of urea to ammonia and carbon dioxide, while proteases catalyze hydrolysis of proteins (to polypeptides and amino acids).

Unlike in microbial fermentation processes (discussed earlier), in enzyme-catalysed processes, the entire substrate could get converted to products (the entire substrate could be utilised for product formation). The scheme may be represented as:

$$\text{Substrate (s)} + [\text{enzyme}] \longrightarrow \text{Products} + [\text{enzyme}] \tag{1.11}$$

$$S + E \longrightarrow P + E \tag{1.12}$$

However, enzyme-catalysed processes are not autocatalytic in nature and adequate concentration of enzyme must have to be maintained in the reaction mixture to attain the desired degree of bioconversion. Also, the cost of synthesis and purification enzyme catalyst is additional in this case.

One of the important industrial applications of lipase enzyme is as catalyst in the manufacture of biodiesels through trans-esterification of vegetable oils [12]. All vegetable oils are esters of glycerol and a high molecular weight carboxylic acid (accordingly, they are called *triglycerides*) and when the oil is reacted with a low molecular weight alcohol (methanol, ethanol or propanol), we obtain the methyl, ethyl or propyl ester of the high molecular weight carboxylic acid, which is biodiesel. Glycerol is released as the byproduct:

$$\begin{array}{l} CH_2OCOR \\ | \\ CHOCOR \\ | \\ CH_2OCOR \end{array} + 3CH_3OH \longrightarrow \underset{\text{(Biodiesel)}}{3RCOOCH_3} + \begin{array}{l} CH_2OH \\ | \\ CHOH \\ | \\ CH_2OH \end{array} \tag{1.13}$$

Biodiesel is a green fuel since it is much more environment-friendly than the diesel oil manufactured from petroleum (petrodiesel), as it does not contain sulfur, neither aromatic hydrocarbons. Biodiesel also does not contribute to CO_2 pollution and consequent global warming, since CO_2 evolved due to the combustion of biodiesel gets reabsorbed by the green plants from which the vegetable oil has been derived (thus maintaining a carbon cycle in the nature). Biodiesel is also *photosynthetically* renewable and it is already being used as an excellent substitute to petrodiesel in all automobiles (cars, busses, trucks, tractors), railway engines and also in engines used for pumping water (irrigation purposes). Manufacture of biodiesels has been discussed in detail in Chapter-5.

By using lipase enzyme as the catalyst, we get a pure form (uncontaminated form) of biodiesel as compared to when a chemical catalyst (alkali catalyst, acid catalyst) is being employed. The overall rate of reaction is, however, lower.

Some of the proteases are used as additives in the manufacture of detergents (they are added to detergents to improve their cleansing power), in meat processing (tenderization) and also in cheese making.

Use of enzymes such as lignin peroxidase, manganese peroxidase and laccases for the delignification of wood to produce cellulosic pulp (which is further used for the manufacture of paper) has also been successful. The process takes place at ordinary temperature (35^0C) and 1 atm and simultaneous bleaching of the pulp could be affected, unlike chemical pulping which demands high temperature (180^0C), use of chemicals (caustic soda, sodium sulfate, sodium sulfite) and a separate bleaching operation.

As in the case of microbial fermentation processes, in enzyme-catalyzed processes also, a high concentration of substrate in the feed solution could cause damage to the activity of the biocatalyst (namely, enzyme). It is postulated that substrate inhibition to enzyme activity is due to the fact that the enzyme-substrate complex (ES) formed, could react with the excess of substrate present to form a non-reactive complex, ESS. This hypothesis is based on the proposed mechanism that in enzyme-catalyzed reactions, the enzyme forms an activated complex (ES) with the substrate which, being unstable, shall decompose fast to produce the desired products. Since a part of ES is converted to the non-reactive ESS, the rate of product formation obviously goes down. In immobilized enzyme reactors however, this problem shall not be acute.

It is also possible that any of the products formed acts as a catalyst poison and thereby tends to bring down the enzyme activity (product inhibition). Once again in immobilized enzyme bioreactors, operating with continuous discharge of product solution, this problem could be overcome to a large extent.

Inhibitors are often purposefully added to the reaction mixture to regulate the enzyme activity. Such additives are also called *modifiers*. These inhibitors could cause competitive, uncompetitive, or non-competitive inhibition to enzyme activity. Use of such inhibitors is discussed in detail in Chapter-2.

References

1. Narayanan, C.M., Biotechnological Education – Urgent Need for Renovation and Reappraisal, Proc. National Convention of Indian Society for Technical Education, Mangalore, December, 2002.
2. Narayanan, C.M., Modeling and Simulation of Fluidized Bed Bioreactors with Liquid Phase Oxygen Utilisation, Int. J. Transport Phenomena, **11**(2), 127-132, 2009.
3. Narayanan, C.M., *Biotechnology and Bioprocess Engineering*, Galgotia Publishers, New Delhi, 2011.
4. Bailey, J.E. and Ollis, D.F., *Biochemical Engineering Fundamentals*, McGraw Hill, Singapore, 1986.

5. Narayanan, C.M. and Vikas Narayan, *Advances in Biogas Manufacture, Bioreactor Design and Biogas Utilisation*, Lambert Academic Publishing, Mauritius, 2020.
6. Narayanan, C.M. and Vikas Narayan, Studies on Synthesis of Lactic Acid and Xanthan Gum from Cheese Whey Permeate in Two Phase and Three Phase Moving Bed Biofilm Reactors, Korean J. Chem. Eng., **38**(9), 1888-902, 2021. **DOI:10.1007/s11814-021-0821-5.**
7. Narayanan, C.M., Performance Analysis of Anaerobic Downflow Stationary Fixed Film (DSFF) Bioreactors and Software Development, Int.J.Chem. Reactor Eng.,**7**(1), Article-61, November, 2009. **DOI:**10.2202/1542-6580.1959.
8. Narayanan, C.M. and Vikas Narayan, Multiparameter Models for Performance Analysis of Upflow Anaerobic Sludge Blanket (UASB) Bioreactors, J. Chemical Tech. and Biotech., **83**, 1170-1176, 2008.
9. Ohshiro,T. and Izumi,Y., Microibial Desulfurization of Organic Sulfur Compounds in Petroleum, Biosci. Biotechnol. Biochem., **63**(1), 1-9, 1999.
10. Maass, D., de Oliveira, D., de Souza, AAU and Souza, SMAGU, Biodesulfurization of a system containing synthetic fuel using Rhodococcus erythropolis ATCC 4277, Appl. Biochem. Biotechnol., **174**(6), 2079-86, 2014.
11. Manjula, P. and Dawn, S.S., *Bio & Enzyme Engineering*, Scitech Publications, Chennai, 2004.
12. De, T., Sikder, J. and Narayanan, C. M., Biodiesel Synthesis and Bioreactor Design – An Overview, Indian J. Chem.Technol., **24**, 575-92, 2017.

Chapter - 2

Kinetics of Biochemical Reactions / Bioconversion Processes

As stated in Chapter – 1, kinetics of biochemical reactions (bioprocesses) are relatively more complex as compared to those of conventional chemical reactions. We hardly encounter simple pseudo-first order or pseudo-second order reactions under biokinetics. Most of the kinetic equations representing biochemical reactions / bioconversion processes are nonlinear and even the simplest one would contain at least two kinetic constants.

Kinetic equations can be developed only experimentally and consequently, a degree of empiricism is inherent in all of the kinetic equations reported. The reliability of each reported kinetic equation thus weighs heavily on the accuracy of the experimental data based on which it has been developed.

2.1 Kinetic experiments

Kinetic equations can be developed based on experiments conducted in

(a) a laboratory batch reactor (may be assumed to be an ideal batch reactor),

(b) a chemostat (a laboratory CSTR- Continuous Flow Stirred Tank Reactor).

The laboratory batch reactor being of low volume and being well agitated, could be assumed to exhibit ideal behavior, in the sense that the composition of the reaction mixture shall not change from one point to another inside the bioreactor at any time (t). Similarly, a chemostat (a laboratory CSTR), since of low volume and is well-agitated, could be assumed to be completely back-mixed (ideal CSTR). It is a flow reactor and once the steady state has been attained, the concentration of any species (substrate, product, cell mass) in the exit stream (C_{Se}, C_{Pe}, x_e) shall be essentially equal to that inside the reactor (see figure 2.1). This is what we mean by *complete backmixing.*

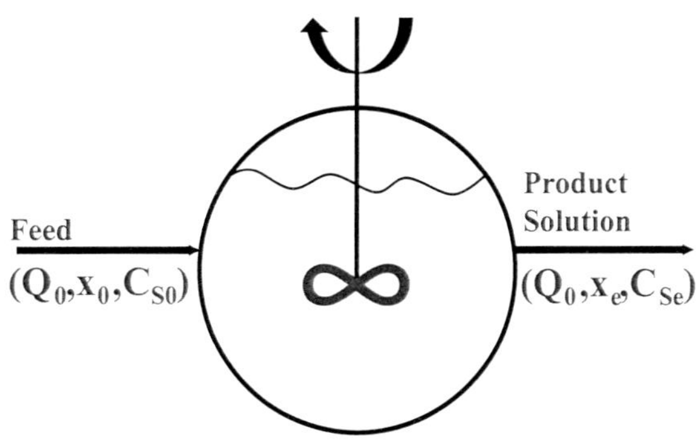

Figure 2.1: Schematic of a chemostat (ideal CSTR)

Batch reactor experiments are relatively easier to conduct, while chemostat experiments demand longer period of time. However, chemostat data are easier to fit into the proposed (assumed) kinetic equations (for the evaluation of kinetic constants). We have illustrated this through examples in the subsequent sections.

The laboratory batch reactor is operated for a particular period of time, samples of reaction mixture are withdrawn at different intervals of time and analyzed to determine the substrate concentration (C_S), product concentration (C_P) and cell mass concentration (x) in each sample (using a spectrophotometer or an HPLC - high pressure liquid chromatograph). In the case of enzyme-catalysed reactions, only C_S and C_P are required to be recorded. We thus obtain an elaborate table that lists the values of C_S, C_P and x at different intervals of time (t).The kinetic equation into which this data fits most accurately is thus located by trial. The batch reactor data could be fitted into any proposed (assumed) kinetic equation, using two different approaches :

(a) Integral analysis,
(b) Differential analysis.

The integral analysis is applicable only if the proposed (assumed) kinetic equation is one that can be integrated analytically. If the assumed kinetic equation is too nonlinear and cannot be integrated analytically (only numerical integration is possible), then this approach becomes non- applicable. The compiled experimental data are to be fitted into the integrated form of the rate equation, to evaluate the kinetic constants. Here again, the integrated form of the assumed kinetic equation must be explicit in the kinetic constants, to permit their evaluation.

In the differential analysis, no integration of the assumed rate equation is necessary. However, the values of reaction rate at different intervals of time (the values of (dC_S / dt) at different t) would have to be determined by plotting substrate concentration (C_S) versus time (t). The slope of the tangent to the curve at a particular time (t) provides the value of (dC_S / dt) at that time (t). An additional table that lists the values of reaction rate, (dC_S / dt), at different values of (t) is then kept prepared.

The performance equation for a chemostat, which, as stated earlier, could be assumed to be equivalent to an ideal CSTR (completely backmixed reactor) is as follows (with reference to Figure 2.1):

$$Q_0\left[C_{SO}-C_{Se}\right]=\left(-r_S\right)_e V$$

or

$$\tau=\left(V / Q_0\right)=\left[C_{SO}-C_{Se}\right] /\left(-r_S\right)_e \tag{2.1.1}$$

where τ = space time (2.1.2)

$(-r_S)_e = (-dC_S / dt)_e$

= rate of reaction (rate of substrate consumption) at $C_S = C_{Se}$ (2.1.3)

V = volume of reactor

Q_0 = volumetric flow rate of feed solution, m^3/s

We have assumed a *constant density system* (the density of the reaction mixture does not change materially due to bioconversion/ biochemical reaction) and accordingly, volume flow rate of feed solution = that of exit solution = Q_0 .

The chemostat is operated by admitting the feed solution (substrate concentration = C_{S0}) at a specified flow rate of Q_0 m^3/s and sufficient time is provided for the reactor operation to attain steady state (usually, steady state is attained within 10-15 minutes). Once steady state has been attained, the concentration of substrate / product / cell mass in the exist stream (C_{Se}, C_{Pe}, x_e) shall not change with time and these are recorded and tabulated. The experiment is then repeated at a different feed flow rate, Q_0. We thus obtain an elaborate table that lists the steady state concentrations such as (C_{Se}, C_{Pe}, x_e) at different values of Q_0. This data is now fitted into equation (2.1), after substituting the assumed rate equation (kinetic equation) into it. If the data fit satisfactorily, then the kinetic constants are evaluated. Otherwise, another kinetic equation is assumed and the trial repeated. The mathematical complexity involved in this analysis is relatively lower, as compared to analysis based on batch reactor experiments. However, as stated earlier, chemostat experiments are more time-consuming.

2.2 Kinetics of Microbial Fermentation Processes

As stated in Chapter-1, in a microbial fermentation process, the substrate is consumed for three different purposes such as for growth and multiplication of microbes, for cell maintenance, and finally, for product formation. As a result, for a precise analysis, we have to write a separate kinetic equation for each of these and then solve them simultaneously. Thus,

(a) Substrate consumption for growth and multiplication of microbes:

This shall be a function of substrate concentration, cell mass concentration and often, the product concentration as well. Thus,

$$\left(-dC_S / dt\right)_1 = \left(-r_S\right)_1 = f_n\left[C_S, x, C_P\right] \tag{2.2.1}$$

(b) Substrate consumption for cell maintenance:

$$\left(-dC_S / dt\right)_2 = \left(-r_S\right)_2 = f_n\left[C_S, x\right] \tag{2.2.2}$$

(c) Substrate utilisation for product formation:

$$\left(-dC_S / dt\right)_3 = \left(-r_S\right)_3 = f_n\left[C_S, x, C_P\right] \tag{2.2.3}$$

We have thus three differential equations that are coupled with each other and are to be solved simultaneously. The procedure shall be invariably laborious.

The task can be simplified by defining what are called the *yield coefficients* (Y, Y_P) and by further assuming that their magnitudes remain more or less constant throughout the process. Such an assumption/ simplification (found to be causing little serious error in many situations) would relax the inter-coupling of these three equations and we shall be left with only one differential equation, based on which the kinetic equation could be developed.

These yield coefficients are defined as follows:

Y = overall yield coefficient for cell mass production

$$= \frac{\text{mass of microbial cells produced (on dry basis)}}{[\text{Total mass of substrate consumed}]}$$

$$= \left(x - x_0\right) / \left(C_{S0} - C_S\right) \tag{2.2.4}$$

Y_P = overall yeild coefficient for product formation

= (mass of product formed)/(total mass of substrate consumed)

$$=\left(C_P - C_{P0}\right)/\left(C_{S0} - C_S\right)$$

$$= C_P/\left(C_{S0} - C_S\right) \qquad (2.2.5)$$

To note that $C_{p0} = 0$ (as no product is formed at $t = 0$).

The earliest kinetic equations representing microbial fermentation processes are those based on the studies reported by biologists. These biologists (botanists, mainly) studied the growth and multiplication characteristics of various microbial species and the equations reported by them were subsequently modified by biochemical engineers / biotechnologists and extended to predict the kinetics of biochemical reactions / bioconversion processes catalyzed by them. A few prominent examples are discussed below.

2.2.1 Monod's kinetic model

Monod, being a botanist, studied the growth kinetics of selected microbes such as *E.coli* and came forward with a rate equation of the following form [1] :

$$\left(dx/dt\right) = \mu x \qquad (2.2.6)$$

where μ = specific growth rate, s^{-1}

Though the above equation appears to be first order in x, the specific growth rate (μ) is not a constant, but is reported to be a function of substrate concentration (C_S) as given below :

$$\mu = \left(\mu_m C_S\right)/\left(K_S + C_S\right) \qquad (2.2.7)$$

where μ_m = maximum specific growth rate, s^{-1}

K_S = saturation constant, g/L

Combining the above two equations together, we get

$$\left(dx/dt\right) = \left(\mu_m C_S\right)x/\left(K_S + C_S\right) \qquad (2.2.8)$$

The above equation contains two independent variables such as the cell mass concentration (x) and the substrate concentration (C_S). To convert it into a more workable form, we may make use of the definition of the overall yield coefficient for cell mass production (Y), given in equation (2.2.4). Thus, assuming that the value of Y does remain more or less constant throughout the bioconversion process under consideration, from equation (2.2.4),

$$x = x_0 + Y\left(C_{S0} - C_S\right) \qquad (2.2.9)$$

$$\left(\frac{x}{Y}\right) = \left[\left(\frac{x_0}{Y} + C_{S0}\right) - C_S\right] \qquad (2.2.10)$$

$$= \left(b - C_S\right) \qquad (2.2.11)$$

where $b = C_{S0} + \left(x_0/Y\right)$ (2.2.12)

Also, $\left(-dC_S/dt\right) = \left(1/Y\right)\left(dx/dt\right)$ (2.2.13)

If we substitute the above two equations such as equations (2.2.11) and (2.2.13) in equation (2.2.8), we get

$$\left(-dC_S/dt\right)=\mu_m C_S\left(b-C_S\right)/\left(K_S+C_S\right) \tag{2.2.14}$$

This is the final form of *Monod's kinetic equation*. It is composed of two kinetic constants, such as μ_m and K_S. Though this equation has been found to be applicable to a good number of bioconversion processes, it does have limitations such as it does not account for

(i) substrate inhibition to microbial growth,

(ii) product inhibition to microbial growth,

(iii) endogenous decay and consequent reduction in cell mass concentration.

It also assumes a constant value of overall yield coefficient for cell mass production (Y). All these limitations must be kept in mind while utilising this kinetic equation for the design of bioprocesses / bioreactors.

Let us assume that experimental data collected using a laboratory batch reactor are available. Then, to determine whether the biochemical reaction (bioconversion process) under consideration follows Monod's kinetic model or not, we may adopt either the integral analysis or the differential analysis. In the case of integral analysis, as stated earlier, we need the integrated form of the kinetic equation. Thus equation (2.2.14) on rearrangement, yields

$$\int_0^t dt=\left(1/\mu_m\right)\int_{C_S}^{C_{S0}}\left(K_S+C_S\right)/C_S\left(b-C_S\right)dC_S \tag{2.2.15}$$

On integration using the method of partial fractions, we get

$$\ln\left[\frac{b-C_S}{b-C_{S0}}\right]+\left(K_S/b\right)\ln\left[\frac{1-b/C_S}{1-b/C_{S0}}\right]=(\mu_m t) \tag{2.2.16}$$

This is the integrated form of Monod's kinetic equation. The value of C_S at any time (t) could be estimated from this equation (though it is implicit in C_S).

The above equation could be rearranged to the following form:

$$(1/t)\ln\left[\frac{b-C_S}{b-C_{S0}}\right]=-\left(K_S/b\right)(1/t)\ln\left[\frac{1-b/C_S}{1-b/C_{S0}}\right]+\mu_m \tag{2.2.17}$$

or,

$$Y=-\left(\frac{K_S}{b}\right)X+\mu_m \tag{2.2.18}$$

where, $Y=y-$ parameter

$$=(1/t)\ln\left[b-C_S/b-C_{S0}\right] \tag{2.2.19}$$

$X=x-$ parameter

$$=(1/t)\ln\left[\left(1-b/C_S\right)/\left(1-b/C_{S0}\right)\right] \tag{2.2.20}$$

Therefore, if the given biochemical reaction does follow Monod's kinetic equation, then a plot of above y-parameter (Y) versus the x-parameter (X) must yield a straight line of

$$\text{slope} = -\left(K_S/b\right) \tag{2.2.21}$$

$$y\text{-intercept} = \mu_m \tag{2.2.22}$$

If a straight line plot (linear plot) does result, then the kinetic constants (μ_m, K_S) could be evaluated from the slope and the y-intercept of the plot.

Let us now consider differential analysis. In this case, integration of the rate equation is not required. For example, equation (2.2.14) could be rearranged as

$$\left(b-C_S\right)/\left(-r_S\right)=(K_S/\mu_m)\,(1/C_S)+(1/\mu_m) \tag{2.2.23}$$

where $(-r_s)=(-dC_S/dt)=$ rate of reaction, g/(L.s).

To note that the minus sign has been used only to indicate that the substrate concentration (C_S) decreases with time (t) and numerical values of reaction rate (without any sign convention) are to be used in the calculations.

From the above equation, it is clear that if the biochemical reaction under consideration follows Monod's kinetics, then a plot of $(b-C_S)/(-r_s)$ versus $(1/C_S)$ must yield a straight line with

$$\text{slope} = (K_S/\mu_m) \tag{2.2.24}$$

$$y\text{-intercept} = (1/\mu_m) \tag{2.2.25}$$

If the plot is found to be far from linear, then it means that the bioprocess does not follow Monod's kinetic equation and an alternate kinetic equation is to be tried. Equation (2.2.23) is simpler to handle than equation (2.2.17) of integral analysis. However, here we need to estimate the values of reaction rate, $(-r_S)$ or $(-dC_S/dt)$, at different values of C_S. This is to be done, as explained in the earlier section, by plotting substrate concentration (C_S) versus time (t) and finding the slope of the tangent to the curve at different values of t. A typical plot is shown in figure (2.2). From the figure, at t = 15 min, the slope of the tangent to the curve is 0.15 mg/(L.min) and hence, at t =15 min, $(-dC_S/dt) = (-r_S) = 0.15$ mg/(L.min) and the corresponding value of C_S= 1.5 mg/L. In this way, an elaborate table containing values of ($-r_S$) at different values of C_S would have to be prepared in advance. This step is additional for differential analysis.

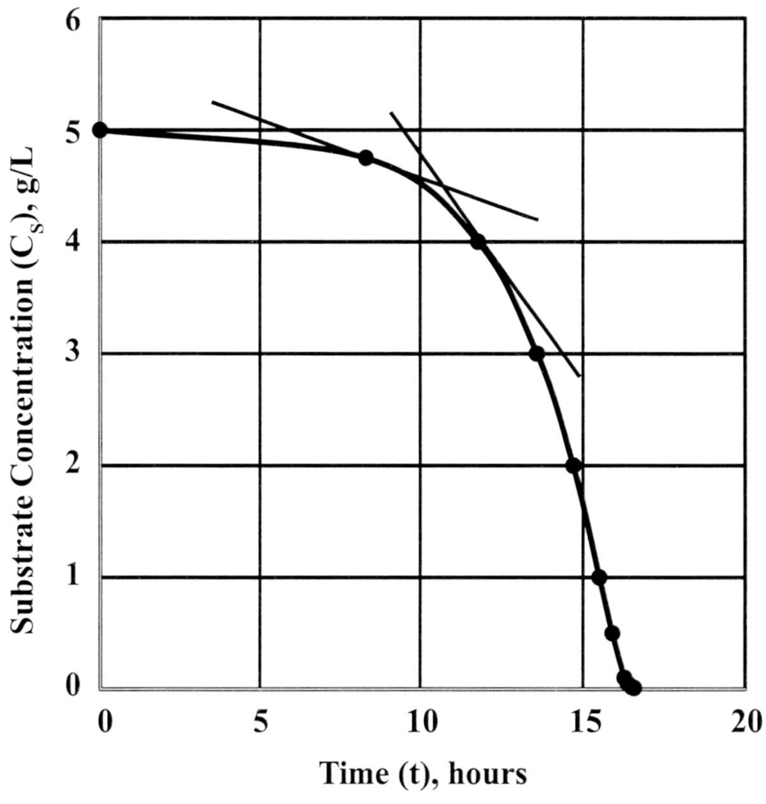

Figure 2.2: Variation of substrate concentration with time for a typical microbial fermentation process

A simplified differential analysis is possible by incorporating a few approximations. Incidentally, it has been observed that even today, a number of investigators do prefer to use this approximate procedure, mainly due to the fact that the computational load shall be distinctly lower. For example, if it is assumed that $\left(\frac{dx}{dt}\right) \approx \left(\frac{\Delta x}{\Delta t}\right)$, then equation (2.2.6) may be rewritten as

$$\mu = \left(\frac{1}{x}\right)\left(\frac{\Delta x}{\Delta t}\right) \tag{2.2.26}$$

Therefore, values of μ at different values of x (at different values of t) could be computed from available batch reactor experimental data. Then, from equation (2.2.7),

$$\left(\frac{1}{\mu}\right) = \left(\frac{K_S}{\mu_m}\right)\left(\frac{1}{C_S}\right) + \left(\frac{1}{\mu_m}\right) \tag{2.2.27}$$

Thus, if a plot of $(1/\mu)$ versus $(1/C_S)$ yields a straight line, then it can be concluded that the bioprocess under consideration follows Monod's kinetic equation. And from the slope and y- intercept of the plot, the values of the kinetic constants (μ_m, K_S) could be determined.

The argument in support to the above approximate analysis is that within the exponential growth phase (within which most laboratory/ industrial bioreactors operate), the variation of x with t is close to linear and the approximation, $\left(\frac{dx}{dt}\right) \approx \left(\frac{\Delta x}{\Delta t}\right)$ shall not necessarily cause serious error in the analysis. However, it is needless to comment that approximate methods must always be used with due caution.

Let us now consider the case when chemostat data are available. If experimental data compiled using a chemostat (laboratory CSTR) are available, then the analysis is relatively simpler and demands lesser computational load. For example, if we substitute Monod's kinetic equation into equation (2.1.1) and rearrange, we get

$$\frac{\left[\tau\left(b - C_{Se}\right)\right]}{\left(C_{S0} - C_{Se}\right)} = \left(K_S/\mu_m\right)\left(1/C_{Se}\right) + \left(1/\mu_m\right) \tag{2.2.28}$$

Thus, if the biochemical reaction does follow Monod's kinetic equation, then a plot of $[\tau \ (b - C_{Se})/(C_{S0} - C_{Se})]$ versus $(1/C_{Se})$ must yield a straight line and from its slope and y- intercept, the kinetic constants (μ_m, K_S) could be evaluated.

The feed solution to the chemostat could often be sterile. In such a case (if experiments were conducted with a sterile feed), then $x_0 = 0$, $b = C_{SO}$. Equation (2.2.28) then gets reduced to

$$\tau = \left(K_S / \mu_m\right)(1/C_{Se}) + (1/\mu_m) \tag{2.2.29}$$

It can be seen that kinetic analysis employing chemostat data is computationally simpler and easier. However, as stated earlier, chemostat experiments demand larger amount of time. In each run, we have to wait for 10 to 15 minutes to make sure steady state has been attained. Also, each run would have to be repeated at least 2-3 times to ensure experimental accuracy.

It maybe noted that the performance equation for a chemostat could be written in terms of cell mass concentration (x) as well. For example, equation (2.1.1) may be rewritten as

$$\tau = (x_e - x_0)/\left(r_{xe}\right) \tag{2.2.30}$$

where $\left(r_{xe}\right) = (dx/dt)_e$

$= \text{rate of cell mass formation at } x = x_e \qquad (2.2.31)$

If the process follows Monod's kinetics, then

$$\left(\frac{dx}{dt}\right)_e = \mu x_e \qquad (2.2.32)$$

Equation (2.2.30) then becomes,

$$\tau = \frac{(x_e - x_0)}{\left(\mu x_e\right)} \qquad (2.2.33)$$

If the chemostat is operated with a sterile feed, then $x_0 = 0$. In such a case, the above equation reduces to

$$\tau = \left(1/\mu\right) \qquad (2.2.34)$$

The reciprocal of the space time (τ) is termed as the *dilution rate* (*D*). In other words,

$$D = \left(1/\tau\right) \qquad (2.2.35)$$

Equation (2.2.34) can be, therefore, rewritten as

$$D = \mu \qquad (2.2.36)$$

Agreed, the terminology *dilution rate* is a bit misleading (it is expressed in s^{-1}). Readers must exercise caution to avoid getting confused with this terminology.

2.2.2 Modified forms of Monod's kinetic equation

In the earlier section, we have considered bioprocesses that deal with a single substrate (S). However, investigations have demonstrated that Monod's kinetic model, discussed above, could be extended to biochemical processes involving multiple substrates as well. For example, let the process be represented as

$$S_1 + S_2 + \text{(Microbes)} \longrightarrow \text{Product(s)} + \text{More microbes} \qquad (2.2.37)$$

Then, equation (2.2.8) could be modified as

$$\left(\frac{dx}{dt}\right) = \left[\frac{\left(\mu_m C_{S1} x\right)}{\left(K_{S1} + C_{S1}\right)}\right]\left[\frac{C_{S2}}{\left(K_{S2} + C_{S2}\right)}\right] \qquad (2.2.38)$$

The above kinetic equation fits to a process in which a single microbial culture is being used, but the substrate is composed of more than one carbon source (S_1, S_2). In such a case, the microbes would have a tendency to attack the more assimilable substrate (say S_1) first and thereafter, they feed on S_1. The rate of consumption of S_1 and that of S_1 would be different and are to be represented by different kinetic constants (K_{S1} and K_{S2}).

Incidentally, even if a mixed substrate (containing multiple carbon sources) is being handled and a complex culture of microbes is employed, we could still use a kinetic equation of the form of equation (2.2.14), except that in such a case, C_S would represent the concentration of the total dissolved substrate (total dissolved organic matter) in the reaction mixture (that could have come from different carbon sources). Also, if the microbial action is such that the products from the reaction catalyzed by microbe -1 form the food for microbe –2 (in other words, a *commensalistic* relationship exists between microbes, see, for example, equations 1.6 to 1.10 of Chapter – 1), then, x would represent the total mass of microbial cells present in the reaction mixture per unit volume.

As stated earlier, in the case of aerobic bioconversion processes (that are catalyzed by aerobic microbes), a specific concentration of dissolved oxygen is a must in the substrate solution (in the reaction mixture). Usually, this concentration of dissolved oxygen does not affect the kinetics of the process, so long as it is at the prescribed value (this value is also quite small in most cases). However, if dissolved oxygen concentration (C_S) does influence the rate of the biochemical reaction materially, then its influence could also be incorporated into Monod's kinetic equation by modifying the expression for μ (equation – 2.2.7) as

$$\mu = \mu_m \left[\frac{C_S}{K_S + C_S} \right] \left[\frac{C_g}{K_g + C_g} \right] \tag{2.2.39}$$

where C_g = concentration of dissolved oxygen, g/L

K_g = oxygen utilisation coefficient, g/L

As discussed in Chapter-1, instead of using external air supply (by means of huge air compressors), LPO technology could be utilised to meet the oxygen requirement of aerobic processes / aerobic microbes. Here, a calculated amount of hydrogen peroxide (H_2O_2) is added to the reaction mixture (substrate solution) which releases nascent oxygen in solution and this nascent oxygen meets the oxygen requirement of microbes and that of substrate oxidation. For aerobic processes employing liquid phase oxygen (LPO) utilisation, the kinetic equation (equation 2.2.7) gets modified as

$$\mu = \mu_m \left[\frac{C_S}{K_S + C_S} \right] \left[\frac{C_{hp}}{K_{Lp} + C_{hp}} \right] \tag{2.2.40}$$

where C_{hp} = concentration of hydrogen peroxide in solution, g/L

K_{Lp} = LPO utilisation coefficient, g/L

We have stated earlier that one of the limitations of Monod's kinetic model is that it does not account for cell death or cell decay. We have discussed in Chapter-1 on endogenous decay of microbial cells, which occurs at excessively high values of cell mass concentration (x).

When the cell population is too high, they shall suffer from food shortage and many microbial cells undergo lysis or rupture and the nutrients leaking out from the ruptured cells are used as food by the remaining living cells. Once this cannibalistic attitude of microbial cells to ingest each other gets developed, the effective cell mass concentration in the reaction mixture (x) tends to decline. Endogenous decay of microbial cells could be incorporated into Monod's kinetic equation by defining an endogenous decay coefficient, k_d and by assuming that the rate of endogenous decay is first order in cell mass concentration (x). Experimental investigations have demonstrated that such an assumption is not too inaccurate. Thus, equation (2.2.6) gets modified to,

$$\left(\frac{dx}{dt}\right) = \mu x - k_d x$$

$$= (\mu - k_d)x \tag{2.2.41}$$

It is important to keep in mind that no substrate consumption occurs during endogenous decay; substrate consumption occurs only during cell growth and multiplication and during product formation. Accordingly, it becomes necessary to define a *true yield coefficient* (Y_e), which is different from the overall yield coefficient (Y) defined earlier. Thus,

$$Y_e = \text{(actual mass of cells produced)/(mass of substrate consumed)} \tag{2.2.42}$$

This differs from the overall yield coefficient, Y (also called observed yield coefficient), since Y is defined as

Y = (net mass of cells produced)/(mass of substrate consumed)

$$= \frac{(\text{Actual mass of cells produced}) - (\text{mass of cells destroyed due to endogenous decay})}{(\text{Mass of substrate consumed})} \tag{2.2.43}$$

Obviously, the magnitude of Y shall be lower than Y_e. Also, unlike Y whose magnitude can be experimentally determined, Y_e is a kinetic constant and its value can be evaluated only through kinetic analysis. To note further, that though Y_e can be treated as a constant, the value of Y does change with time in cases like this. Based on the definition of Y_e, therefore,

$$\left(\frac{-dC_S}{dt}\right) = \left(\frac{1}{Y_e}\right)(\mu x)$$

$$= (\mu_m / Y_e)C_S\, x/(K_S + C_S) \tag{2.2.44}$$

Equations (2.2.41) and (2.2.44) represent Monod's kinetic equation that incorporates endogenous decay of microbes as well. To note that these equations contain four kinetic constants such as μ_m, K_S, k_d and Y_e. We may use equation (2.2.11) as the relationship between the two independent parameters (x and C_S), but the overall yield coefficient (Y) cannot be treated as a constant and is a function of time. Due to these constraints, equations (2.2.41) and (2.2.44) cannot be integrated analytically, they can be solved only numerically (numerical solution of two simultaneous differential equations using any of the available numerical algorithms such as the Gill's method). Kinetic analysis based on batch reactor experimental data therefore becomes inapplicable here.

However, kinetic analysis could be conveniently performed based on experimental data collected on a chemostat. Thus, if we substitute equation (2.2.44) in the performance equation of the chemostat (namely, equation 2.1.1) and rearrange, we get

$$\frac{(\tau x_e)}{(C_{S0} - C_{Se})} = \left(\frac{K_S Y_e}{\mu_m}\right)\left(\frac{1}{C_{Se}}\right) + \left(\frac{Y_e}{\mu_m}\right) \tag{2.2.45}$$

Thus, if the modified Monod's kinetic equation does describe the kinetics of the bioprocess under consideration, then a plot of $\left[\frac{(\tau x_e)}{(C_{S0} - C_{Se})}\right]$ versus $\left(\frac{1}{C_{Se}}\right)$ must yield a straight line, with

$$\text{slope} = (K_S Y_e / \mu_m) \tag{2.2.46}$$

$$y\text{- intercept} = (Y_e / \mu_m) \tag{2.2.47}$$

From the slope and y- intercept of the straight line plot therefore, we could deduce the value of the saturation constant (K_S) and that of the ratio $\left(\frac{Y_e}{\mu_m}\right)$.

Now, if we write the performance equation of the chemostat in terms of cell mass concentrations (see equation 2.2.30), then

$$\tau = (x_e - x_0) / (\mu - k_d) x_e \tag{2.2.48}$$

Incidentally, equation (2.1.1) could also be written as

$$\tau = \frac{(C_{S0} - C_{Se})}{[(1/Y_e)(\mu x_e)]} \tag{2.2.49}$$

or

$$(\mu x_e) = \frac{(C_{S0} - C_{Se}) Y_e}{\tau} \tag{2.2.50}$$

If we substitute the above expression for (μx_e) in equation (2.2.48) and rearrange, we get

$$\frac{(x_e - x_0)}{(\tau x_e)} = Y_e \left[\frac{(C_{S0} - C_{Se})}{(\tau x_e)}\right] - k_d \tag{2.2.51}$$

This once again is equation to a straight line with slope = Y_e and y- intercept = $(-k_d)$. All the four kinetic constants (μ_m, K_S, k_d, Y_e) could be thus evaluated from these two straight line plots.

Incidentally, if the chemostat is operated with a sterile feed $(x_0 = 0)$, then the above equation gets simplified to

$$\left(\frac{1}{\tau}\right) = Y_e \left[\frac{(C_{S0} - C_{Se})}{(\tau x_e)}\right] - k_d \tag{2.2.52}$$

As stated in Chapter – 1, endogenous decay of microbes could be avoided by optimizing the cell mass concentration in the bioreactor. Also, this problem occurs predominantly in suspended growth systems, hardly occurs in attached growth bioreactors.

2.2.3 Haldane – Andrews kinetic equation (incorporating substrate inhibition)

We have discussed in Chapter-1 on the ill-effect of high substrate concentration on microbial activity. High concentrations of substrate tend to impede microbial growth and multiplication and often lead to microbial death or decay. This is what is termed as *substrate inhibition*. Monod's kinetic equation, discussed in the earlier sections, does not account for substrate inhibition to microbial activity.

Haldane – Andrews kinetic equation [2] provides an improvement over Monod's equation, in the sense that it accounts for substrate inhibition to microbial growth and multiplication as well, by incorporating an additional kinetic parameter called *substrate inhibition coefficient* (K_{Si}).

$(dx / dt) = \mu x$

where $\mu = (\mu_m C_S) / \left[K_S + C_S + (C_S^2 / K_{Si}) \right]$ (2.2.53)

A large magnitude of K_{Si} indicates substrate inhibition to microbial activity is insignificant and vice versa.

Based on a constant value of overall yield coefficient for cell mass production (Y), we could use equations (2.2.11) to (2.2.13) for obtaining the expression for rate of substrate consumption, $\left(\frac{-dC_S}{dt} \right)$ in terms of substrate concentration (C_S). Thus,

$$(-dC_S / dt) = (1 / Y)(dx / dt) \tag{2.2.54}$$

$$(x / Y) = (b - C_S) \tag{2.2.55}$$

$$b = C_{S0} + (x_0 / Y) \tag{2.2.56}$$

Accordingly, based on equation (2.2.53),

$$(-dC_S / dt) = (\mu_m C_S)(b - C_S) / (K_S + C_S + C_S^2 / K_{Si}) \tag{2.2.57}$$

This is the final form of *Haldane – Andrews kinetic equation*. It does consider substrate inhibition to microbial growth, but does not account for endogenous decay of microbial cells and product inhibition to microbial activity. It also assumes a constant value of overall yield coefficient (Y).

If kinetic analysis is performed using integral analysis based on batch reactor experimental data, then we need the integrated form of above equation. On integrating equation (2.2.57). using the method of partial fractions, we get

$$\left[1 + b / K_{Si}\right] \ln \left[\frac{b - C_S}{b - C_{S0}} \right] + (K_S / b) \ln \left[\frac{1 - b / C_S}{1 - b / C_{S0}} \right] - (1 / K_{Si})[C_{S0} - C_S] = \mu_m t \tag{2.2.58}$$

This may be re-arranged as

$$[C_{S0} - C_S] / t = [b + K_{Si}] \left(\frac{1}{t} \right) \ln \left[\frac{b - C_S}{b - C_{S0}} \right] + (K_S K_{Si} / b) \left(\frac{1}{t} \right) \ln \left[\frac{1 - b / C_S}{1 - b / C_{S0}} \right] - (\mu_m K_{Si}) \tag{2.2.59}$$

The above equation is of the form.

$$y = a_0 + a_1 x_1 + a_2 x_2 \tag{2.2.60}$$

where

$$a_0 = (-\mu_m K_{Si}) \tag{2.2.61}$$

$$a_1 = (b + K_{Si}) \tag{2.2.62}$$

$$a_2 = (K_S K_{Si} / b) \tag{2.2.63}$$

$$y = [C_{S0} - C_S] / t \tag{2.2.64}$$

$$x_1 = (1/t)\ln\left(\frac{b - C_S}{b - C_{S0}}\right) \tag{2.2.65}$$

$$x_2 = (1/t)\ln\left[\frac{1 - b/C_S}{1 - b/C_{S0}}\right] \tag{2.2.66}$$

The batch reactor experimental data could be fitted into the above equation (polynomial) and then solved for the three unknowns (a_0, a_1, a_2)) using matrix algebra as,

$$\begin{bmatrix} \sum y \\ \sum x_1 y \\ \sum x_2 y \end{bmatrix} = \begin{bmatrix} n & \sum x_1 & \sum x_2 \\ \sum x_1 & \sum x_1^2 & \sum x_1 x_2 \\ \sum x_2 & \sum x_1 x_2 & \sum x_2^2 \end{bmatrix} \begin{bmatrix} a_0 \\ a_1 \\ a_2 \end{bmatrix} \tag{2.2.67}$$

where n = total number of data points compiled.

Equation (2.2.67) can be solved either by matrix inversion (Kramer's rule) or by successive elimination to obtain the values of a_0, a_1, a_2 and from them, the kinetic constants (μ_m, K_S, K_{Si}) could be evaluated.

If the value of any of the kinetic constants so computed appears anomalous (either too large, too low or less than zero), then it means that the bioprocess under consideration does not follow Haldane – Andrews kinetics. Analysis with an alternate kinetic equation is to be attempted.

Let us now consider differential analysis. In this case, as discussed earlier, a plot of C_S versus time (t) is to be prepared first and the values of reaction rate, $(-r_S)$ or $(-dC_S/dt)$, at different values of C_S (at different t) are to be determined from the slope of the tangent to the curve (at the specific C_S or t) and listed. Then, equation (2.2.57) could be re-arranged as,

$$(b - C_S)/(-r_S) = (1/\mu_m) + (1/\mu_m K_{Si})C_S + (K_S/\mu_m)(1/C_S) \tag{2.2.68}$$

This is also of the form,

$$y = a_0 + a_1 x_1 + a_2 x_2 \tag{2.2.69}$$

where $y = \dfrac{(b - C_S)}{(-r_S)}$ (2.2.70)

$$x_1 = C_S, \quad x_2 = (1/C_S) \tag{2.2.71}$$

$$a_0 = (1/\mu_m) \tag{2.2.72}$$

$$a_1 = 1/(\mu_m K_{Si}), \quad a_2 = (K_S/\mu_m) \tag{2.2.73}$$

Once again, the experimental data could be fitted into the above polynomial (equation – 2.2.69) and the kinetic constants (μ_m, K_S, K_{Si}) evaluated using matrix algebra, as discussed earlier.

It can be seen that computation of the independent variables (x_1, x_2) is much simpler in this case (as compared to integral analysis). However, the first step of computing the values of ($-r_S$) from C_S versus t plot shall be additional.

With due caution, we may utilize the *approximate differential analysis* also here. For example, assuming $\left(\frac{dx}{dt}\right) \approx (\Delta x / \Delta t)$,

$$\mu = \left(\frac{1}{x}\right)\left(\frac{\Delta x}{\Delta t}\right) \tag{2.2.74}$$

Thus, the values of μ at different values of x (at different t) can be computed from the above equation, using the available experimental data. Equation (2.2.53) may be then rearranged as,

$$(1/\mu) = (1/\mu_m) + \left[1/(\mu_m K_{Si})\right] C_S + (K_S/\mu_m)(1/C_S) \tag{2.2.75}$$

which, once again, is of the form,

$$y = a_0 + a_1 x_1 + a_2 x_2 \tag{2.2.76}$$

Though computationally simpler, the approximation involved in this analysis must be kept in mind, before utilising this method for kinetic analysis.

Kinetic analysis using chemostat experimental data is equally possible. The procedure shall be then computationally simpler. Thus, if we substitute equation (2.2.57) into the performance equation of chemostat (equation – 2.1.1) and rearrange, we get

$$\left[\tau(b - C_{Se})/(C_{S0} - C_{Se})\right] = (1/\mu_m) + \left[1/(\mu_m K_{Si})\right] C_{Se} + (K_S/\mu_m)(1/C_{Se}) \tag{2.2.77}$$

This also is of the form,

$$y = a_0 + a_1 x_1 + a_2 x_2 \tag{2.2.78}$$

When the number of kinetic constants is more than two (here, there are three kinetic constants) and we have to go for a polynomial fit, then analysis based on chemostat experimental data can be seen to be more handy and convenient.

If the chemostat is operated with sterile feed ($x_0 = 0$), then the above equation shall get further simplified to

$$\tau = (1/\mu_m) + (K_S/\mu_m)(1/C_{Se}) + \left[1/(\mu_m K_{Si})\right] C_{Se} \tag{2.2.79}$$

In all the above cases, it is essential to ensure that the computed values of polynomial coefficients (a_0, a_1, a_2) predict realistic values of kinetic constants. If any of the computed values of kinetic constants is observed to be negative or abnormally large, then it can be concluded that the assumed kinetic equation does not predicts the kinetics of the bioprocess under consideration. Trials with alternate kinetic equations would have to be then attempted.

2.2.4 Kinetic Equations incorporating Product Inhibition

We have discussed on product inhibition to microbial growth and multiplication also in Chapter - 1. One or more of products formed could act as a toxin or poison to the microbial cells. Alternately, the product formed could disturb one of the influencing operating parameter such as the pH of the medium and this, in turn, would tend to disturb the microbial activity.

Many of the investigations in this connection have demonstrated that product inhibition to microbial growth could be accounted for by incorporating an additional kinetic constant called *product inhibition coefficient* into the Monod-type kinetic equation. For example, Aiba and coworkers [3] have recommended the following modification in the expression for μ :

$$\mu=\mu_m\left[C_S/(K_S+C_S)\right]\left[C_P/(K_P+C_P)\right] \tag{2.2.80}$$

where C_P = concentration of product (P) in the reaction mixture, g/L

K_P = product inhibition coefficient, g/L

They also have assumed a constant value of overall yield coefficient (Y) and have developed the expression for $(-dC_S/dt)$ based on equations (2.2.11) to (2.2.13). Thus,

$$(-dC_S/dt)=\mu_m\left[C_S(b-C_S)/(K_S+C_S)\right]\left[K_P/(K_P+C_P)\right] \tag{2.2.81}$$

By defining a yield coefficient for product formation (Y_P) as given in equation (2.2.5) and assuming that it remains more or less constant during the process, we could express the product concentration (C_P) in terms of substrate concentration (C_S). Thus, based on equation (2.2.5),

$$C_P=Y_P\left(C_{S0}-C_S\right) \tag{2.2.82}$$

Substituting the above expression for C_P in equation (2.2.81) and re-arranging, we get

$$(-dC_S/dt)=(\mu_m K_P/Y_P)\left[C_S(b-C_S)/\{(K_S+C_S)(p-C_S)\}\right] \tag{2.2.83}$$

where $p=C_{S0}+(K_P/Y_P)$ (2.2.84)

This kinetic equation also contains three kinetic constants $\left(\mu_m, K_S, K_P\right)$.

If kinetic experiments have been conducted using a chemostat, then for the evaluation of the kinetic constants (for testing whether the bioprocess satisfies Aiba et.al's kinetic equation given above), we could substitute above equation (2.2.83) in the performance equation of chemostat (namely, equation 2.1.1) and rearrange it, to get

$$\left[\tau(b-C_{Se})/(C_{S0}-C_{Se})\right]=\left[Y_P(p-K_S)/(K_P\mu_m)\right]+\left[pK_SY_P/(K_P\mu_m)\right](1/C_{Se}) -\left[Y_P/(K_P\mu_m)\right]C_{Se} \tag{2.2.85}$$

This is of the form,

$$y=a_0+a_1x_1+a_2x_2 \tag{2.2.86}$$

where $x_1=(1/C_{Se})$, $x_2=C_{Se}$. (2.2.87)

The experimental data could be now fitted into equation (2.2.85) and it could be solved for the polynomial coefficients (a_0, a_1, a_2) using matrix algebra (as discussed in the earlier section) and the kinetic constants computed there-from.

For the differential analysis using batch reactor experimental data (assuming the values of reaction rate $(-r_S)$, at different values of C_S have already been estimated from the C_S versus t plot), we could once again rearrange equation (2.2.83) into a similar form as,

$$\left[\frac{(b-C_S)}{(-r_S)}\right]=\left[\frac{Y_P(p-K_S)}{(K_P\mu_m)}\right]+\left[\frac{pK_SY_P}{(K_P\mu_m)}\right]\left(\frac{1}{C_S}\right)-\left[\frac{Y_P}{(K_P\mu_m)}\right]C_S \tag{2.2.88}$$

This once again forms a polynomial fit of the form,

$$y=a_0+a_1x_1+a_2x_2 \tag{2.2.89}$$

For an *approximate* differential analysis, μ - values are to be first computed from equation (2.2.74) and then, equation (2.2.80) could be re-arranged after substituting for C_P from equation (2.2.82) as

$$(1/\mu)=\left[Y_P(p-K_S)/(K_P\mu_m)\right]+\left[pK_SY_P/(K_P\mu_m)\right](1/C_S)-\left[Y_P/(K_P\mu_m)\right]C_S \tag{2.2.90}$$

This is also of the form,

$$y=a_0+a_1x_1+a_2x_2 \tag{2.2.91}$$

Integration of equation (2.2.83) and development of the integrated form of this kinetic equation, we leave to the readers to perform as an interesting exercise (see Ex.2.E at the end of this Chapter).

Product inhibition to microbial growth (and thereby to the rate of substrate utilisation) could often be an exponential function in product concentration (C_P). No doubt, in such situations, the kinetic equation shall become more nonlinear (more complex). For example, for the microbial fermentation of molasses (to convert its sucrose content to lactic acid) using a culture of *Enterococcus faecalis*, Anjana and Kumar [4] have reported a kinetic equation of the form

$$(-dC_S/dt)=[(\mu_m/Y)C_Sx/(K_S+C_S)]\exp(-C_P/K_P) \tag{2.2.92}$$

Typical values of kinetic parameters reported by them are, (μ_m/Y) = 3.33 g/(g.hr), K_S = 0.1 g/L and K_P = 29.17 g/L. They have fitted the experimental data into the above equation and evaluated the kinetic constants using genetic algorithm (GA).

2.2.5 Alternate Kinetic Models (Tessier's, Contois' and Ollis' kinetic models)

A number of alternate kinetic equations have been proposed by different authors to describe microbial fermentation processes, but each has distinct limitations of its own. For example, Tessier [5] has proposed a kinetic equation in which the specific growth rate (μ) is reported to be an exponential function in substrate concentration (C_S). Thus

$$\mu=\mu_m\left[1-exp(-C_S/K_S)\right] \tag{2.2.93}$$

Based on a constant value of overall yield coefficient for cell mass production (Y), the expression for the rate of substrate utilisation shall be

$$(-dC_S/dt)=\mu_m\left[1-exp(-C_S/K_S)\right](b-C_S) \tag{2.2.94}$$

Except for the fact that μ has been expressed as an exponential function in C_S, the above equation differs little from Monod's kinetic equation (equation – 2.2.14). All the inherent limitations of Monod's equation remain behind such as substrate inhibition and product inhibition to microbial growth are not accounted for, neither is endogenous decay of microbes.

Due to the presence of an exponential function, Tessier's kinetic equation is nonlinear and evaluation of kinetic constants is more laborious, mostly demanding a trial and error procedure. For example, integration of equation (2.2.94) yields

$$(\mu_m t)= \int_{C_S}^{C_{S0}} \frac{dC_S}{\left[1-exp\left(-C_S / K_S\right)\right]\left(b-C_S\right)} \tag{2.2.95}$$

The above integral can be evaluated only numerically using any of the popular algorithms such as Simpson's rule or trapezoidal rule. Evidently, the kinetic constants (μ_m , K_S) can be evaluated only by trial.

Contois [6] has proposed a kinetic equation which is also basically analogous to Monod's kinetic model, except that it proposes μ to be a function of not only the substrate concentratin (C_S), but also the cell mass concentration (x). Thus,

$$\mu= \left(\mu_m C_S\right) / \left(K_C x+C_S\right) \tag{2.2.96}$$

where K_C = Contois' kinetic constant, g substrate / g cell mass.

It can be seen that as the cell mass concentration (x) increases, the specific growth rate (μ) tends to decrease. This could appear anomalous, particularly in suspended growth systems and this forms the major limitation of this kinetic model. Zabot et.al [7] have reported that aerobic fermentation of lactose (present in cheese whey) using *Xanthomonas campestris* culture to produce Xanthan gum, follows Contois kinetics with μ_m = 2.9 hr^{-1} and K_C = 2.0 g/g.

The kinetic model proposed by Ollis [8] does present some improvements over Monod's conventional approach in the sense that cell growth and multiplication, substrate utilisation and product formation are dealt as three different steps and separate equation or correlation has been proposed for each step. Substrate consumption for cell maintenance has also been separately considered and cell death or cell decay due to presence of toxins has also been accounted for. However, the analysis suffers from a serious limitation in that cell growth and multiplication has been assumed to follow logistic equation, which considers the specific growth rate (μ) as constant, independent of substrate concentration (C_S). Treating μ to be a constant is a serious approximation and this significantly lowers the applicability of Ollis' kinetic model to real – life systems. The model is outlined below :

(a) Cell growth and multiplication :

As stated above, cell growth and multiplication is assumed to follow the *logistic equation*. This equation assumes μ to be a constant, but considers cell death or cell decay due to the presence of toxins as well. It is being assumed that the rate of such inhibition to cell growth is second order in cell mass concentration (x). Thus,

$$\left(dx / dt\right)=\mu x-kx^2 \tag{2.2.97}$$

$$=\mu x\left[1-\alpha_0 x\right] \tag{2.2.98}$$

where $\alpha_0 =\left(k / \mu\right)$ (2.2.99)

k = second order decay coefficient, L/(g.s) (2.2.100)

On integration using the method of partial fractions,

$$(\mu t)=\ln\left[x(1-\alpha_0 x_0)/\{x_0(1-\alpha_0 x)\}\right] \tag{2.2.101}$$

or,

$$x=x_0\exp(\mu t)/\{1-(\alpha_0 x_0)[1-\exp(\mu t)]\} \tag{2.2.102}$$

This is called the *logistic equation*. It predicts that a plot of x versus t [or, $(\alpha_0 x)$ versus t] shall be sigmoidal in nature (see figure 2.3). It demonstrates an exponential growth phase and also a stationary phase at which $(\alpha_0 x)$ = $(\alpha_0 x_S)$ = 1.0. It does not show any death phase.

Since, $\alpha_0=\left(\dfrac{1}{x_S}\right)$, where x_S is the cell mass concentration in the stationary phase, equation (2.2.102) may be rewritten as

$$x=\frac{x_0\exp(\mu t)}{\left\{1-\left(\dfrac{x_0}{x_S}\right)[1-\exp(\mu t)]\right\}} \tag{2.2.103}$$

As stated earlier, the gross approximation associated with logistic equation is that it assumes the specific growth rate (μ) to be a constant (independent of substrate concentration, C_S).

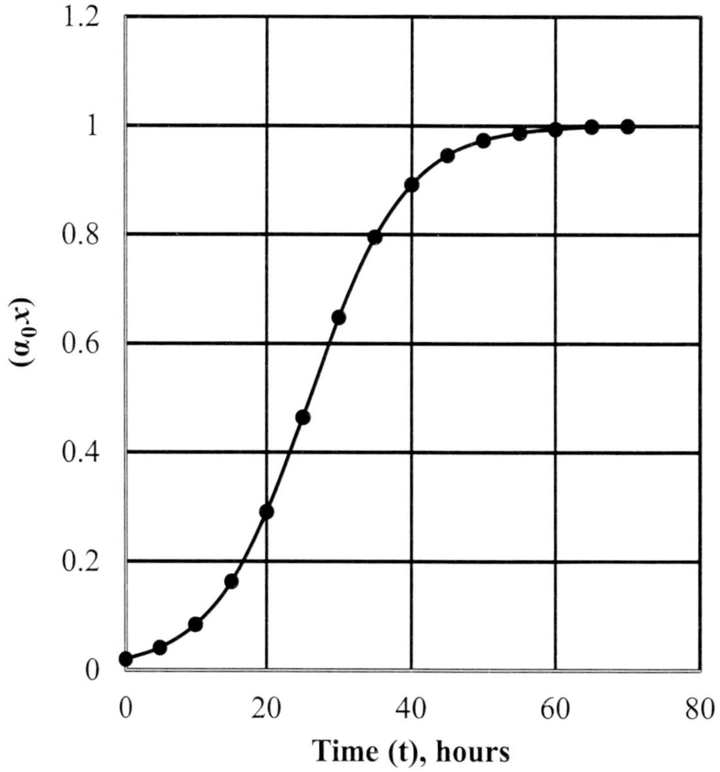

Figure 2.3a: Logistic curve (sigmoidal)

(b) Substrate Consumption / Utilisation:

$$(-dC_S/dt)=(1/Y_S)(dx/dt)+(1/Y_{SP})(dC_P/dt)+(k_e x) \tag{2.2.104}$$

where Y_S = *selective* yield coefficient for cell mass production

$$= \frac{\text{(net mass of cells formed on dry basis)}}{\text{(mass of substrate consumed for cell mass production)}} \tag{2.2.105}$$

Y_{SP} = selective yield coefficient for product formation

$$= \frac{\left(\text{mass of product formed}\right)}{\left(\text{mass of substrate consumed for product formation}\right)} \tag{2.2.106}$$

k_e = cell maintenance coefficient, s^{-1}

It has been assumed that the values of selective yield coefficients (Y_S or Y_{SP}) remain more or less unaltered during the entire process (this once again forms an approximation, discussed below). To note that the first term on the right hand side of the above equation depicts the rate of substrate consumption for cell mass production, the second term represents the rate of substrate consumption for product formation and the last term accounts for substrate consumption for cell maintenance.

The difference between Y_S and Y (and similarly, that between Y_{SP} and Y_P) is not difficult to comprehend. Y (the overall yield coefficient for cell mass production) is the ratio of net mass of dry cells formed to the total mass of substrate consumed, while Y_S represents the ratio of net mass of dry cells formed to the mass of substrate consumed for cell mass production only. Evidently, the selective yield coefficient (Y_S) shall be larger in magnitude than the overall yield coefficient (Y). Also, when cell formation (cell production) and cell death or cell decay occur simultaneously as is in the present case considered, the net mass of cells in the reaction mixture shall change with time and it would be erroneous to assume a constant value of Y_S or Y. We have already discussed on this in Section (2.2.2). When both growth and multiplication of cells and cell death or cell decay due to presence of toxins or any other reason occur simultaneously, neither the overall yield coefficient (Y) nor the selective yield coefficient (Y_S) could be treated as a constant, though the *true* yield coefficient (Y_S) could be.

(c) Product formation:

The product formation is assumed to follow Leudeking – Piret – type kinetics [9]. Thus,

$$\left(\frac{dC_P}{dt}\right) = \alpha_1\left(\frac{dx}{dt}\right) + \alpha_2 x \tag{2.2.107}$$

If the above equation for product formation is substituted in equation (2.2.104) and rearrange, we get

$$\begin{aligned} -(dC_S/dt) &= [(1/Y_S) + (\alpha_1/Y_{SP})](dx/dt) + [(\alpha_2/Y_{SP}) + k_e]x \\ &= \alpha_3(dx/dt) + \alpha_4 x \end{aligned} \tag{2.2.108}$$

Thus, the equation for product formation (equation 2.2.107) and that for substrate utilisation (equation 2.2.108) are similar in nature, each consisting of two kinetic constants.

If we substitute the expression for $\left(\frac{dx}{dt}\right)$ from equation (2.2.98) and that for x from equation (2.2.103) in the above two equations and integrate them, we obtain,

$$\left(C_{S0}-C_S\right)-\left(\frac{\alpha_4 x_S}{\mu}\right)\ln\left[1-\left(\frac{x_0}{x_S}\right)\left\{1-\exp\left(\mu_t\right)\right\}\right]=\alpha_3\left(x-x_0\right) \tag{2.2.109}$$

$$\left(C_{P0}-C_P\right)-\left(\frac{\alpha_2 x_S}{\mu}\right)\ln\left[1-\left(\frac{x_0}{x_S}\right)\left\{1-\exp\left(\mu_t\right)\right\}\right]=\alpha_1\left(x-x_0\right) \tag{2.2.110}$$

In most processes, $C_{P0} = 0$ (unless a partial product recycle is employed).

Ollis' kinetic model is thus composed of six kinetic constants $\left(\alpha_1, \alpha_2, \alpha_3, \alpha_4, \mu_t, x_S\right)$. Its most serious limitation is that μ has been assumed to be a constant. It is also based on constant values of selective yield coefficients (Y_S, Y_{SP}).

Based on an elaborate batch reactor experimental data that lists values of C_S, x and C_P at different intervals of time (t), all the six kinetic constants of of Ollis' kinetic model could be evaluated. The procedure is summarized below:

(i) First, plot x versus t. If the process does follow Ollis' kinetic model, then this plot shall be a logistic curve as shown in figure (2.3). It can be seen that x increases exponentially with time (t) and after a particular period of time (at $t \geq t_S$), it attains a constant value (x_S). The value of x_S can be thus read directly from this plot. The value of t_S at which the stationary phase starts, could also be noted down.

(ii) As per equation (2.2.101), a plot of $\ln\left[\dfrac{x\left(\dfrac{1-x_0}{x_S}\right)}{\left\{x_0\left(\dfrac{1-x}{x_S}\right)\right\}}\right]$ versus time (t) must yield a staright line (passing through the origin) with slope = μ. The value of μ can be thus evaluated from this plot.

(iii) From equation (2.2.107), it can be seen that $t \geq t_S$, when $x = x_s$ and $\left(\dfrac{dx}{dt}\right) = 0$,

$$\left(dC_P / dt\right) = \left(\alpha_2 x_S\right) = \text{constant}$$

or

$$C_P = \left(\alpha_2 x_s\right)t + C_1 \tag{2.2.111}$$

where C_1 = constant of integration. Thus, a plot of C_P versus time (t) at all values of t greater than t_S must yield a straight line of slope = $\left(\alpha_2 x_S\right)$. Since x_S has already been determined in step (i), the value of α_2 could be deduced. To note that only C_P values at $t \geq t_S$ must be used for preparing this plot.

(iv) Equation (2.2.110) may now be expressed as,

$$F\left(C_P, t\right) = \alpha_1\left(x - x_0\right) \tag{2.2.112}$$

To note that the expression on the LHS of equation (2.2.110) contains only two variables such as C_P and t. All others are constants [whose values are already known or have already been determined] such as x_0, x_S, μ, α_2. The above equation (2.2.112) once again tells us that a plot $F(C_p, t)$ versus $\left(x - x_0\right)$ should result in a straight line plot (passing through the origin) with slope = α_1.

(v) The kinetic constants (α_3, α_4) of equation (2.2.107) can be similarly evaluated. Thus at $t \geq t_S$, when $x = x_S$, $(dx/dt) = 0$,

$$(-dC_S / dt) = \alpha_4 x_S \tag{2.2.113}$$

or

$$C_S = -(\alpha_4 x_S)t + C_2 \tag{2.2.114}$$

where C_2 = constant of integration. Thus, a plot of C_S versus time (t) at $t \geq t_S$, shall yield a straight line plot with negative slope of magnitude, $(\alpha_4 x_S)$. Here also, only values of C_S at $t \geq t_S$ must be used for preparing the plot.

(vi) Now, equation (2.2.109) is of the form,

$$F(C_S, t) = \alpha_3 (x - x_0) \tag{2.2.115}$$

Thus, a plot of $F(C_S, t)$ versus $(x - x_0)$ should also yield a straight line (passing through the origin) with slope = α_3.

All the six kinetic constants of Ollis' model can be thus evaluated from the above five straight line plots and the logistic curve.

In spite of the approximations involved, Ollis' kinetic model has been found to be applicable to selected bioconversion processes (See Exercise No. 2.G at the end of this Chapter).

2.2.6 Kinetics of Microbial Processes involving Complex Cultures (Microbial Interactions):

Complex cultures (mixed cultures) of microbes are used in a large number of industrial processes. If the different families of microbes present in the complex culture do not interfere with each other's growth and multiplication or interact constructively with each other, the processes can be handled with relative case. In fact, many microbes multiply faster in mixed cultures (when cultured in presence of other microbes) rather than in single cultures (when cultured alone). This is due to the beneficial inter-relationship existing among the microbes.

However, if destructive interactions (amensalistic relationships) exist among microbes, either mixed cultures are to be avoided or special precautions are to be taken to minimize such interactions.

Microbial interactions may be broadly classified into the following categories:

(a) Constructive interactions
- (i) Neutralism
- (ii) Mutualism
- (iii) Symbiosis
- (iv) Commensalism

(b) Destructive interactions
- (i) Amensalism
- (ii) Competition
- (iii) Parasitism
- (iv) Predator-Prey

Neutralism is exhibited by those microbes who grow and multiply independently even in a mixed culture and do not interfere with the growth and multiplication of other microbes present. The rate of growth and multiplication of these microbes shall be the same when cultured alone or when cultured in presence of other microbes. Unfortunately, very few of the commercial microbes exhibit this characteristic.

Mutualism refers to microbes being beneficial to each other and consequently, these microbes multiply faster in a mixed culture rather than when cultured alone. This happens mostly when microbe-1 releases the nutrients for the growth and multiplication of microbe-2, and the latter, in turn releases the nutrients required for microbe-1. The scheme may represented as

$$S_1 + \text{(microbes -1)} \longrightarrow S_2 + \text{(more microbes -1)} \tag{2.2.116}$$

$$S_2 + \text{(microbes -2)} \longrightarrow S_1 + \text{(more microbes -2)} \tag{2.2.117}$$

For example, aerobic bacteria consume carbohydrates (S_1) and oxygen for their metabolic activities and release $CO_2(S_2)$. This carbon dioxide (S_2) is absorbed by green algae during photosynthesis, which converts it to carbohydrates (S_1) and release oxygen.

Symbiosis can be classified under mutualism itself. Here also, the microbes remain beneficial to each other, by feeding on a substance which otherwise could be a toxin to other microbes. For example, acetogens (acetogenic microbes) convert ethanol to acetates and hydrogen gas. This favors the growth and multiplication of methanogens (since ethanol acts as toxin to their growth). However, hydrogen released tends to inhibit the activity of acetogens (product inhibition) and methanogens absorb this hydrogen (along with CO_2) and convert them to methane. This, in turn, facilitates the growth and multiplication of acetogens:

$$C_2H_5OH + H_2O \xrightarrow[\text{acetogens}]{} CH_3COO^- + (5/2)H_2 \tag{2.2.118}$$

$$4H_2 + CO_2 \xrightarrow[\text{methanogens}]{} CH_4 + 2H_2O \tag{2.2.119}$$

Commensalism is a type of constructive interaction that is exhibited by a large number of microbes. In this case, microbe-2 functions beneficial to microbe-1, but the reverse does not occur. Microbe-2 either consumes (feeds on) a substance that otherwise is toxin to microbe-1 or releases nutrients for the growth and multiplication of microbe-1. For example, *Pseudomonas* feeds on methane and catalyzes its conversion to methanol, but its activity is inhibited by methanol it produces (product inhibition). When cultured together with *hypomicrobia*, the latter consumes methanol by converting it to carbon dioxide and water, and thereby foster the growth and multiplication of *Pseudomonas* cells.

Similarly, when yeast cells (*Saaccharomyces cerevisiae*) and *Acetobactor* are cultured together in a glucose solution (or in a starchy substrate), the alcohol produced by the former by the fermentation of glucose/starch is used as substrate by *Acetobacter* for the production of acetic acid.

In all the above cases, the microbial activity is better and superior in a mixed culture rather than when cultivated alone (in a single culture).

Destructive interactions among microbes are not unusual and we have to take suitable precautions to avoid them or to minimize them.

Amensalism occurs when one of the products released by one set of microbes acts as toxin to the other set of microbes and thereby inhibits their growth. Obviously, they should not be cultured together.

Competition, as the term itself implies, refers to different microbes competing for the same carbon source and consequently, the growth and multiplication of one getting hampered in preference to the other. For example, it has been reported that the growth of *Staphylococcus aureus* gets reduced by more than 25% when cultured together with *E.coli.*

Parasitism is exhibited by those microbes (such as *bacteriophages*) which enter other bacterial cells (host cells) and disrupt their growth by consuming most of the nutrients.

In the *predator – prey* relationship, one set of microbes acts as predator to the other set of microbes, as they consume or ingests the cells of the prey. Often, a predator is purposefully added to the culture broth to control or regulate the population of the prey. This then becomes a beneficial relationship.

Kinetic equations discussed in the earlier sections can be easily extended to the systems handling mixed/complex culture of microbes, provided the microbial interactions prevalent is accurately known. For example, consider two sets of microbes that exhibit mutualism. The scheme is that sketched in equations (2.2.116) and (2.2.117). If we assume Haldane- Andrews kinetic equation (equation – 2.2.53) is valid for both microbes, then

$$\left(\frac{dx_1}{dt}\right) = \mu_1 x_1 \qquad (2.2.120)$$

$$\left(\frac{dx_2}{dt}\right) = \mu_2 x_2 \qquad (2.2.121$$

where $\mu_1 = \left(\mu_{m1} C_{S1}\right) / \left[K_{S1} + C_{S1} + \left(C_{S1}{}^2 / K_{Si,1}\right)\right]$ (2.2.122)

$$\mu_2 = \left(\mu_{m2} C_{S2}\right) / \left[K_{S2} + C_{S2} + \left(C_{S2}{}^2 / K_{Si,2}\right)\right] \qquad (2.2.123)$$

To estimate the rate of consumption of S_1 and S_2, we need to define four yield coefficients such as

$Y_{S1} = Y_{S1} / x_1 = (\text{mass of microbe} - 1 \text{ produced}) / (\text{mass of } S_1 \text{ consumed})$

= yield coefficient for cell mass production (microbe – 1) (2.2.124)

$Y_{P1} = Y_{S2} / x_1 = (\text{mass of } S_2 \text{ (product of reaction} - 1) \text{ produced}) / (\text{mass of microbe} - 1 \text{ produced})$

= yield coefficient for product formation (reaction – 1) (2.2.125)

$Y_{S2} = Y_{S2} / x_2 = (\text{mass of microbe} - 2 \text{ produced}) / (\text{mass of S2 consumed})$ (2.2.126)

$Y_{P2} = Y_{S1} / x_2 = (\text{mass of } S_1 \text{ (product of reaction-2) produced}) / (\text{mass of microbe-2 produced})$ (2.2.127)

Accordingly,

$$\left(\frac{-dC_{S1}}{dt}\right)=\left(\frac{1}{Y_{S1}}\right)\left(\frac{dx_1}{dt}\right)-\left(Y_{P2}\right)\left(\frac{dx_2}{dt}\right)$$

$$=\left(\frac{1}{Y_{S1}}\right)\left(\mu_1 x_1\right)-\left(Y_{P2}\right)\left(\mu_2 x_2\right) \quad (2.2.128)$$

$$\left(\frac{-dC_{S2}}{dt}\right)=\left(\frac{1}{Y_{S2}}\right)\left(\frac{dx_2}{dt}\right)-\left(Y_{P1}\right)\left(\frac{dx_1}{dt}\right)$$

$$=\left(\frac{1}{Y_{S2}}\right)\left(\mu_2 x_2\right)-\left(Y_{P1}\right)\left(\mu_1 x_1\right) \quad (2.2.129)$$

Let us now consider a mixed culture of microbes that exhibit predator – prey interaction. For example, microbe-1 is the predator that feeds on the prey (microbe-2). The prey (microbe-2) feeds on the substrate. Let us assume that only the prey consumes the substrate, the predator does not. The predator only feeds on the prey. The scheme shall be as given below:

$$S + \text{(microbe -2)} \longrightarrow P + \text{(more microbes -2)} \quad (2.2.130)$$

$$\text{(Microbes -2)} + \text{(microbes -1)} \longrightarrow \text{(more microbes -1)} \quad (2.2.131)$$

If Monod's kinetic equation (equation – 2.2.6) holds true for both microbes, then

$$\left(\frac{dx_1}{dt}\right)=\mu_1 x_1 \quad (2.2.132)$$

$$\left(\frac{dx_2}{dt}\right)=\mu_2 x_2-\left(\frac{1}{Y_1}\right)\left(\mu_1 x_1\right) \quad (2.2.133)$$

$$\left(\frac{-dC_S}{dt}\right)=\left(\frac{1}{Y_2}\right)\left(\mu_2 x_2\right) \quad (2.2.134)$$

where $\mu_1=\dfrac{\left(\mu_{m1}\, x_2\right)}{\left(K_{S1}+x_2\right)}$ (2.2.135)

$$\mu_2=\frac{\left(\mu_{m2} C_S\right)}{\left(K_{S2}+C_S\right)} \quad (2.2.136)$$

Y_1 = yield coefficient for microbe – 1 (predator)

= (mass of microbe-1 produced)/(mass of prey (microbe-2) consumed) (2.2.137)

Y_2 = yield coefficient for microbe – 2 (prey)

= (mass of prey (microbe- 2) produced)/(mass of substrate consumed) (2.2.138)

The yield coefficients have, nevertheless, been assumed to remain more or less constant throughout the process.

Example 2.1: Anaerobic digestion of a sample of textile mill wastewater using a complex culture of microbes was studied in a laboratory batch reactor and the results obtained are tabulated below. The original feed sample had a BOD of 25 mg/L and it had been inoculated to obtain a cell mass concentration of 0.01 mg/L.

Time, hr	Substrate concentration, mg/L
2.4	22.61
2.6	21.3228
2.8	19.4613
2.9	18.2667
3.0	16.8683
3.1	15.2433
3.2	13.36797
3.3	11.2175
3.4	8.7657

Determine whether the kinetics of the process follow Monod's kinetic equation (and if so, evaluate the kinetic constants) based on

(a) Integral analysis
(b) Differential analysis
(c) Approximate differential analysis

Take overall yield coefficient for cell mass production = 0.57

Solution:

(a) Integral analysis:

The integral form of Monod's kinetic equation (after rearrangement) is that given in equation (2.2.18) :

$$Y = -\left(K_S / b\right)\left(X\right) + \mu_m \tag{i}$$

where $Y = y\text{- parameter} = \left(\frac{1}{t}\right) \ln\left[\frac{\left(b - C_S\right)}{\left(b - C_{S0}\right)}\right]$

$X = x\text{- parameter} = \left(1/t\right)\ln\left[\left(1 - b/C_S\right)/\left(1 - b/C_{S0}\right)\right]$

In the present case,

$b = C_{S0} + \left(x_0 / Y\right) = 25.0 + \left(0.01/0.57\right) = 25.01754$ mg/L

The values of x- parameter and y- parameter are computed from the data provided and are listed in Table (2.1.1).

Table 2.1.1

Time (hr)	x-parameter (X), hr^{-1}	y-parameter (Y), hr^{-1}
2.6	2.11895	2.05776
2.8	2.1459	2.0565
2.9	2.1609	2.05273
3.0	2.1782	2.047
3.1	2.19927	2.03968
3.2	2.2264	2.03079
3.3	2.2634	2.0206
3.4	2.3175	2.0092

The plot of y- parameter (Y) versus x- parameter (X) is shown in figure (2.1.1). It is more or less linear, indicating that the data do fit into Monod's kinetic equation. From the plot,

$$\text{slope} = \left(\frac{K_S}{b}\right) = 0.2688$$

or $$K_S = (0.2688)\,(25.01754) = 6.7247 \text{ mg/L}$$

$$y\text{- intercept} = \mu_m = 2.631 \text{ hr}^{-1}$$

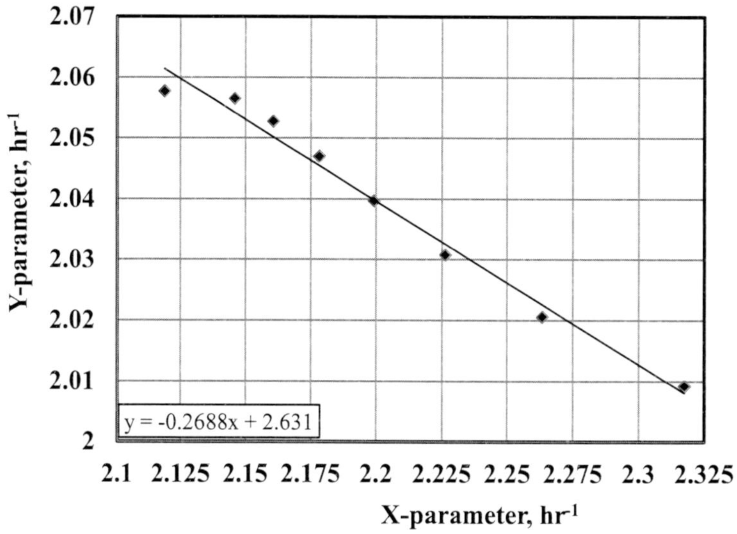

Figure 2.1.1: Monods kinetics (integral analysis)

(b) Differential analysis:

For performing the differential analysis, we have to first estimate the values of reaction rate, $-(r_S)$ at different intervals of time (at different values of substrate concentration, C_S). This can be accomplished, as discussed in Section (2.2), by plotting C_S versus time (t) and then finding the slope of the tangent to the curve at specified values of t. Alternately, we may fit the C_S versus t data into a correlation using any of the curve- fitting algorithms such as regression analysis. In the present case, it is seen that the data fit into a fourth order polynomial of following form (with $R^2 = 0.9999$) :

$$C_S = -0.4866t^4 + 1.9799t^3 - 2.8652t^2 + 1.2078t + 24.989 \tag{ii}$$

Therefore, $$-(r_S) = (-dC_S/dt) = 1.9464t^3 - 5.9397t^2 + 5.7364t - 1.2078 \tag{iii}$$

We have to now check whether the reported data fit into equation (2.2.23):

$$\frac{(b - C_S)}{(-r_S)} = \left(\frac{K_S}{\mu_m}\right)\left(\frac{1}{C_S}\right) + \left(\frac{1}{\mu_m}\right) \tag{iv}$$

The computed values of $(-r_S)$, the y- parameter, $\dfrac{(b-C_S)}{(-r_S)}$, and the x- parameter, $\left(\dfrac{1}{C_S}\right)$, are listed in table (2.1.2).

Table 2.1.2

Time (hr)	C_S, mg/L	$(-r_s)$, mg/(L.hr)	$(b-C_S)/(-r_S)$ (hr)	$(1/C_S)$, L/mg
2.6	21.3228	7.74879	0.4768	0.046898
2.8	19.4613	10.9974	0.5052	0.051384
2.9	18.2667	12.9282	0.52217	0.05474
3.0	16.8683	15.0789	0.5404	0.05928
3.1	15.2433	17.4611	0.55977	0.0656
3.2	13.36797	20.0866	0.5799	0.0748
3.3	11.2175	22.967	0.60	0.08915
3.4	8.7657	26.1139	0.6223	0.11408

The plot of $\left[(b-C_S)/(-r_S)\right]$ versus $(1/C_S)$ is shown in figure (2.1.2). It is more or less linear. From the plot,

$$\text{slope} = (K_S / \mu_m) = 2.0404$$

$$y\text{- intercept} = (1/\mu_m) = 0.409$$

Therefore, $\mu_m = 2.445\ \text{hr}^{-1}$

$K_S = (2.0404)\ (2.445) = 4.9887$ mg/L

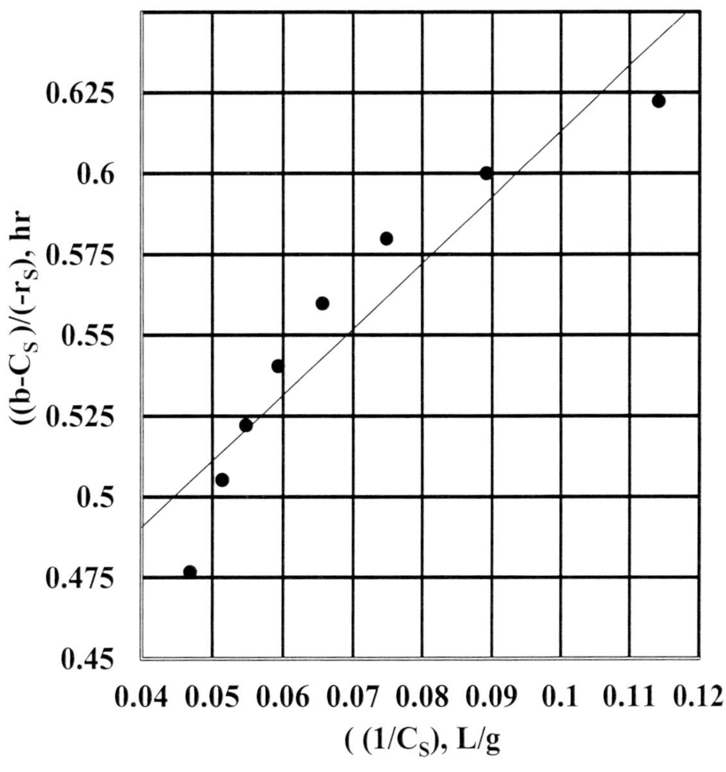

Figure 2.1.2: Monods kinetics (differential analysis)

(c) Approximate Differential analysis :

In this case, we have to check whether the data fit into equation (2.2.27) :

$$\left(\frac{1}{\mu}\right)=\left(\frac{K_S}{\mu_m}\right)\left(\frac{1}{C_S}\right)+\left(\frac{1}{\mu_m}\right) \qquad \text{(v)}$$

The values of specific growth rate (μ) are to be first computed from equation (2.2.26) as,

$$\mu=\left(\frac{1}{x_{avg}}\right)\left(\frac{\Delta x}{\Delta t}\right) \qquad \text{(vi)}$$

where $x=(b-C_S)Y=[25.01754-C_S](0.57)$

The computed values of x, μ, $\left(\frac{1}{\mu}\right)$ and $\left(\frac{1}{C_S}\right)$ are listed in table (2.1.3).

Table 2.1.3

Time (hr)	**C_S, mg/L**	***x*, mg/L**	**$\left(\frac{\Delta x}{\Delta t}\right)$ mg/(L.hr)**	***x*(avg), mg/L**	**μ, hr^{-1}**	**$\left(\frac{1}{\mu}\right)$, hr**	**$\left(\frac{1}{C_S}\right)$, L/mg**
2.4	22.61	1.3723	-	-	-	-	-
2.6	21.3228	2.106	3.6685	1.73915	2.10936	0.474	0.04869
2.8	19.4613	3.167	5.305	2.6365	2.01214	0.497	0.05138
2.9	18.2667	3.8479	6.809	3.50745	1.9413	0.5151	0.05474
3.0	16.8683	4.645	7.971	4.2464	1.877	0.5327	0.05928
3.1	15.2433	5.5713	9.263	5.10815	1.81337	0.55146	0.0656
3.2	13.36797	6.64	10.687	6.1056	1.7503	0.5713	0.0748

The plot of $\left(\frac{1}{\mu}\right)$ versus $\left(\frac{1}{C_S}\right)$ is shown in figure (2.1.3). This plot is also more or less linear with slope = 2.1342 units and y- intercept = 0.3923 units.

Thus, $\left(\frac{K_S}{\mu_m}\right)=2.1342$, $\left(\frac{1}{\mu_m}\right)=0.3923$

Therefore, $\mu_m=\left(\frac{1}{0.3923}\right)=2.549\ hr^{-1}$

K_S = (2.1342) (2.549) = 5.44 mg/L

Interestingly, results from this approximate method do agree closely with those obtained through earlier two methods.

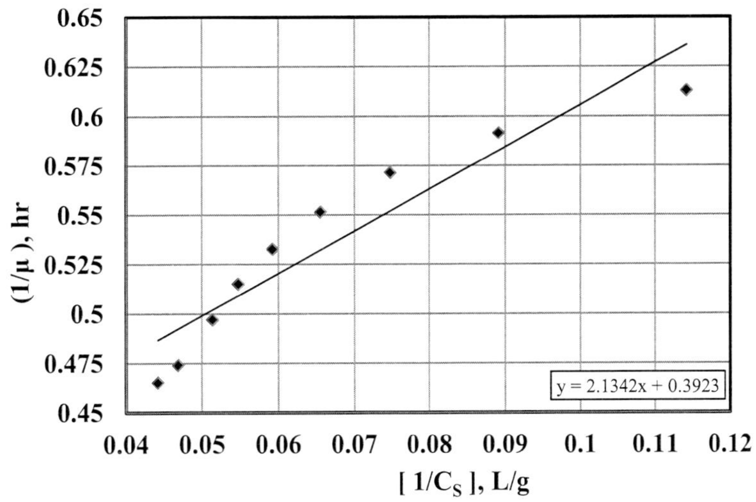

Figure 2.1.3: Monods kinetics (Approximate differential analysis)

Example 2.2 :Aerobic dephenolization of a sample of coke oven wastewater was studied in a 1.0 L chemostat and the data compiled are given below. The concentration of phenolics in the feed solution was 80 mg/L and it was essentially sterile.

Feed flow rate, L/hr	Steady state concentration of phenolics in exit stream, mg/L
0.3352	70.0696
0.3338	65.04
0.3317	59.867
0.3289	54.897
0.3252	50.00
0.3202	44.995
0.3136	39.9766
0.305	35.01
0.2935	29.995
0.278	24.961

Determine whether the process follows Haldane – Andrews kinetic equation (that accounts for substrate inhibition to microbial activity as well) and if so, evaluate the kinetic constants. The overall yield coefficient for cell mass production may be taken as constant and equal to 0.40.

Solution:

The performance equation of a chemostat that receives a sterile feed and which handles a bioconversion process that follows Haldane – Andrews kinetic equation is that given in equation (2.2.79) :

$$\tau = \left(\frac{1}{\mu_m}\right) + \left[\frac{1}{(\mu_m K_{Si})}\right] C_{Se} + \left(\frac{K_S}{\mu_m}\right)\left(\frac{1}{C_{Se}}\right) \qquad \text{(i)}$$

This is of the form

$$y = a_0 + a_1 x_1 + a_2 x_2 \quad \text{(ii)}$$

where $y = \tau = \left(\frac{V}{Q_0}\right) = \left(\frac{1.0}{Q_0}\right)$

$$x_1 = C_{Se}, \quad x_2 = \left(\frac{1}{C_{Se}}\right)$$

$$a_0 = \left(\frac{1}{\mu_m}\right), \; a_1 = \frac{1}{(\mu_m K_{Si})}, \; a_2 = \left(\frac{K_S}{\mu_m}\right).$$

Let us now check whether the given data fit into the above polynomial satisfactorily using matrix algebra. Thus, from equation (2.2.67),

$$\begin{bmatrix} \sum y \\ \sum x_1 y \\ \sum x_2 y \end{bmatrix} = \begin{bmatrix} N & \sum x_1 & \sum x_2 \\ \sum x_1 & \sum x_1^2 & \sum x_1 x_2 \\ \sum x_2 & \sum x_1 x_2 & \sum x_2^2 \end{bmatrix} \begin{bmatrix} a_0 \\ a_1 \\ a_2 \end{bmatrix} \quad \text{(iii)}$$

where N = no. of data points = 10. The elements of the above coefficient matrix are now computed and are listed in table (2.2.1). To note that,

$$\sum x_1 x_2 = 10 \left(\text{since } x_1 = C_{Se} \text{ and } x_2 = \frac{1}{C_{Se}}\right) \quad \text{(iv)}$$

Table 2.2.1A

Q_0, L/hr	$x_1 = C_{Se}$, mg/L	$y = \tau$, hr	$x_2 = \left(\frac{1}{C_{Se}}\right)$, L/mg	$x_1^2 = (C_{Se})^2$
0.3352	70.0696	2.9833	0.01427	4909.7488
0.3338	65.04	2.9958	0.015375	4230.2616
0.3317	59.867	3.0147	0.0167	3584.0577
0.3289	54.897	3.04	0.018216	3013.68
0.3252	50.00	3.075	0.02	2500.00
0.3202	44.995	3.123	0.02222	2024.55
0.3136	39.9766	3.1887	0.02501	1598.1285
0.305	35.01	3.2787	0.02856	1225.70
0.2935	29.995	3.4071	0.03334	899.70
0.2780	24.961	3.597	0.04	623.0515
	$\sum x_1 = 474.8112$	$\sum y = 31.7033$	$\sum x_2 = 0.23369$	$\sum x_1^2 = 24608.81875$

Table 2.2.1B

$\left(10^6\right)x_2^2 = 10^6\left(\frac{1}{C_{Se}}\right)^2$	$x_1y = \tau C_{Se}$, (mg.hr)/L	$x_2y = \left(\tau / C_{Se}\right)$,(L.hr)/mg
203.676	209.038637	0.042576
236.395	194.8468	0.04606
279.013	180.481	0.050356
331.820	166.887	0.055376
400.000	153.75	0.0615
493.937	140.5194	0.0694077
625.730	127.47338	0.07976
815.860	114.7873	0.09365
1111.480	102.19596	0.1135889
1605.00	89.7847	0.1441
$\sum x_2^2 = 61.0291 \times 10^{-4}$	$\sum x_1y = 1479.76417$	$\sum x_2y = 0.756374$

Equation (iii), therefore, becomes

$$\begin{bmatrix} 31.7033 \\ 1479.76417 \\ 0.756374 \end{bmatrix} = \begin{bmatrix} 10 & 474.8112 & 0.23369 \\ 474.8112 & 24608.818 & 10.0 \\ 0.23369 & 10.0 & 61.0291 X 10^{-4} \end{bmatrix} \begin{bmatrix} a_0 \\ a_1 \\ a_2 \end{bmatrix}$$

On solving,

$$a_0 = 2.18896$$

$$a_1 = 0.00477$$

$$a_2 = 32.2977$$

Therefore, $\mu_m = \left(\frac{1}{a_0}\right) = 0.4568 \text{ hr}^{-1}$

$$K_S = \left(\frac{a_2}{a_0}\right) = 14.7548 \text{ mg/L}, \qquad K_S = \left(\frac{a_0}{a_1}\right) = 458.9 \text{ mg/L}$$

The kinetic equation representing the dephenolization process is, therefore,

$$\left(-dC_S / dt\right) = \frac{(0.4568)C_S\left(b - C_S\right)}{\left[14.7548 + C_S + C_S^2 / 458.9\right]}$$

where $\text{b} = C_{S0} + \left(\frac{x_0}{Y}\right) = \left(C_{S0} + 2.5x_0\right)$.

Example 2.3: Removal of color (dyes) from a sample of textile wastewater through anaerobic fermentation using a complex bacterial culture (sludge) is studied in the laboratory using 2.0 L chemostat. The data collected are presented below. The feed sample had a dye concentration of 500 mg/L and was essentially sterile.

Feed flow rate, L/d	Steady state concentration in effluent, mg/L	
	Substrate (dye)	Active cell mass
2.0	95.988	263.486
2.2	122.00	246.7848
2.4	157.6925	223.6862
2.6	209.6756	189.8616
2.8	292.4168	135.8408
3.0	444.684	36.2185

Determine whether the process follows Monod's kinetic equation (that incorporates the effect of endogenous decay of microbes as well) and if so, evaluate the kinetic constants.

Solution:

If endogenous decay of microbes is anticipated, then the process could follow a kinetic equation of the form shown in equation (2.2.41). Accordingly, we have to first check whether the data reported fit into equation (2.2.45), namely the performance equation of the chemostat. Thus,

$$Y = \left(Y_e K_S / \mu_m\right)\left(1/C_{Se}\right) + \left(Y_e / \mu_m\right) \tag{i}$$

where $Y = y$- parameter $= \dfrac{\left(\tau x_e\right)}{\left(C_{S0} - C_{Se}\right)}$ (ii)

Now, $\tau = \left(V/Q_0\right) = \left(2.0/Q_0\right)$, $C_{S0} = 500$ mg/L.

The values of y- parameter (Y) and those of $\left(1/C_{Se}\right)$ are now computed from the data supplied and are listed in table (2.3.1).

Table 2.3.1

Q_0, L/d	τ, d	y- parameter (Y), d	(1/ C_{Se}), L/mg
2.0	1.0	0.652175	0.0104179
2.2	0.909	0.593525	0.0081967
2.4	0.8333	0.54455	0.0063414
2.6	0.7692	0.503049	0.0047693
2.8	0.71428	0.46742	0.0034197
3.0	0.6666	0.43651	0.0022488

The plot of *y*- parameter (*Y*) versus $\left(1/C_{Se}\right)$ is shown in figure (2.3.1). It is a straight line plot, with

$$\text{slope} = (Y_e K_S / \mu_m) = 26.399$$

and $$y\text{- intercept} = (Y_e / \mu_m) = 0.3771$$

Therefore, K_S = (26.399) / (0.3771) = 70 mg/L

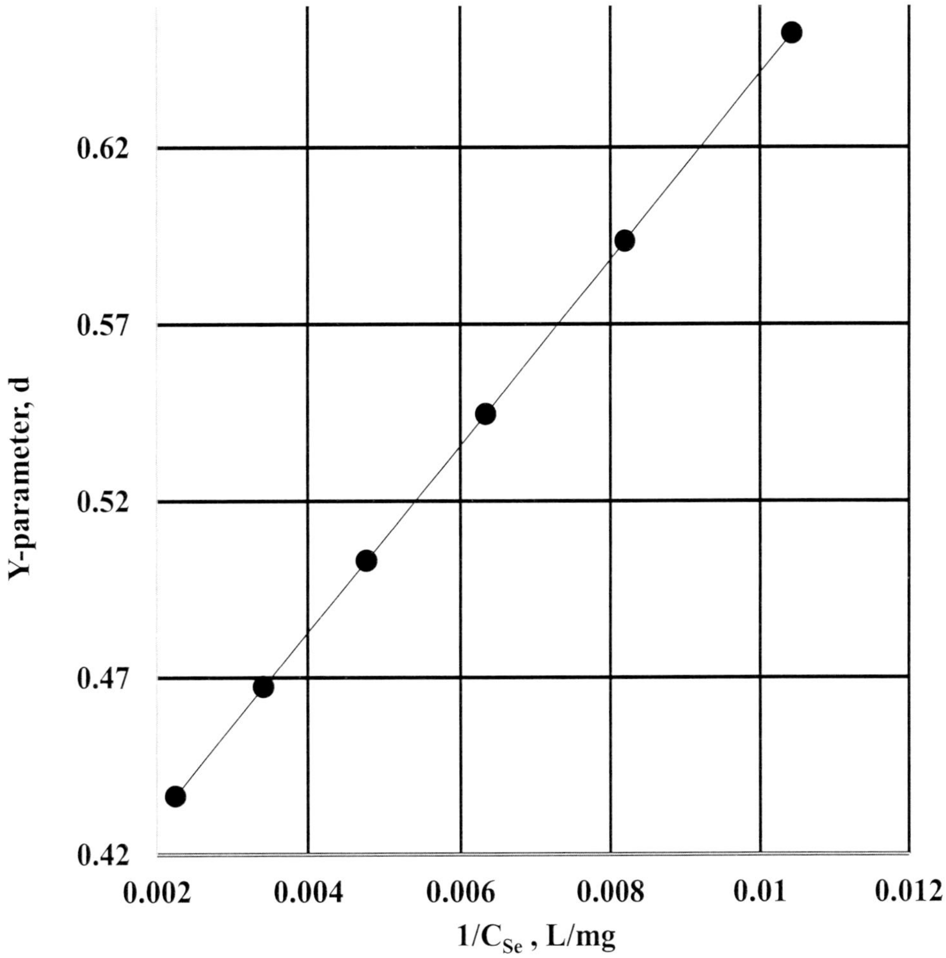

Figure 2.3.1

Let us now check whether the data fit into equation (2.2.52) as well (which is once again the performance equation of a chemostat, that receives a sterile feed). Thus,

$$\left(1/\tau\right) = Y_e\left(X\right) - k_d \tag{iii}$$

$$\text{where } X = x\text{- parameter} = (C_{S0} - C_{Se}) / \left(\tau x_e\right) \tag{iv}$$

It can be seen that the above *x*- parameter is nothing but the reciprocal of the *y* - parameter defined in equation (ii). The values of *x*- parameter (*X*) and those of $\left(1/\tau\right)$ are listed in table (2.3.2).

Table 2.3.2

$(1/\tau)$, d^{-1}	*x*- parameter (*X*), d^{-1}
1.0	1.53333
1.1	1.684848
1.2	1.836363
1.3	1.987878
1.4	2.13939
1.5	2.2909

The plot of $(1/\tau)$ versus the *x*- parameter (*X*) is shown in figure (2.3.2). It is also a straight line plot, with

$$\text{slope} = Y_e = 0.66$$

and

$$y\text{- intercept} = k_d = 0.012 \text{ d}^{-1}$$

Since, $(Y_e / \mu_m) = 0.3771$ (determined earlier),

$$\mu_m = (0.66 / 0.3771) = 1.7502 \text{ d}^{-1}$$

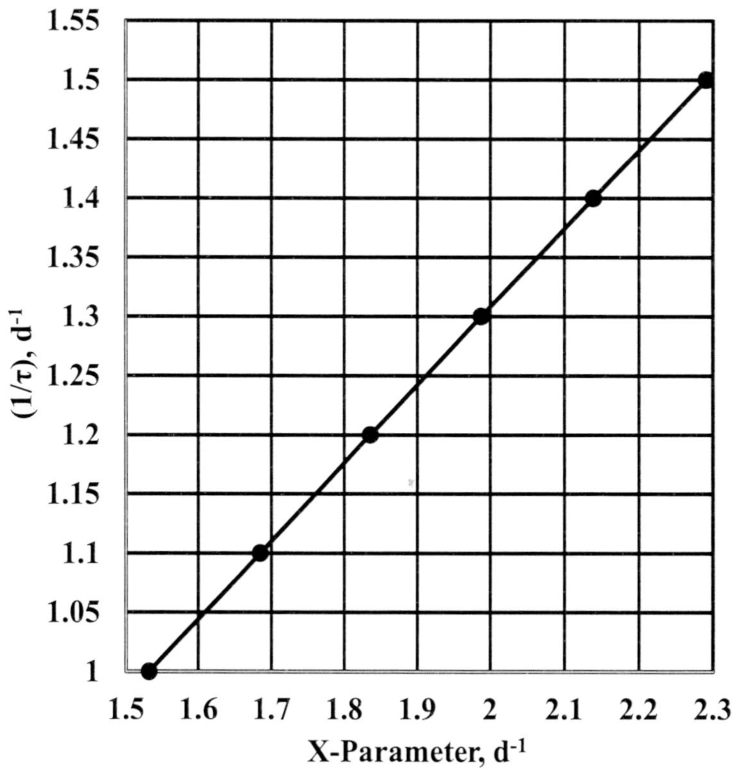

Figure 2.3.2

The process thus follows Monod's kinetic equation (with allowance for endogenous decay of microbes) and the kinetic constants associated with are,

$$\mu_m = 1.7502 \text{ d}^{-1}, \quad K_S = 70.0 \text{ mg/L}, \quad Y_e = 0.66, \quad k_d = 0.012 \text{ d}^{-1}$$

Example 2.4:
Microbial fermentation of a starchy substrate using a recombinant culture of Aerobacter was studied in the laboratory and the process was observed to follow Tessier's kinetic equation with kinetic constants, μ_m = 0.5 hr^{-1} and K_S = 0.25 g/L. If this process is conducted in an industrial batch reactor with a feed solution containing 10 g of substrate per liter and a cell mass concentration of 0.01 g/L, estimate the time required to attain 70% conversion of substrate. The overall yield coefficient for cell mass production = 0.25. Assume that the performance of the industrial batch reactor is close to ideal.

Solution:
The performance equation of an ideal batch reactor is nothing but the integrated form of the kinetic equation (namely, equation – 2.2.95):

$$(\mu_m t) = \int_{C_{Se}}^{C_{S0}} dC_S / \{[1-\exp(-C_s/K_S)](b-C_S)\} \qquad \text{(i)}$$

In the present case,

$$C_{S0} = 10.0 \text{ g/L}, \quad K_S = 0.25 \text{ g/L}, \quad \mu_m = 0.50 \text{ hr}^{-1}$$

Since 70% substrate conversion is being demanded,

$$C_{Se} = (0.3)C_{S0} = 3.0 \text{ g/L}$$

Also, $b = C_{S0} + (x_0/Y)$ = (10.0) + (0.01/0.25) = 10.04 g /L

Therefore, equation (i) reduces to,

$$(0.5t) = \int_{3.0}^{10.0} F(C_S)\,dC_S \qquad \text{(ii)}$$

where $F(C_S) = 1/\{[1-\exp(-4C_S)](10.04-C_S)\}$ (iii)

The integral of equation (ii) is to be evaluated numerically, say by Simpson's rule. For this, the values of $F(C_S)$ at different values of C_S are computed and are listed in table (2.4.1).

Table – 2.4.1

C_S, g/L	$F(C_S)$, L/g
3.0	0.1420463
3.5	0.1529053
4.0	0.16556
4.5	0.180505
5.0	0.1984127
5.5	0.220264
6.0	0.247524
6.5	0.282486
7.0	0.328947

C_S , g/L	$F(C_S)$, L/g
7.5	0.3937
8.0	0.490196
8.5	0.64935
9.0	0.961538
9.5	1.85185
10.0	25.0

Therefore, from Simpson's rule,

$$(0.5t) = I = (h/3)\left[f_0 + 4(f_1 + f_3 + \ldots f_{n-1}) + 2(f_2 + f_4 + \ldots f_{n-2}) + f_n\right]$$

where h = step size = 0.5 g/L, f_0 = 0.1420463 L/g, f_1 = 0.1529053 L/g, f_2 = 0.16556 L/g and so on. Finally, $f_n = f_{14}$ = 25.0 L/g. Therefore,

$$(0.5t) = I = (0.5\ t) = I = 7.4751$$

or $$t = (7.4751/0.5) = \mathbf{14.9502\ hr.}$$

This is the batch time required to achieve 70 % substrate conversion.

Example 2.5:

Microbial fermentation of a glucose – limited substrate is being conducted in a 1.2 m^3 CSTR using a mixed culture of Dictyostelium discoideum (predator) and E.coli (prey). The feed solution with a glucose concentration of 0.5 g/L and that is essentially sterile is admitted to the bioreactor at 75.0 L/hr. If both microbes follow Monod – type kinetics and the kinetic constants are those listed below, determine the fractional conversion of glucose attained. What shall be the steady state concentration of microbe – 1 (predator) in the exit stream? Assume that the performance of the bioreactor is close to ideal and the density of the reaction mixture remains more or less constant throughout the process.

Parameter	Microbe – 1 (predator)	Microbe – 2 (prey)
μ_m, hr^{-1}	0.24	1.8
K_S, g/L	0.004	0.0045
Y	0.10	0.20

Solution:

Let us assume that the predator feeds only on the prey and the substrate (glucose) is consumed only by the prey. Accordingly, (see equations – 2.2.132 to 2.2.136 of Section – 2.2),

$$(dx_1/dt)_e = \mu_1 (x_1)_e$$

where $\mu_1 = \mu_{m1}(x_2)_e / \left[K_{S1} + (x_2)_e\right]$

$$= (0.24)(x_2)_e / \left[0.004 + (x_2)_e\right] \quad \text{(i)}$$

Since the feed is sterile,

$$\mu_1 = D = (1/\tau) = (75/1200) = (1/16)\ hr^{-1}$$

Substituting the above value of μ_1 in equation (i) and solving, we get

$$(x_2)_e = 0.00140845 \text{ g/L} = 1.40845 \text{ mg/L}$$

Since substrate is consumed only by the prey,

$$(-dC_S / dt)_e = (1/Y_2)\mu_2 (x_2)_e$$
$$= (\mu_{m2} / Y_2) C_{Se} (x_2)_e / (K_{S2} + C_{Se})$$

The performance equation of the CSTR is, therefore,

$$\tau = (C_{S0} - C_{Se}) / (-dC_S / dt)_e$$
$$= (C_{S0} - C_{Se})(K_{S2} + C_{Se}) / [(\mu_{m2} / Y_2) C_{Se} (x_2)_e] \quad \text{(ii)}$$

Since τ = 16 hr, C_{S0} = 0.5 g/L, K_{S2} = 0.0045 g/L, μ_{m2} = 1.8 hr^{-1}, Y_2 = 0.2 and $(x_2)_e$ = 1.40845 × 10^{-3} g/L,

$$C_{Se} = 0.3 \text{ g/L}$$

The fractional conversion (α) of glucose attained is,

$$\alpha = (C_{S0} - C_{Se}) / C_{S0}$$
$$= (0.5 - 0.3) / (0.5) = \mathbf{40\%}$$

Now, for microbe – 2 (prey), the net rate of production shall be,

$$(dx_2 / dt)_e = \mu_2 (x_2)_e - (1/Y_1)\mu_1 (x_1)_e \quad \text{(iii)}$$

where $\mu_2 = \mu_2 C_{Se} / (K_{S2} + C_{Se})$

$= (1.8)(0.3) / (0.0045 + 0.3) = 1.7734 \text{ hr}^{-1}$

Therefore, $(dx_2 / dt)_e$ = (1.7734) (1.40845 × 10^{-3}) – (1/1.6) $(x_1)_e$

$$= 0.0024977 - 0.625\,(x_1)_e \quad \text{(iv)}$$

Now, the performance equation of the CSTR (written in terms of the concentration of microbe – 2),

$$\tau = [(x_2)_e - (x_2)_0] / (dx_2 / dt)_e$$
$$= (x_2)_e / (dx_2 / dt)_e, \quad \text{since the feed is sterile}$$

or, $\quad 16 = (1.40845 \times 10^{-3}) / [0.0024977 - 0.625\,(x_1)_e]$

Therefore, $(x_1)_e$ = 0.0038554 g/L = **3.8554 mg/L**

This is the concentration of microbe – 1 (predator) in the exit stream.

2.3 Kinetics of Enzyme Catalyzed Reactions

As discussed in Chapter-1, enzymes are selective catalysts. They more or less preferentially activate a specific reaction pathway and consequently, little side reactions occur, little side products are formed. The product solution shall be thus composed of mainly the primary product (desired product) and this brings down the cost of downstream processing (product recovery).

Earliest attempts to determine the kinetics of enzyme – catalyzed reactions were based on the hypothesis that the enzyme forms an intermediate complex (activated complex, ES) with the substrate and this complex, being enormously reactive (and thereby unstable), decomposes fast to yield the product(s). The scheme may be represented as:

$$S + E \rightleftharpoons ES \tag{2.3.1}$$

$$ES \rightarrow P + E \tag{2.3.2}$$

Each of these steps is assumed to be an elementary reaction. In other words, the reaction is assumed to be taking place in two steps, each step being an elementary reaction.

Once a probable mechanism of the reaction has been selected, then the rate equation (kinetic equation) can be developed based on two different approaches, as given below :

(a) approach based on a chosen rate-controlling step,
(b) that based on a steady-state approximation.

Michaelis and Menten [10] were the first to develop a kinetic equation for simple enzyme- catalysed reactions based on the rate-controlling step approach, while Briggs and Haldane [11] were the first to employ the steady state approximation to derive the kinetics of relatively simple enzyme-catalyzed biochemical reactions. Nevertheless, the kinetic equations developed by both of them are practically identical.

2.3.1 Michaelis – Menten and Briggs – Haldane kinetic equations

Let us start with the Michaelis- Menten scheme. The mechanism considered by them is same as that presented in equations (2.3.1) and (2.3.2), Michaelis and Menten assumed that the second step (namely, dissociation of enzyme-substrate complex, ES) is relatively slow and could be treated as the rate-controlling step, while the first step (namely, the formation of enzyme – substrate complex, ES) is relatively fast and attains equilibrium within a short duration of time. Thus,

$$S + E \rightleftharpoons ES \quad (K = k_1 / k_2) \tag{2.3.3}$$

$$ES \longrightarrow P + E \ \text{(rate constant} = k_3\text{)} \tag{2.3.4}$$

where k_1 is the rate constant of forward reaction and k_2 that of the backward reaction of equation (2.3.3). The overall rate of the reaction shall be, therefore, more or less equal to that of the second step (rate-controlling step). Thus,

$$(-r_S) = (-dC_S / dt) = k_3 C_{ES} \tag{2.3.5}$$

To note that since each step has been postulated to be an elementary reaction, the kinetic equation for each step can be deduced directly from the stoichiometric equation.

Since, the first step has attained equilibrium, we may define an equilibrium constant (K) for the same as,

$$K = (k_1 / k_2) = C_{ES} / (C_E C_S) \tag{2.3.6}$$

or,

$$C_{ES} = K C_E C_S \tag{2.3.7}$$

Enzyme, being the biocatalyst, is not consumed during the process. In other words, the total concentration of enzyme in the reaction mixture remains unaltered (and remains equal to the initial value, C_{E0}). Thus,

$$C_{E0} = C_E + C_{ES} \tag{2.3.8}$$

where C_{E0} = concentration of enzyme in the reaction mixture at $t = 0$

C_E, C_{ES} = concentration of free enzyme and that of combined enzyme (enzyme in the combined state) in the reaction mixture at any time, t.

Eliminating C_E between the above two equations, such as equations (2.3.7) and (2.3.8), we get

$$C_{ES} = C_{E0}\, C_S / \left[(1/K) + C_S\right] \tag{2.3.9}$$

Substituting this in equation (2.3.5), we get the kinetic equation as

$$\begin{aligned}(-dC_S / dt) &= (k_3 C_{EO}) C_S / \left[(1/K) + C_S\right] \\ &= (r_{max}\, C_S) / (K_m + C_S)\end{aligned} \tag{2.3.10}$$

where $r_{max} = (k_3 C_{EO})$ = maximum reaction rate, g/(L.s)

$K_m = (1/K)$ = Michaelis – Menten constant, g/L

This is called the *Michaelis – Menten kinetic equation.* To note that the reaction rate becomes maximum when the substrate concentration (C_S) is very large (during the initial stages of the process) and at such large values of C_S, we may safely assume

$$(K_m + C_S) \approx C_S \tag{2.3.11}$$

Accordingly, equation (2.3.10) would reduce to

$$(-dC_S / dt) = r_{max} \tag{2.3.12}$$

Similarly, at very low values of C_S, $(K_m + C_S) \approx K_m$ and equation (2.3.10) gets simplified to

$$(-dC_S / dt) = (r_{max} / K_m) C_S \tag{2.3.13}$$

The reaction shall be then first order in C_S. In other words, at large values of C_S ($C_S >> K_m$), the reaction shall be of zero order (equation - 2.3.12) and at low values of C_S ($K_m >> C_S$), it shall be first order in C_S (equation - 2.3.13). Processes of this kind are termed as *reactions of switching order.*

Briggs and Haldane [11] based their analysis on the same mechanism, but used the steady state approach rather than the approach based on a rate - controlling step. This approach is advantageous when we are not in a position to designate any one of the steps of the proposed mechanism as the rate controlling step. Now, based on the mechanism postulated in equations (2.3.3) and (2.3.4),

$$(dC_{ES} / dt) = k_1 C_S C_E - k_2 C_{ES} - k_3 C_{ES} \tag{2.3.14}$$

What is meant by the steady state approach is that we utilise the fact that under steady state conditions, the net rate of production of the enzyme – substrate complex (*ES*) shall be zero (whatever *ES* is produced in the first step, gets consumed in the second step). In other words, under steady-state conditions,

$$(dC_{ES} / dt) = 0 \tag{2.3.15}$$

Putting $(dC_{ES} / dt) = 0$ and solving for C_{ES}, we get

$$C_{ES} = [k_1 / (k_2 + k_3)] C_S \, C_E \tag{2.3.16}$$

$$= k_5 \, C_S \, C_E \tag{2.3.17}$$

Now, from equation (2.3.4),

$$(dC_P / dt) = \text{rate of product formation} = k_3 C_{ES} \tag{2.3.18}$$

Since one mole of substrate on complete conversion yields one mole of product,

$$(-dC_S / dt) = (dC_P / dt) = k_3 C_{ES} \tag{2.3.19}$$

Substituting for C_{ES} from equation (2.3.17), we get,

$$(-dC_S / dt) = k_3 \, k_5 \, C_S \, C_E \tag{2.3.20}$$

Since enzyme, being the biocatalyst, is not consumed during the process, from equation (2.3.8),

$$C_{E0} = C_E + C_{ES}$$

$$= C_E (1 + k_5 \, C_S)$$

or

$$C_E = C_{E0} / (1 + k_5 \, C_S) \tag{2.3.21}$$

Substituting the above expression for C_E in equation (2.3.20) and rearranging, we get

$$(-dC_S / dt) = [k_3 C_{E0}] C_S / [(1 / k_5) + C_S] \tag{2.3.22}$$

where $(1 / k_5) = (k_2 + k_3) / k_1$. (2.3.23)

This is called *Briggs – Haldane kinetic equation.* It can be seen that this kinetic equation is practically same as that developed by Michaelis and Menten (equation 2.3.10), with $K_m = (1 / k_5) = (k_2 + k_3) / k_1$.

Since both kinetic equations are identical, hereafter we shall be using the terminology, *Michaelis - Menten kinetic equation,* only.

This kinetic equation has been developed basically for single substrate reaction (though it could be extended to multiple substrates). Substrate inhibition or product inhibition to enzyme activity or inhibition caused by an externally added inhibitor has not been accounted for. Since, as discussed in Chapter – 1, enzymes are mostly used in commercial bioreactors in the immobilized form, neither substrate inhibition nor product inhibition to enzyme activity shall be significant (since neither the substrate nor the product tends to accumulate within the support particle, in which the enzyme has been immobilized). If an inhibitor has been added to the reaction mixture from an external source to regulate the enzyme activity, then the inhibition caused by it would have to be accounted for (this is discussed subsequently in this Section).

Michaelis – Menten scheme has received wide acceptability for describing the kinetics of a number of bioprocesses (though it is the simplest form of kinetic equation describing enzyme – catalyzed reactions).

Equation (2.3.10) can be easily integrated to yield,

$$\left(C_{S0}-C_S\right)+K_m \ln\left(C_{S0}/C_S\right)=r_{max}t \tag{2.3.23}$$

or,

$$\left[\left(C_{S0}-C_S\right)/t\right]=-K_m\left[\left(1/t\right)\ln\left(C_{S0}/C_S\right)\right]+r_{max} \tag{2.3.24}$$

Thus, if the enzyme – catalyzed reaction under consideration does follow Michaelis – Menten kinetics, a plot of $\left[\left(C_{S0}-C_S\right)/t\right]$ versus $\left[\left(1/t\right)\ln\left(C_{S0}/C_S\right)\right]$ must yield a straight line with slope = $-K_m$ and y- intercept = r_{max}. This forms the integral analysis based on batch reactor experimental data.

We may very well employ the differential analysis, for determining the kinetics of the reaction. For this, as discussed in the earlier section, the values of reaction rate, $\left(-r_S\right)$ or $\left(-dC_S/dt\right)$ at different values of C_S are to be first determined by plotting C_S versus time (t) and finding the slope of the tangent to the curve at different values of t (see figure 2.4). If a degree of approximation is admissible, then the values of $\left(-r_S\right)$ may also be computed as

$$\left(-r_S\right)=\left(-dC_S/dt\right)\approx\left(\Delta C_S/\Delta t\right) \tag{2.3.25}$$

Now, equation (2.3.10) may be rearranged as

$$\left[1/\left(-r_S\right)\right]=\left(K_m/r_{max}\right)\left(1/C_S\right)+\left(1/r_{max}\right) \tag{2.3.26}$$

Thus, a plot of $\left[1/\left(-r_S\right)\right]$ versus $\left(1/C_S\right)$ would yield a straight line with slope = $\left(K_m/r_{max}\right)$ and y – intercept = $\left(1/r_{max}\right)$, if the process under consideration does follow Michaelis – Menten kinetics. Such a plot is called *Lineweaver – Burk plot.*

Equation (2.3.26) may also be written as,

$$\left[C_S/\left(-r_S\right)\right]=\left(1/r_{max}\right)\left(C_S\right)+\left(K_m/r_{max}\right) \tag{2.3.27}$$

This once again forms equation to a straight line with slope = $\left(1/r_{max}\right)$, and y- intercept = $\left(K_m/r_{max}\right)$. This is called the *Hanes- Woolf plot.*

A still alternate rearrangement of equation (2.3.26) is,

$$r=-K_m\left(r/C_S\right)+r_{max} \tag{2.3.28}$$

where, $r=\left(-r_S\right)$. This predicts that a plot of reaction rate, r or $\left(-r_S\right)$ versus $\left(r/C_S\right)$ should yield a straight line of slope = $-K_m$ and y- intercept = r_{max}, if the process follows Michaelis – Menten kinetics. Such a plot is called *Eadie – Hofstee plot.*

The analysis may be performed using any one of these plots / equations. The values of kinetic constants determined must necessarily be the same irrespective of the plot / equation selected. If they do differ, that must be attributed to the experimental inaccuracy associated with the data collected.

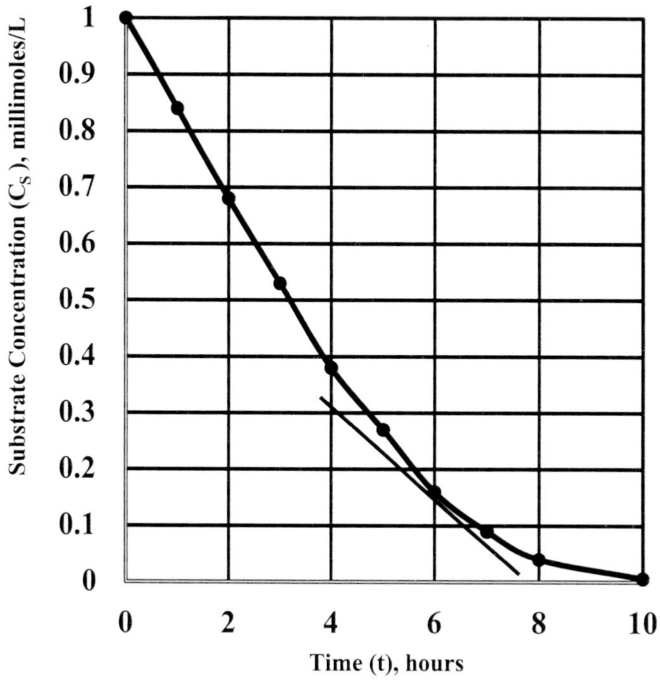

Figure 2.4: Variation of substrate concentration with time for a typical enzyme catalysed reaction

An alternate approach often used in kinetic analysis is the *initial rate approach.* The reaction is conducted for a small interval of time (Δt) and the change in substrate concentration (ΔC_S) is recorded. Then,

$$r_0 = \text{initial rate of reaction} = (\Delta C_S / \Delta t) \tag{2.3.29}$$

The procedure is repeated for different values of C_{S0} (initial substrate concentration). We thus obtain an elaborate table that lists the r_0 - values at different values of C_{S0}. Since the change in substrate concentration (C_S) within a small interval of time (Δt) is small, it shall not be too erroneous for us to substitute $C_S = C_{S0}$ at $(-r_S) = r_0$. Agreed, this does involve an approximation. Equations (2.3.26) to (2.3.28) could be, therefore, rewritten as,

$$(1/r_0) = (K_m / r_{max})(1/C_{S0}) + (1/r_{max}) \tag{2.3.30}$$

$$(C_{S0} / r_0) = (1/r_{max}) C_{S0} + (K_m / r_{max}) \tag{2.3.31}$$

$$(r_0) = -K_m (r_0 / C_{S0}) + (r_{max}) \tag{2.3.32}$$

Each of the above equations predicts a straight line plot, from the slope and y – intercept of which the kinetic constants (K_m , r_{max}) could be evaluated. The procedure is relatively simpler, since it demands only one straight line plot. Computation of reaction rate $(-r_S)$, at different values of t shall not be required (this eliminates a good amount of computational load). However, the approach involves approximation (as mentioned above). Also, accurate determination of r_0 - values often causes concern. Since biochemical reactions are relatively slow, to obtain a measurable change in C_S, we would have to wait for a long period of time and accordingly, it would not be possible to maintain Δt very low.

If the enzyme – catalyzed reaction under consideration has been studied in a laboratory chemostat and steady state values of substrate concentration (namely, C_{Se}) at different values of space time, $\tau\left(\text{where } \tau = V / Q_0\right)$ or feed flow rate (Q_0) have been collected, then we may use the performance equation of chemostat (namely, equation – 2.1.1) for the kinetic analysis. Thus, from equation (2.1.1),

$$\left[\tau / \left(C_{S0} - C_{Se}\right)\right] = 1 / \left(-r_{Se}\right) \tag{2.3.33}$$

Comparing the above equation with equation (2.3.26), we get,

$$\left[\tau / \left(C_{S0} - C_{Se}\right)\right] = \left(K_m / r_{max}\right)\left(1 / C_{Se}\right) + \left(1 / r_{max}\right) \tag{2.3.34}$$

This also is equation to a straight line with slope = $\left(K_m / r_{max}\right)$ and y- intercept = $\left(1 / r_{max}\right)$. Since demands only one straight line plot, the analysis is computationally simpler.

2.3.2. Modified Michaelis – Menten Schemes (incorporating inhibition to enzyme activity)

Substrate inhibition to enzyme activity is not unusual. What has been postulated is that at large substrate concentrations, the enzyme – substrate complex (*ES*) tends to react with excess substrate present to form a non- reactive complex (ES_2). Being non – reactive, ES_2 shall not decompose further to form the product (s). The yield of product (s) thus goes down. A part of enzyme also gets lost in the form of this unreactive complex, ES_2. The mechanism may be represented as,

$$E + S \rightleftharpoons ES \tag{2.3.35}$$

$$ES + S \rightleftharpoons ES_2 \tag{2.3.36}$$

$$ES \longrightarrow P + E \tag{2.3.37}$$

Based on the above mechanism and considering the last step (equation - 2.3.37) to be rate – controlling and by following the same procedure as discussed in the earlier section, the rate equation could be derived as,

$$\left(-r_S\right) = \left(-dC_S / dt\right) = \left(r_{max} C_S\right) / \left[K_m + C_S + \left(C_S^{\,2} / K_{Si}\right)\right] \tag{2.3.38}$$

where, K_{Si} = substrate inhibition coefficient. g/L

It can be seen that the above kinetic equation is analogous to Haldane – Andrews kinetic equation for microbial fermentation (as discussed in Section 2.2.3)

It is important to note that in the present case, r_{max} represents the maximum reaction rate that would be attained *in the absence of substrate inhibition.* When the substrate inhibition to enzyme activity does exist, $\left(-r_S\right) = r_{max}$ shall never be attained. The effect of substrate inhibition shall be highest at t = 0 (at the start of the process) when substrate concentration is highest at C_{S0}. As the reaction proceeds, the substrate concentration (C_S) starts decreasing and consequently, the effect of substrate inhibition also decreases. Due to decrease in the inhibitory effect to enzyme activity, the rate of reaction tends to increase with time (with decrease in C_S). The rate ultimately reaches a maximum value (at $C_S = C_{Sm}$) and thereafter, it starts decreasing with further decrease in C_{Se} (see figure 2.5). At $C_S = C_{Sm}$, the substrate concentration has reduced to a fairly low magnitude and as a result, the effect of substrate inhibition to enzyme activity has also reduced to a minimum. The rate of reaction, therefore, starts decreasing with time (with decrease in C_S) thereafter.

To keep in mind that in the absence of substrate inhibition, when the process follows classical Michaelis – Menten kinetics (equation 2.3.10), the rate of reaction decreases continuously with time (with decrease in substrate concentration), as shown earlier in figure (2.4).

The value of C_{Sm} at which the rate of reaction shall be the highest can be analytically determined by putting $d(-r_S)/dC_S = 0$ and then solving it for C_{Sm}. This yields,

$$C_{Sm} = \sqrt{K_m K_{Si}} \tag{2.3.39}$$

The maximum rate so achieved (at $C_S = C_{Se}$) shall be less than r_{max}.

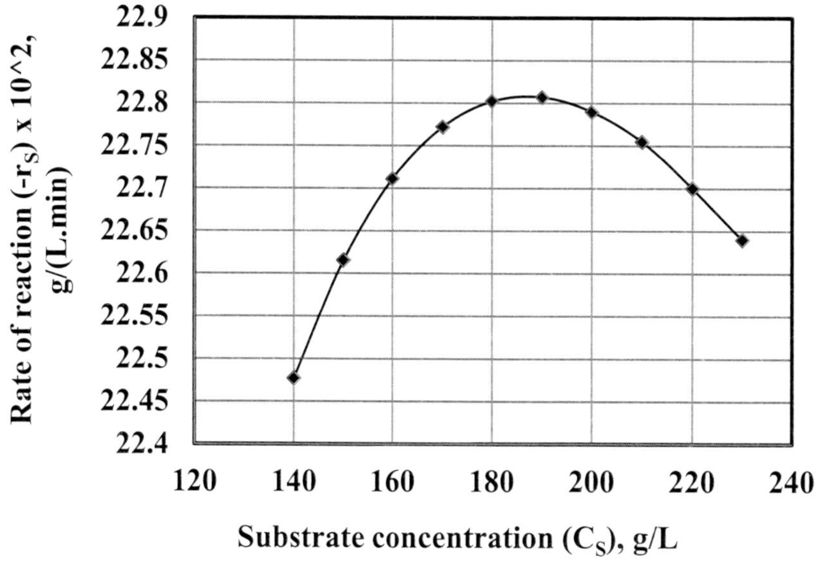

Figure 2.5: Typical rate curve for enzyme-catalyzed reaction with substrate inhibition

Equation (2.3.38) can be easily integrated and rearranged to the following form:

$$\left[\frac{t}{(C_{S0}-C_S)}\right] = \left(\frac{1}{r_{max}}\right) + \left(\frac{K_m}{r_{max}}\right)\left[\frac{\ln(C_{S0}/C_S)}{(C_{S0}-C_S)}\right] + \left[\frac{1}{(2K_{Si}r_{max})}\right](C_{S0}+C_S) \tag{2.3.40}$$

This is of the form,

$$y = a_0 + a_1 x_1 + a_2 x_2 \tag{2.3.41}$$

where $a_0 = (1/r_{max})$, $a_1 = (K_m/r_{max})$, $a_2 = 1/(2K_{Si}\,r_{max})$ (2.3.42)

$$x_1 = [\ln(C_{S0}/C_S)]/(C_{S0}-C_S), x_2 = (C_{S0}+C_S) \tag{2.3.43}$$

The kinetic constants (K_m, r_{max}, K_{Si}) can be evaluated by fitting the experimental data into the above equation and solving the resultant matrix equation (see equation – 2.2.67) as discussed in the earlier Section.

Alternately, if we substitute for K_{Si} from equation (2.3.39) as, $K_{Si} = \left(C_{Sm}^{\ 2} / K_m\right)$, then equation (2.3.40) could be further rearranged to

$$\left[t/\left(C_{S0}-C_S\right)\right]=\left(1/r_{max}\right)+\left(K_m/r_{max}\right)\left[\{\ln\left(C_{S0}/C_S\right)\}/\left(C_{S0}-C_S\right)+\left(C_{S0}+C_S\right)/(2C_{Sm}^{\ 2})\right] \qquad (2.3.44)$$

It can be seen that this is equation to a straight line and is of the form,

$$Y=\left(K_m/r_{max}\right)X+\left(1/r_{max}\right) \qquad (2.3.45)$$

where, $X = x$- parameter $= \left[\{\ln\left(C_{S0}/C_S\right)\}/\left(C_{S0}-C_S\right)+\left(C_{S0}+C_S\right)/(2C_{Sm}^{\ 2})\right]$ (2.3.46)

$Y = y$- parameter $= \left[t/\left(C_{S0}-C_S\right)\right]$ (2.3.47)

Thus, from the slope and intercept of the above straight line plot, the values of the kinetic constants (K_m, r_{max}) could be estimated. Once the value of K_m is known, the value of K_{Si} is obtained from equation (2.3.39).

It must be however noted that a plot of reaction rate ($-r_S$) versus substrate concentration (C_S) is to be prepared (as shown in figure – 2.5) at the outset and from this plot, the value of C_{Sm} is to be noted down.

For differential analysis, equation (2.3.38) could be rearranged to

$$\left[1/\left(-r_S\right)\right]=\left(1/r_{max}\right)+\left(K_m/r_{max}\right)\left(1/C_S\right)+\left[1/\left(K_{Si}\,r_{max}\right)\right]C_S \qquad (2.3.48)$$

This is also of the form,

$$y=a_0+a_1x_1+a_2x_2 \qquad (2.3.49)$$

where $x_1=\left(1/C_S\right)$, $x_2=\left(C_S\right)$.

It goes without saying that values of reaction rate, ($-r_S$) at different values of C_S must have to be kept computed in advance.

If experimental data on a chemostat are available, we could fit them into equation (2.1.1), which, for the present case, could be rearranged to,

$$\left[\tau/\left(C_{S0}-C_{Se}\right)\right]=\left(1/r_{max}\right)+\left(K_m/r_{max}\right)\left(1/C_{Se}\right)+\left[1/\left(K_{Si}\,r_{max}\right)\right]C_{Se} \qquad (2.3.50)$$

This once again is of the form

$$y=a_0+a_1x_1+a_2x_2 \qquad (2.3.51)$$

Inhibitors (also called as *reversible inhibitors*) are often added to the reaction mixture to regulate the activity of the enzyme used. In such situations, they could better be termed as *modifiers* or *additives* . The enzyme activity can be restored back to the original value if found necessary, by adding suitable precursors. That is why they are called *reversible inhibitors*. These inhibitors (modifiers or additives) may be classified into three categories:

(a) Competitive inhibitors
(b) Uncompetitive inhibitors
(c) Mixed inhibitors

Competitive inhibitors compete with the substrate for the active sites of the enzyme. In other words, these inhibitors supposedly react with the enzyme forming a non – reactive enzyme – inhibitor complex, *EI*. A part of the active sites of the enzyme is thus occupied by this inhibitor. The activity of the enzyme obviously declines. The rate of product formation also decreases. The proposed mechanism may be represented as,

$$E + S \rightleftharpoons ES \tag{2.3.52}$$

$$E + I \rightleftharpoons EI \text{ (non-reactive)} \tag{2.3.53}$$

$$ES \longrightarrow P + E \tag{2.3.54}$$

Based on the above – proposed mechanism and considering the last step (equation – 2.3.54) as the rate controlling step, the kinetic equation could be derived as,

$$(-r_S) = (-dC_S / dt) = (r_{max}\, C_S) / (K_m(app) + C_S) \tag{2.3.55}$$

where $K_m(app) = K_m\left[1 + (C_I / K_i)\right]$ (2.3.56)

C_I = concentration of the inhibitor in the reaction mixture, g/L
K_i = competitive inhibition coefficient, g/L

It can be seen that the above kinetic equation is quite similar to the classical Michaelis – Menten equation (equation – 2.3.10), except that the kinetic constant K_m has been replaced by $K_m(app)$, which is a function of inhibitor concentration, C_I.

The above kinetic equation is composed of two independent variables such as C_S and C_I (both decreasing with time). The inter – relationship between the two is not known. Also, since I is added in small amounts, accurate determination of inhibitor concentration (C_I) in the reaction mixtures at different intervals of time (*t*) has been observed to be quite troublesome. It has, therefore, become a common practice to use the initial rate approach for analyzing the kinetics of biochemical processes (enzyme catalyzed processes) like this that involve a reversible inhibitor (competitive, uncompetitive or mixed).

Experiments are conducted in a batch reactor without any inhibitor and r_0- values at different values of C_{S0} are recorded. The data are then fitted into the equation (2.3.26) and the kinetic constants (K_m, r_{max}) are evaluated. The reaction is now conducted in presence of a specific amount of the competitive inhibitor (initial concentration of inhibitor in the reaction mixture = C_{I0}) and once again, r_0- values at different values of C_{S0} are recorded. The value of C_{I0} is maintained the same in all the experimental runs. This data is now fitted into the equation (2.3.55) by rearranging it to the following form:

$$\left[1/r_0\right] = \left[K_m(app)/r_{max}\right](1/C_{S0}) + (1/r_{max}) \tag{2.3.57}$$

The value of $K_m(app)$ can be obtained from the above straight line plot. To note that this plot shall have the same *y*- intercept as the Lineweaver – Burk plot based on equation (2.3.26). The value of K_i can be now obtained from equation (2.3.56) as

$$K_m(app) = K_m\left[1 + (C_{I0} / K_i)\right]$$

or

$$K_i = C_{I0} / \left[(K_m(app)/K_m) - 1\right] \tag{2.3.58}$$

In the case of *uncompetitive inhibitors*, it has been postulated that they react with a part of the enzyme-substrate complex (*ES*) forming a non-reactive complex, *ESI*. Since a part of *ES* and a part of enzyme are thus lost as *ESI*, the rate of product formation decreases. The mechanism can be represented as

$$E + S \rightleftharpoons ES \tag{2.3.59}$$

$$ES + I \rightleftharpoons ESI \tag{2.3.60}$$

$$ES \longrightarrow P + E \tag{2.3.61}$$

Here also, considering the last step (equation 2.3.61) to be rate-controlling, the kinetic equation could be derived as,

$$(-r_S) = (-dC_S / dt) = \left[r_{max}(app) C_S\right] / [K_m(app) + C_S] \tag{2.3.62}$$

where $r_{max}(app) = r_{max} / \left[1 + C_I / K_i^{'}\right]$ (2.3.63)

$$K_m(app) = K_m / \left[1 + C_I / K_i^{'}\right] \tag{2.3.64}$$

$K_i^{'}$ = uncompetitive inhibition coefficient, g/L

The procedure of evaluating the kinetic constants is the same as that for competitive inhibition. The reaction is first conducted without the inhibitor and the data compiled (r_0- values at different values of C_{S0}) are fitted into equation (2.3.26) to evaluate the kinetic constants (r_{max}, K_m). The reaction is now conducted with a specific amount of inhibitor having been added to the reaction mixture (such that the initial concentration of inhibitor in the reaction mixture = C_{I0}) and once again r_0- values at different values of C_{S0} are recorded. This data is now fitted into equation (2.3.62), after rearranging into the following form:

$$(1 / r_0) = [K_m(app) / r_{max}(app)](1 / C_{S0}) + \left[1 / r_{max}(app)\right] \tag{2.3.65}$$

$$= (K_m / r_{max})(1 / C_{S0}) + \left[1 / r_{max}(app)\right] \tag{2.3.66}$$

The *y*- intercept of above straight line plot provides the value of $r_{max}(app)$ and thereafter, the value of $K_i^{'}$ could be retrieved from equation (2.3.63) as,

$$K_i^{'} = C_{IO} / \left\{\left[r_{max} / r_{max}(app)\right] - 1\right\} \tag{2.3.67}$$

It can be seen that the straight line plot based on equation (2.3.66) shall have the same slope as the Lineweaver- Burk plot based on equation (2.3.26), but a larger y- intercept.

In the case of *mixed inhibition*, the inhibitor could block some of the active sites of the enzyme, but also could further retard the product formation by converting a part of the enzyme – substrate complex (*ES*) into non – reactive *ESI*. The scheme could be as follows:

$$E + S \rightleftharpoons ES \tag{2.3.68}$$

$$E + I \rightleftharpoons EI \tag{2.3.69}$$

$$ES + I \rightleftharpoons ESI \tag{2.3.70}$$

$$ES \longrightarrow P + E \tag{2.3.71}$$

The kinetic equation based on the above mechanism shall be

$$(-r_S)=(-dC_S/dt)=\left[r_{max}(app)C_S\right]/[K_m(app)+C_S] \tag{2.3.72}$$

where $r_{max}(app)=r_{max}/\left[1+C_I/K_i^{'}\right]$ (2.3.73)

$$K_m(app) = K_m\left[1+C_I/K_i\right]/\left[1+C_I/K_i^{'}\right] \tag{2.3.74}$$

As before, the kinetic constants (r_{max}, K_m) are determined from the Lineweaver-Burk plot based on equation (2.3.26), using experimental data collected without inhibitor. Experimental data compiled with a specific initial inhibitor concentration of C_{IO} are then fitted into equation (2.3.72) after rearranging it to the following form:

$$(1/r_0)= [K_m(app)/r_{max}(app)](1/C_{S0})+\left[1/r_{max}(app)\right] \tag{2.3.75}$$

The values of $K_m(app)$ and $r_{max}(app)$ could be computed from the slope and y- intercept of above straight line plot. The values of inhibition coefficients $(K_i, K_i^{'})$ could be then deduced from (2.3.73) and (2.3.74) as

$$K_i^{'} = C_{I0}/\left\{\left[r_{max}/r_{max}(app)\right]-1\right\} \tag{2.3.76}$$

$$K_i = C_{I0}/\left\{\left[K_m(app)(1+C_{I0}/K_i^{'})/K_m\right]-1\right\} \tag{2.3.77}$$

A specific case of above scheme is one in which the inhibition coefficients are more or less equal in magnitude. In other words, $K_i = K_i^{'}$. In such a case, the kinetic equation gets reduced to

$$(-r_S)=(-dC_S/dt)=\left[r_{max}(app)C_S\right]/(K_m+C_S) \tag{2.3.78}$$

Lineweaver – Burke plots for the above – discussed three categories of reversible inhibition are sketched and illustrated in figure (2.6).

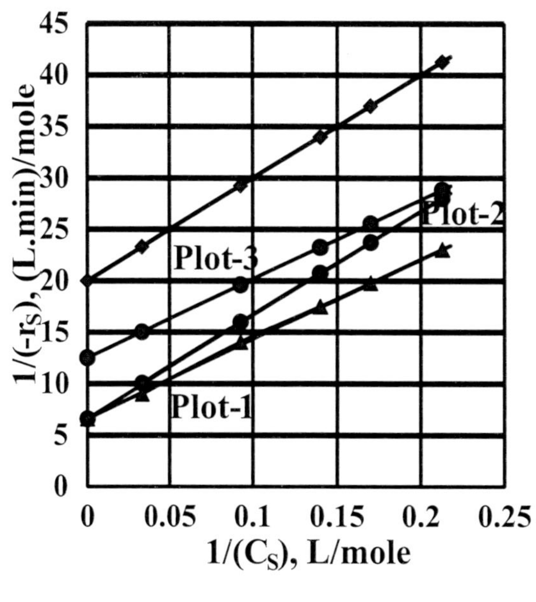

Figure 2.6: Typical Lineweaver – Burk plots for Reversible Inhibition to Enzyme Activity. Plot 1: Without inhibitor. Plot 2: With competitive reversible inhibition. Plot 3: With uncompetitive reversible inhibition. Topmost plot: With Mixed inhibition.

2.3.3 Kinetics of Enzyme – catalyzed reactions with Multiple Substrates

Enzyme catalyzed bioprocesses with multiple substrates are not uncommon. An excellent example is the manufacture of biodiesels from vegetable oils by reacting them with a low molecular weight alcohol (mostly, methanol) in presence of lipase enzyme as the catalyst. The reaction stoichiometry is shown in equation (1.13) of Chapter – 1. The vegetable oil (which is an ester of glycerol and a high molecular weight carboxylic acid) reacts with methanol to yield biodiesel (methyl ester of the high molecular weight carboxylic acid) and glycerol as products. The process may be represented as,

$$S_1 + 3S_2 + E \longrightarrow 3P_1 + P_2 + E \tag{2.3.79}$$

Michaelis - Menten kinetic model (equation 2.3.10) could very well be extended to enzyme catalysed reactions with multiple substrates. For example, let us consider a two substrate reaction, but yielding only one product (P). Thus,

$$S_1 + S_2 + E \longrightarrow P + E \tag{2.3.80}$$

Based on the postulated mechanism that both S_1 and S_2 combine with the enzyme forming enzyme - substrate complexes (ES_1 and ES_2) and each of them decomposes to yield the product P, the kinetic equation could be derived as

$$(dC_P / dt) = (r_{max} C_{S1} C_{S2}) / (K_1 + K_{m1} C_{S1} + K_{m2} C_{S2} + C_{S1} C_{S2}) \tag{2.3.81}$$

where K_1, K_{m1}, K_{m2} = kinetic constants, g/L

r_{max} = maximum reaction rate, g/L

We may utilise differential analysis for the evaluation of the kinetic constants (and for verifying whether the experimental data do fit into the above – proposed kinetic equation). Thus, since one mole of S_1 on complete conversion yields one mole of P,

$$(-r_{S1}) = (-dC_{S1} / dt) = (dC_P / dt) \tag{2.3.81a}$$

Therefore, the above equation (equation - 2.3.81) may be rearranged as

$$(C_{S1} C_{S2})(-r_{S1}) = (K_1/r_{max}) + (K_{m1}/r_{max})\, C_{S1} + (K_{m2}/r_{max})\, C_{S2} + (1/r_{max})\, C_{S1} C_{S2} \tag{2.3.82}$$

This is of the form,

$$y = a_0 + a_1 x_1 + a_2 x_2 + a_3 x_3 \tag{2.3.83}$$

where $x_1 = C_{S1}$, $x_2 = C_{S2}$ and $x_3 = (C_{S1} C_{S2})$.

The values of $(-r_{S1})$ at different values of (C_{S1}) are to be first determined by plotting C_{S1} versus time (t). The polynomial coefficients (a_0 to a_3) can be then evaluated by solving the matrix equation (see equation (2.2.67) of earlier section) and the kinetic constants computed therefrom.

If experimental data have been compiled using a chemostat, then we may utilize the performance equation of chemostat (namely equation – 2.1.1) written in terms of C_{S1}. Thus,

$$\tau = [(C_{S1})_0 - (C_{S1})_e] / (-r_{S1})_e$$

or

$$\tau / [(C_{S1})_0 - (C_{S1})_e] = 1 / (-r_{S1})_e \tag{2.3.84}$$

On comparing the above equation with equation (2.3.82), we get

$$\frac{(C_{S1})_e (C_{S2})_e \tau}{[(C_{S1})_0 - (C_{S1})_e]} = \left(\frac{K_1}{r_{max}}\right) + \left(\frac{K_{m1}}{r_{max}}\right)(C_{S1})_e + \left(\frac{K_{m2}}{r_{max}}\right)(C_{S2})_e + \left(\frac{1}{r_{max}}\right)(C_{S1})_e (C_{S2})_e \quad (2.3.85)$$

This is once again of the form,

$$y = a_0 + a_1 x_1 + a_2 x_2 + a_3 x_3 \quad (2.3.86)$$

The analysis could be made simpler, if we employ the initial rate approach (the approximation associated with this approach having been born with). Experiments are to be conducted at a specific value of $(C_{S2})_0$ and r_0- values at different values of $(C_{S1})_0$ are to be recorded. Experiments are to be repeated at least at five values of $(C_{S2})_0$ and in each case, r_0 versus $(C_{S1})_0$ values be recorded.

We thus obtain at least five sets of values of r_0 and $(C_{S1})_0$, each at a specified value $(C_{S2})_0$. Now, rearrangement of equation (2.3.81) after writing it in terms of initial rate r_0) and initial concentrations of S_1 and S_2, yields

$$(1/r_0) = (1/r_{max})\left[K_{m2} + K_1/(C_{S2})_0\right]\left[1/(C_{S1})_0\right] + (1/r_{max})\left[1 + K_{m1}/(C_{S2})_0\right] \quad (2.3.87)$$

This shows that a plot of $(1/r_0)$ versus $[1/(C_{S1})_0]$ at any specific value of $(C_{S2})_0$ would yield a straight line plot with

$$\text{slope} = (K_{m2}/r_{max}) + (K_1/r_{max})\left[1/(C_{S2})_0\right] \quad (2.3.88)$$

and

$$y\text{- intercept} = (1/r_{max}) + (K_{m1}/r_{max})\left[1/(C_{S2})_0\right] \quad (2.3.89)$$

We could thus prepare a family of plots (at least five in number) of $(1/r_0)$ versus $[1/(C_{S1})_0]$ each at a specific value of $(C_{S2})_0$. The slope and y – intercept of each are to be noted down.

Now, from equation (2.3.89), a plot of y- intercept versus $[1/(C_{S2})_0]$ would yield a straight line with slope = (K_{m1}/r_{max}) and y – intercept = $(1/r_{max})$. The values of K_{m1} and r_{max} can be thus determined from this straight line plot.

Further, from equation (2.3.88), a plot of slope versus $[1/(C_{S2})_0]$ would also yield a straight line with slope = (K_1/r_{max}) and y- intercept = (K_{m2}/r_{max}). This plot would thus provide the values of K_1 and K_{m2} as well.

It must be kept in mind that the above procedure based on initial rate of reaction, is suitable only when both the substrates (S_1 and S_2) are aqueous solutions.

For the manufacture of biodiesel from vegetable oil and methyl acetate using lipase enzyme as the catalyst, Du et.al [12] and De et.al [13-15] have reported that a kinetic equation of the following form is applicable:

$$(-r_{S1}) = \left(\frac{-dC_{S1}}{dt}\right) = \left(\frac{(r_{max} C_{S1} C_{S2})}{C_{S1}C_{S2} + K_{m1}C_{S1} + K_{m2}C_{S2} + K_{m2}\left(\frac{C_{S2}^{\ 2}}{K_{Si}}\right)}\right) \quad (2.3.90)$$

where C_{S1} = concentration of vegetable oil, mol/L

C_{S2} = concentration of methyl acetate, mol/L

K_{Si} = substrate inhibition coefficient, mol/L

The above equation predicts that excess of methyl acetate causes inhibition to enzyme activity. It is, therefore, recommended that the vegetable oil and methyl acetate be used in the stoichiometric proportion of 1:3 (by mole) and no excess methyl acetate be used.

The above equation may be rearranged to

$$\frac{1}{(-r_{S1})}=\left(\frac{1}{r_{max}}\right)+\left(\frac{K_{m1}}{r_{max}}\right)\left(\frac{1}{C_{S2}}\right)+\left(\frac{K_{m2}}{r_{max}}\right)\left(\frac{1}{C_{S1}}\right)+\left[\frac{K_{m2}}{K_{si}r_{max}}\right]\left(\frac{C_{S2}}{C_{S1}}\right) \quad (2.3.91)$$

This is also of the form,

$$y=a_0+a_1x_1+a_2x_2+a_3x_3 \quad (2.3.92)$$

where $x_1=(1/C_{S2})$, $x_2=(1/C_{S1})$, $x_3=(C_{S2}/C_{S1})$.

Use of methyl acetate in place of methanol has been recommended whenever immobilized lipase enzyme is being employed as the catalyst [12]. This is because the byproduct glycerol (formed when methanol is being used for trans-esterification) tends to form a layer around the immobilized lipase and thereby tends to deactivate it. With methyl acetate, the byproduct shall be glycerol triacetate, which is reportedly harmless to the enzyme catalyst.

Example 2.6: Enzymatic hydrolysis of a polysaccharide (used as food additive) is studied in a laboratory batch reactor and the results obtained are given below:-

Time (hr)	Substrate concentration , millimol/L
0.0	1.0
1.0	0.8409
2.0	0.6838
3.0	0.5359
4.0	0.40
5.0	0.2797
6.0	0.1774
6.5	0.1340

Determine whether the kinetics of the process follow Michaelis – Menten scheme (and if so, evaluate the kinetic constants) using

(a) Integral analysis

(b) Differential analysis based on

  (i) Lineweaver – Burk equation

  (ii) Hanes – Woolf equation

  (iii) Eadie – Hofstee equation

(c) Approximate differential analysis

Solution:

(a) Integral Analysis:

The integrated form of Michaelis – Menten equation (after a rearrangement) is that given in equation (2.3.24) :

$$\left[\left(C_{S0}-C_S\right)/t\right]=-K_m\left[(1/t)\ln\left(C_{S0}/C_S\right)\right]+r_{max}$$

or
$$Y=-K_m\left(X\right)+r_{max} \qquad \text{(i)}$$

where $Y = y$- parameter $= \left(C_{S0}-C_S\right)/t$ (ii)

$X = x$- parameter $= (1/t)\ln\left(C_{S0}/C_S\right)$ (iii)

From the data reported, the values of y- parameter and those of the x- parameter are computed and are listed in table (2.6.1).

Table – 2.6.1

Time (hr)	y- parameter(Y), mM/hr	x- parameter (X), hr^{-1}
1.0	0.1591	0.17328
2.0	0.1581	0.19
3.0	0.1547	0.2079
4.0	0.150	0.2291
5.0	0.144	0.2548
6.0	0.1371	0.2882
6.5	0.13323	0.3092

The plot of y- parameter (Y) versus x- parameter (X) is shown in figure (2.6.1). It is a straight line plot, with

$$\text{slope} = K_m = 0.2015 \text{ mM}$$

$$y\text{- parameter} = r_{max} = 0.1956 \text{ mM/hr}$$

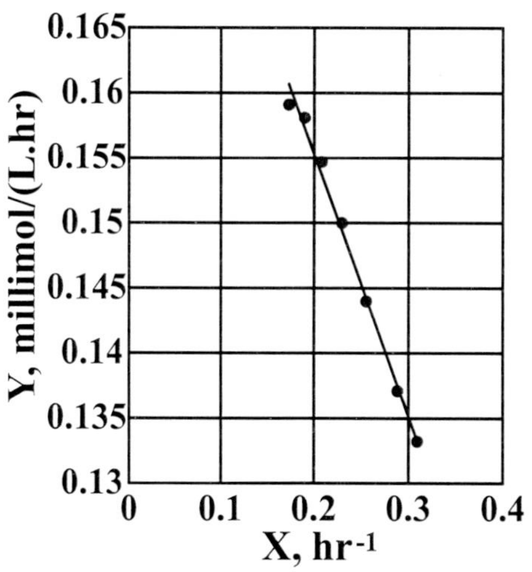

Figure 2.6.1: Michaelis – Menten kinetics (Integral analysis)

Since the data fit into a linear plot, it could be concluded that the process follows Michaelis – Menten kinetics.

(b) Differential Analysis:

For performing the differential analysis, we have to first estimate the values of the reaction rate, , either graphically (by plotting versus time and then finding the slope of the tangent to the curve at specified values of time) or by fitting the data into a correlation. It is seen that the reported data of versus fit into a third degree polynomial of the following form (with $R^2 = 0.9999$):

$$C_S = 1.0042 - 0.1654 + 0.0016\,t^2 + 0.005\,t^3 \quad \text{(iv)}$$

Therefore, $(-r_S) = (-dC_S / dt) = 0.1654 - 0.0032\,t - 0.0015\,t^2$ (v)

Using the above correlation, the C_S versus t data is made more elaborate by computing additional values of substrate concentration (C_S) at additional values of time (t). The results are tabulated in table (2.6.2). Computed values of reaction rate, $(-r_S)$, are also listed in this table.

Table – 2.6.2

Time, hr	**C_S, mM**	**$(-r_S)$, mM/hr**	**$(1/C_S)$, mM^{-1}**	**$1/(-r_S)$, hr/mM**
1.0	0.8409	0.1607	1.1892	6.22277
1.5	0.7614	0.1572	1.3134	6.3603
2.0	0.6838	0.153	1.4624	6.5359
2.5	0.6085	0.148	1.64335	6.7556
3.0	0.5359	0.1423	1.866	7.0274
3.5	0.4663	0.1358	2.144	7.3624
4.0	0.40	0.1286	2.50	7.776
4.5	0.33786	0.1206	2.9598	8.2901
5.0	0.2797	0.1119	3.5752	8.9365
5.5	0.2261	0.1024	4.4228	9.7632
6.0	0.1774	0.0922	5.6369	10.846
6.5	0.134	0.08122	7.4627	12.3122

(i) Let us start with Lineweaver – Burk equation / plot (equation 2.3.26) :

$$1/(-r_S) = (K_m / r_{max})(1/C_S) + (1/r_{max}) \quad \text{(vi)}$$

The values of $[1/(-r_S)]$ and those of $(1/C_S)$ are also listed in table (2.6.2). The plot of $[1/(-r_S)]$ versus $(1/C_S)$, namely the Lineweaver – Burk plot, is shown in figure (2.6.2). It is a linear plot, with

$$\text{slope} = (K_m / r_{max}) = 0.9908$$

$$y\text{- intercept} = (1/r_{max}) = 5.1954$$

Therefore, $r_{max} = (1/5.1954) = 0.1925$ mM/hr

$K_m = (0.9908/5.1954) = 0.1907$ mM

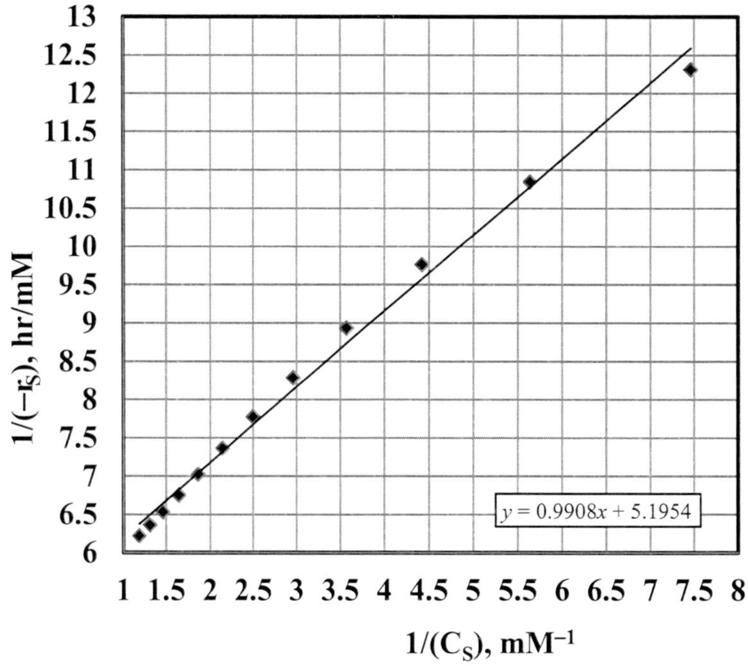

Figure 2.6.2: Lineweaver – Burk plot

(ii) Let us now consider Hanes – Woolf equation (2.3.27) :

$$[C_S/(-r_S)] = (1/r_{max})(C_S) + (K_m/r_{max}) \quad \text{(vii)}$$

The values of $[C_S/(-r_S)]$ at different values of C_S are listed in table (2.6.3) and the plot of $[C_S/(-r_S)]$ versus C_S is shown in figure (2.6.3).

Table – 2.6.3

Time, hr	C_S, mM	$(-r_S)$, mM/hr	$[C_S/(-r_S)]$, hr	$(-r_S)/C_S$, hr^{-1}
1.0	0.8409	0.1607	5.2327	0.1911
1.5	0.7614	0.1572	4.8427	0.2065
2.0	0.6838	0.153	4.4692	0.2237
2.5	0.6085	0.148	4.1107	0.24326
3.0	0.5359	0.1423	3.766	0.2655
3.5	0.4663	0.1358	3.4331	0.29128
4.0	0.40	0.1286	3.1104	0.3215
4.5	0.33786	0.1206	2.801	0.3570
5.0	0.2797	0.1119	2.50	0.40

Time, hr	C_S, mM	$(-r_S)$, mM/hr	$[C_S/(-r_S)]$, hr	$(-r_S)/C_S$, hr^{-1}
5.5	0.2261	0.1024	2.2074	0.453
6.0	0.1774	0.0922	1.9241	0.5197
6.5	0.134	0.08122	1.6498	0.6061

As can be seen from figure (2.6.3), the plot is linear with slope = 4.9902 hr/mM and y-intercept = 1.0701 hr.

Therefore, $(1/r_{max}) = 4.9902$

Or $r_{max} = 0.2$mM/hr

$$K_m = \left(\frac{1.0701}{4.9902}\right) = 0.214 \text{ mM}$$

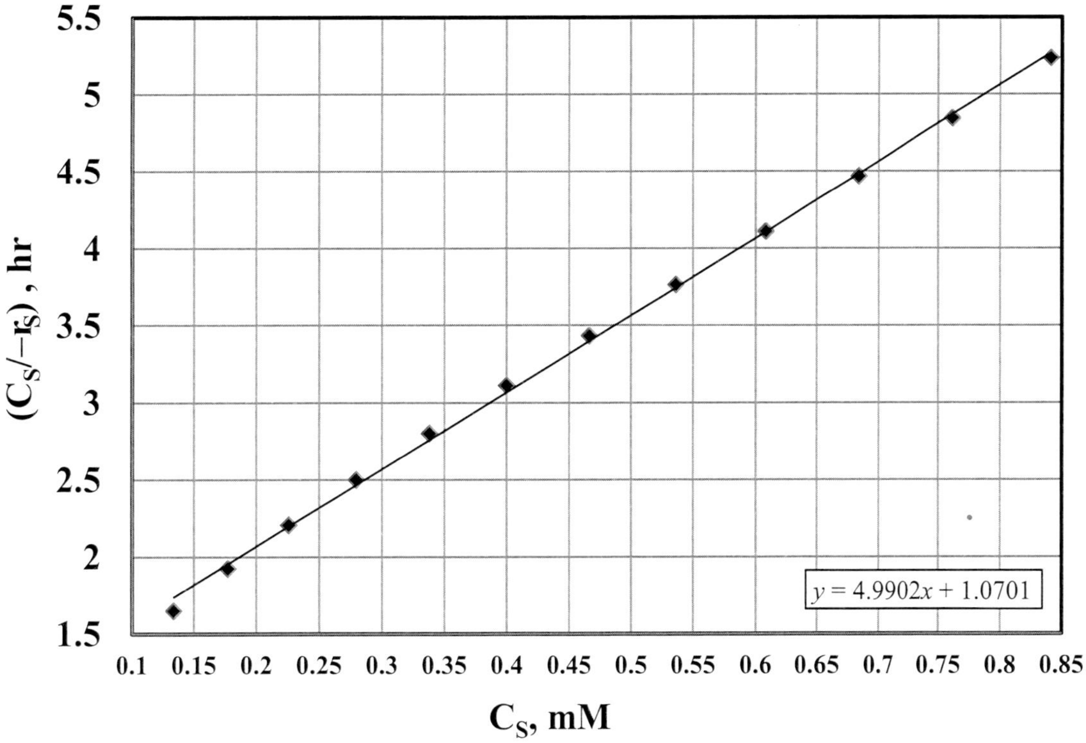

Figure 2.6.3: Hanes - Woolf plot

(iii) As per Eadie – Hofstee equation (equation 2.3.28),

$$(-r_S) = -K_m\ [(-r_S)/C_S] + r_{max} \qquad \text{(viii)}$$

The computed values of $(-r_S)$ and those of $[(-r_S)/C_S]$ are also listed in table (2.6.3). The plot between the two is shown in figure (2.6.4). Once again, the plot is linear, with

$$\text{slope} = K_m = 0.1987 \text{ mM}$$

$$y\text{- intercept} = r_{max} = 0.1954 \text{ mM/hr}$$

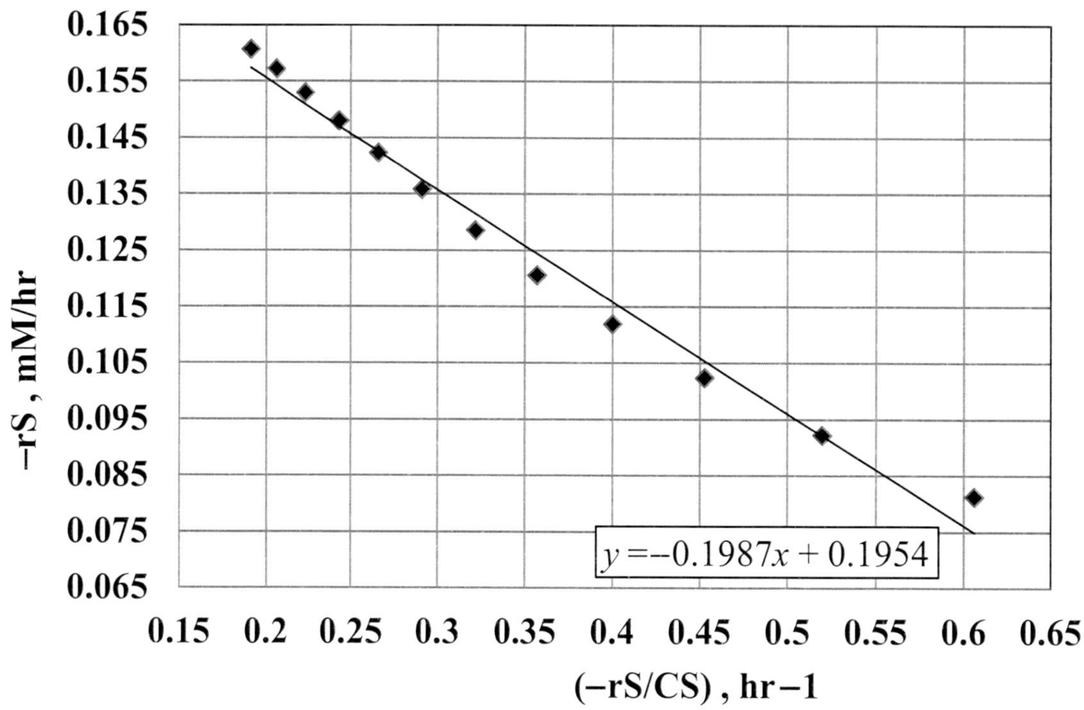

Figure 2.6.4: Eadie - Hofstee plot

(c) Approximate Differential Analysis:

For an approximate differential analysis, which avoids computation of $(-r_S)$ values either graphically or through curve fitting, we have to check whether the data compiled fit into the approximate form of equation (2.3.26):

$$\left(\Delta t / \Delta C_S\right) = \left(K_m / r_{max}\right)[1 / C_S\left(avg\right)] + \left(1 / r_{max}\right) \tag{ix}$$

The values of y- parameter, namely $\left(\Delta t / \Delta C_S\right)$ and those of the x- parameter $\left[1 / C_S\left(avg\right)\right]$ are computed from the data supplied and are presented in table (2.6.4). The plot between the two is shown in figure (2.6.5).

Table – 2.6.4

Time, hr	C_S, mM	$(\Delta C_S/\Delta t)$, mM/hr	$C_S(avg)$, mM	$(\Delta t/\Delta C_S)$, hr/mM	$[1/ C_S(avg)]$, mM^{-1}
1.0	0.8409				
1.5	0.7614	0.159	0.8011	6.2893	1.2483
2.0	0.6838	0.1552	0.7226	6.4433	1.384
2.5	0.6085	0.1506	0.64615	6.6401	1.5476
3.0	0.5359	0.1452	0.5722	6.887	1.7476
3.5	0.4663	0.1392	0.5011	7.1839	1.9956
4.0	0.40	0.1326	0.43315	7.5415	2.3087
4.5	0.33786	0.1243	0.3689	8.046	2.7107
5.0	0.2797	0.1163	0.3088	8.5984	3.2383
5.5	0.2261	0.1072	0.2529	9.3283	3.9541

Time, hr	C_S, mM	$(\Delta C_S/\Delta t)$, mM/hr	$C_S(avg)$, mM	$(\Delta t/\Delta C_S)$, hr/mM	$[1/C_S(avg)]$, mM^{-1}
6.0	0.1774	0.0974	0.20175	10.2669	4.9566
6.5	0.1340	0.0868	0.1557	11.5207	6.4226

It can be seen that figure (2.6.5) also displays a straight line plot with slope = 1.0283 hr and y- intercept = 5.122 hr/mM. Thus,

$$K_m = (1.0283)/(5.1222) = 0.2$$

$$r_{max} = (1/5.1222) = 0.19523 \text{ mM/h}$$

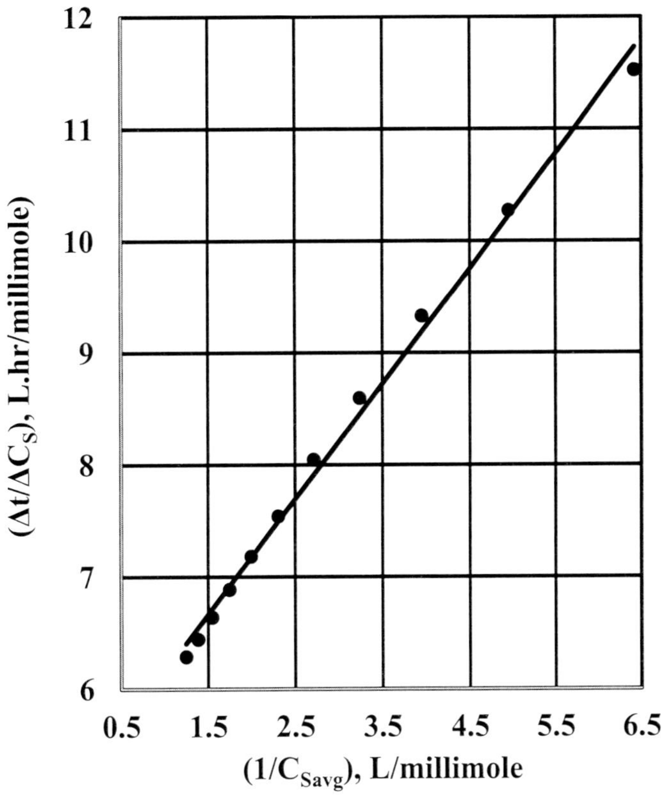

Figure 2.6.5: Michaelis – Menten kinetics (Approximate differential analysis)

Example 2.7 : Hydrolysis of benzoyl ethyl ester using the enzyme papain is studied in the laboratory and it is presumed that the substrate inhibition to enzyme activity could exist. The experimental data compiled are given below:

Initial substrate concentration, g/L	Substrate converted in first 10 min, g/L	Initial substrate concentration: 240g/L	
		Time, min	Substrate concentration, g/L
230	2.264	49.77	228.75
220	2.27	93.65	218.80
210	2.2755	133.86	209.66
200	2.2790	181.33	198.85

Initial substrate concentration, g/L	Substrate converted in first 10 min, g/L	Initial substrate concentration: 240g/L	
		Time, min	Substrate concentration, g/L
190	2.2807	214.30	191.33
180	2.2802	268.70	178.93
170	2.277	318.0	167.67
160	2.27115	356.8	158.88
150	2.2615	397.0	149.75
140	2.2477	446.7	138.56

Derive a kinetic equation for the process.

Solution:

From the first set of experimental data, that lists the values of initial rate of reaction (r_0) at different values of initial substrate concentration (C_{S0}), it can be observed that as the initial substrate concentration increases, the rate of reaction (r_0) first increases, reaches a maximum value and thereafter decreases. This indicates possible substrate inhibition to enzyme activity. It can be further seen that if we plot the initial rate of reaction (r_0) versus initial substrate concentration (C_{S0}) based on the given data, we shall obtain the same curve as that shown in the figure (2.5) of Section (2.3.2). Thus, from figure (2.5), the substrate concentration at which the initial rate of reaction is maximum can be read as close to 187 g/L. With allowable error therefore, we can take

$$C_{Sm} = 187.0 \text{ g/L}$$

Now, consider the data reported in the last two columns of the table (C_S versus t data at C_{S0} = 240 g/L). We could try to fit this data into the integrated form of Michaelis – Menten kinetic equation (that incorporates substrate inhibition to enzyme activity as well), namely equation (2.3.45). Thus,

$$Y = (K_m / r_{max})(X) + (1/r_{max}) \quad \text{(i)}$$

where $Y = y\text{- parameter} = t/(C_{S0} - C_S)$ (ii)

$$X = x\text{- parameter} = \left[\{\ln(C_{S0}/C_S)\}/(C_{S0} - C_S) + (C_{S0} + C_S)/(2C_{Sm}^{\ 2})\right] \quad \text{(iii)}$$

The values of x- parameter (X) and y- parameter (Y) are now computed and are listed in table (2.7.1).

Table – 2.7.1

Time, min	x-parameter (X), L/g	y- parameter (Y), (L.min/g)
49.77	0.01096985	4.4240
93.65	0.0109224	4.41745
133.86	0.010884	4.41199

Time, min	*x*-parameter (*X*), L/g	*y*- parameter (*Y*), (L.min/g)
181.33	0.0108456	4.40656
214.30	0.0108240	4.403123
268.70	0.0107980	4.39987
318.0	0.0107870	4.3965
356.8	0.0107880	4.3984
397.0	0.0107990	4.39889
446.7	0.0108280	4.403588

The plot of *y*- parameter (*Y*) versus *x*- parameter (*X*) is shown in figure (2.7.1). It can be seen that the data fit closely to a straight line plot. From the plot,

$$\text{slope} = \left(K_m / r_{max}\right) = 145.79$$

$$y\text{- intercept} = \left(1 / r_{max}\right) = 2.825$$

Therefore, $r_{max} = 0.354$ g/(L.min)

$K_m = (145.79/2.825) = 51.609$ g/L

Also, $K_{Si} = \left(C_{Sm}{}^2 / K_m\right) = (187.0)^2/(51.609) = 677.5756$ g/L

The kinetic equation representing the process is, therefore,

$$\left(-dC_S / dt\right) = \left(0.354 C_S\right) / (51.609 + C_S + C_S{}^2 / 677.57)$$

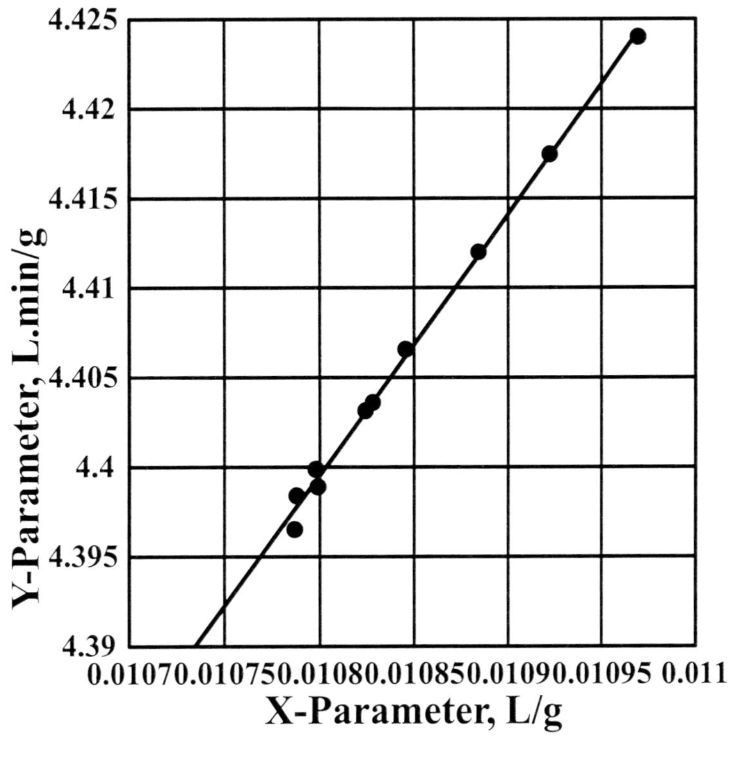

Figure 2.7.1: Enzyme – catalysed reaction with substrate inhibition (integral analysis)

Example 2.8 : The following data are collected when an enzyme – catalyzed reaction was studied in a laboratory bench scale reactor:

Initial substrate concentration, μM	**Initial rate of reaction, μM/ min**			
	C_{I0} **= 1.0 mM**	C_{I0} **= 2.0 mM**	C_{I0} **= 3.0 mM**	C_{I0} **= 4.0 mM**
3.0	0.1013	0.07824	0.0637	0.05377
4.0	0.1350	0.1042	0.085	0.0716
5.0	0.1684	0.1300	0.1060	0.0894
10.0	0.3350	0.2585	0.2105	0.1775
30.0	0.9818	0.7565	0.6154	0.5186

Determine the kinetic equation of the reaction. What is the nature of inhibition (competitive, uncompetitive or mixed)?

Solution:

The determine the nature of inhibition and thereby to develop the kinetic equation, let us first fit the data into Lineweaver – Burk equation for each value of C_{I0} (initial inhibitor concentration). For this, the value of $(1/r_0)$ and $(1/\ C_{S0})$ are first computed and are listed in table (2.8.1)

Table – 2.8.1

$(1/\ C_{S0})$**, μM**$^{-1}$	$(1/r_0)$**, min/μM**			
	C_{I0} **= 1.0 mM**	C_{I0} **= 2.0 mM**	C_{I0} **= 3.0 mM**	C_{I0} **= 4.0 mM**
0.3333	9.87167	12.781186	15.698587	18.5977
0.250	7.4074	9.5969	11.7647	13.96648
0.20	5.9382	7.6923	9.43396	11.18568
0.10	2.98507	3.8685	4.7506	5.6338
0.0333	1.0185	1.32187	1.62496	1.92827

The plot of $(1/r_0)$ versus $(1/C_{S0})$ at the four different values of C_{I0} (C_{I0} = 1.0, 2.0, 3.0 and 4.0 mM) are shown in figure (2.8.1) . All the four are straight line plots with different values of slope and y- intercept. This indicates a case of mixed inhibition to enzyme activity. The values of slope and y- intercept are listed below (in table - 2.8.2).

Table – 2.8.2

C_{I0}**, mM**	**Slope, min**	y**- intercept, min/μM**
1.0	29.506	0.0351
2.0	38.197	0.0498
3.0	46.883	0.06
4.0	55.561	0.0769

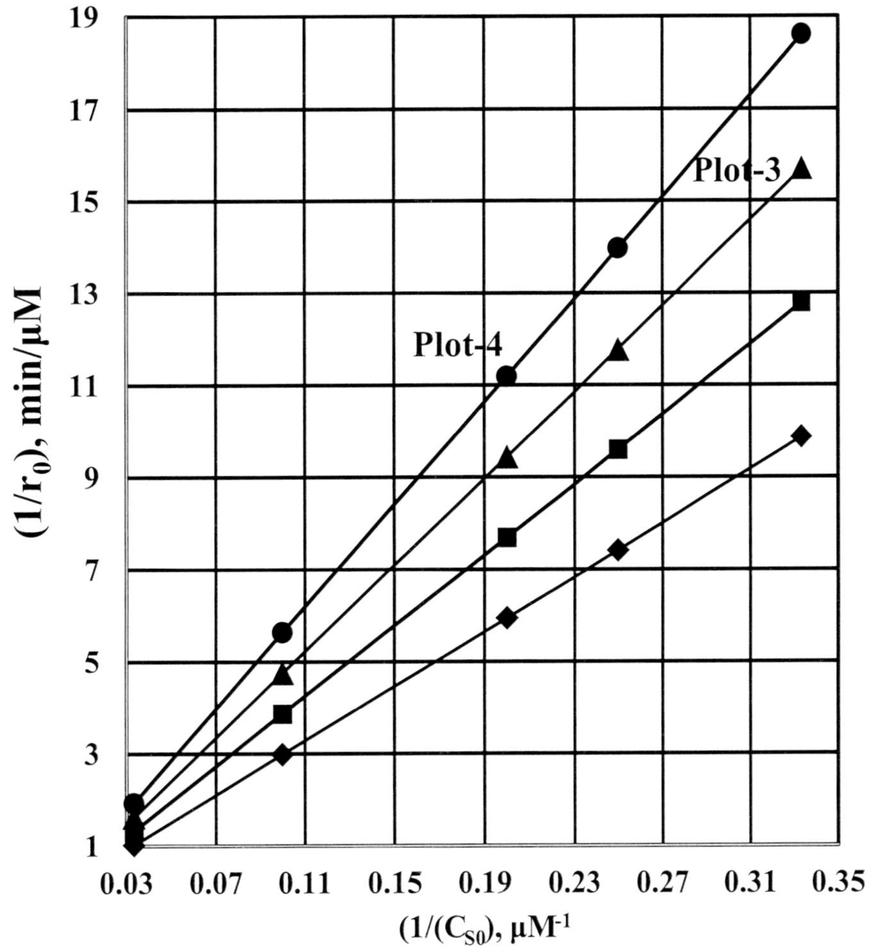

Figure 2.8.1

Now, from equation (2.3.75), the slope of each plot is given by,

$$\text{slope} = K_m(app)/r_{max}(app)$$

$$= \left[K_m/\left(K_i r_{max}\right)\right]C_{I0} + \left(K_m/r_{max}\right) \quad \text{(i)}$$

And, y- intercept $= 1/r_{max}(app)$

$$= \left[1/\left(K_i' r_{max}\right)\right]C_{I0} + (1/r_{max}) \quad \text{(ii)}$$

It is, therefore, apparent that if we plot the slope – values against C_{I0} values (listed in table 2.8.2), we must get a straight line with slope $= \left[K_m/\left(K_i r_{max}\right)\right]$ and y- intercept $= \left(K_m/r_{max}\right)$. This plot is shown in figure (2.8.2). It is a straight line plot with slope = 8.6869 min/mM and y- intercept = 20.818 min. Therefore,

$$\left[K_m/\left(K_i r_{max}\right)\right] = 8.6869 \text{ min/ mM}$$

$$(K_m/r_{max}) = 20.818 \text{ min}$$

and K_i = (20.818/8.6869) = 2.39648 mM.

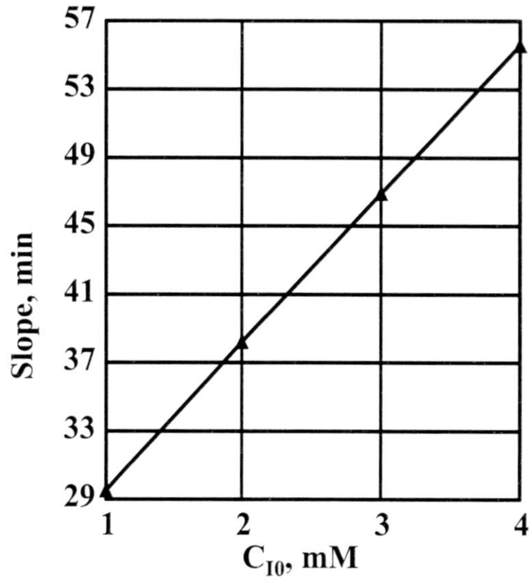

Figure 2.8.2

Similarly, as per equation (ii), the plot of y- intercept values (listed in table 2.8.2) versus C_{I0} values must also yield a straight line with slope = $\left[1/\left(K_i^{'}r_{max}\right)\right]$ and y- intercept = $(1/r_{max})$. This plot is shown in figure (2.8.3), which is also a linear plot and has a slope = 0.0136 min (μM. mM) and y- intercept = 0.0216 min/μM. Therefore,

$$K_i^{'}r_{max} = (1/0.0136)$$

$$r_{max} = (1/0.0216) = 46.2963 \ \mu\text{M/min} = 0.04629 \ \text{mM/min}$$

And $K_i^{'} = (0.0216/\ 0.0136) = 1.5882$ mM

Since $(K_m/r_{max}) = 20.818$ (determined earlier),

$$K_m = (20.818)(0.04629) = 0.96379 \ \text{mM}$$

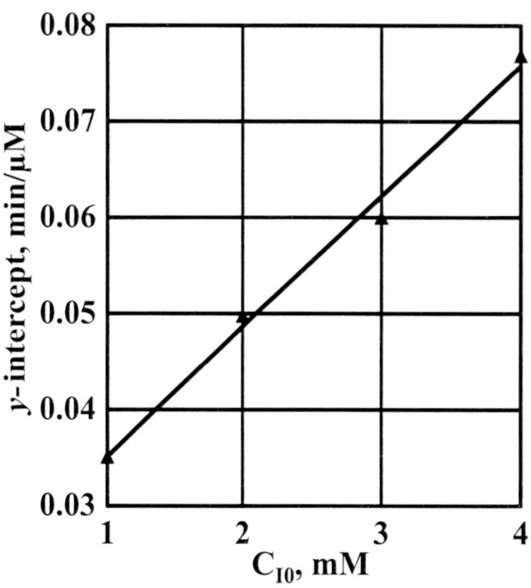

Figure 2.8.3

The kinetic equation representing the process is (as per equation 2.3.72),

$$\left(-dC_S / dt\right) = r_{max}\left(app\right)C_S / \left[K_m\left(app\right) + C_S\right]$$

where $r_{max}\left(app\right) = 0.04629\left[1 + C_I / (1.5882)\right]$

$K_m\left(app\right) = 0.96379\left[1 + C_I / (2.39648)\right] / \left[1 + C_I / (1.5882)\right]$

Example 2.9:
Trans-esterification of jatropha oil using methyl acetate in presence of lipase enzyme as the catalyst (to produce biodiesel, jatropha oil methyl ester) is studied in the laboratory in a 1.0 L chemostat and the data collected are summarized below. The feed mixture was composed of oil and methyl acetate in the molar ratio of 1:5.

Feed flow rate, L/hr	Steady state concentration of oil in the exit stream, mol/L
12.8648	0.645
9.2153	0.615
6.4583	0.573
5.384	0.547
4.6852	0.525
4.1193	0.503
3.445	0.470
3.4079	0.448
2.711	0.422
2.50	0.405

Develop a kinetic equation for the process.

Take density of oil = 907.2 kg/ m^3· and that of methyl acetate = 937 kg/ m^3, molecular weight of jatropha oil = 887.0 g/mol, that of methyl acetate = 74.0 g/ mol.

Solution:

As for trans-esterification of vegetable oil with methyl acetate, we could try to fit the experimental data compiled into a kinetic equation of the form given in the equation (2.3.91). When rewritten for the chemostat, equation (2.3.91) becomes

$$y = a_0 + a_1x_1 + a_2x_2 + a_3x_3 \tag{i}$$

where, y = y-parameter = $\tau / [\left(C_{S1}\right)_O - \left(C_{S1}\right)_e]$

$$a_0 = \left(1 / r_{max}\right),\quad a_1 = \left(K_{m2} / r_{max}\right),\ a_2 = \left(K_{m1} / r_{max}\right),\ a_3 = K_{m2} / (r_{max}K_{Si})$$

Also, $x_1 = \left(1 / C_{S1}\right)$, $x_2 = \left(1 / C_{S2}\right)$, $x_3 = \left(C_{S2} / C_{S1}\right)$.

In the above equations, $C_{S1} = \left(C_{S1}\right)_e$, $C_{S2} = \left(C_{S2}\right)_e$. We have avoided the suffix "*e*" as a matter of convenience. This notation shall be used throughout the solution of this Example.

Since the oil and methyl acetate are used in 1:5 ratio (by mole),

$$\left(C_{S1}\right)_O = 1 / \left[\left(M_A / \rho_A\right) + \left(5M_B / \rho_B\right)\right]$$

where, M_A = molecular weight of oil = 887.0 g/mol

ρ_A = density of oil = 907.2 g/L

M_B = molecular weight of methyl acetate = 74.0 g/mo

ρ_B = density of methyl acetate = 937 g/L

Therefore, $(C_{S1})_O = 0.7285$ mol/L

$(C_{S2})_O = 5(0.7285) = 3.6425$ mol/L

Also, since 3 moles of methyl acetate react with one mole of oil (as per stoichiometry),

$$C_{S2} = (C_{S2})_O - 3\left[(C_{S1})_O - C_{S1}\right]$$

$$= 5\ (C_{S1})_O - 3(C_{S1})_O + 3C_{S1} = 2\ (C_{S1})_O + 3C_{S1}$$

$$= (1.457 + 3C_{S1}) \tag{ii}$$

The experimental data can be fitted into equation (i) using matrix algebra (similar to that shown in equation 2.2.67) as

$$\begin{bmatrix} y \\ x_1 y \\ x_2 y \\ x_3 y \end{bmatrix} = \begin{bmatrix} N & x_1 & x_2 & x_3 \\ x_1 & x_1^2 & x_1 x_2 & x_1 x_3 \\ x_2 & x_1 x_2 & x_2^2 & x_2 x_3 \\ x_3 & x_1 x_3 & x_2 x_3 & x_3^2 \end{bmatrix} \begin{bmatrix} a_0 \\ a_1 \\ a_2 \\ a_3 \end{bmatrix} \tag{iii}$$

where, N = no. of data points = 10.

All the elements of the coefficient matrix as well as those of the LHS matrix are computed from the experimental data presented and are listed in table (2.9.1). To note that the values of C_{S2} have been computed from equation (ii). Also,

$$\tau = (1/Q_0)hr = (60/Q_0)\ \text{min} \tag{iv}$$

where Q_0 = feed flow rate, L/hr

Table – 2.9.1 A

Q_0, L/hr	τ, min	y, min/M	C_{S1}, M	C_{S2}, M	x_1, M^{-1}	x_2, M^{-1}	x_3
12.8648	4.6639	55.85478	0.645	3.392	1.550387	0.2948	5.2589
9.2153	6.5109	57.364968	0.615	3.320	1.6260	0.30284	5.3691
6.4583	9.2903	59.74477	0.573	3.176	1.7452	0.31486	5.54275
5.384	11.144	61.40073	0.547	3.098	1.82815	0.322789	5.66362
4.6852	12.8062	62.92986	0.525	3.032	1.90476	0.329815	5.7752
4.1193	14.5656	64.5926	0.503	2.966	1.98807	0.33715	5.8966
3.445	17.4173	67.3782	0.470	2.867	2.12766	0.34879	6.10
3.0794	19.48437	69.46295	0.448	2.801	2.23214	0.357015	6.2522
2.711	22.13136	72.206719	0.422	2.723	2.36967	0.36724	6.4526

Q_0, L/hr	τ, min	y, min/M	C_{S1}, M	C_{S2}, M	x_1, M^{-1}	x_2, M^{-1}	x_3
2.50	24.0	74.190979	0.405	2.672	2.46913	0.37425	6.5975
		$\Sigma y =$ 645.12658			$\Sigma x_1 =$ 19.84117	$\Sigma x_2 =$ 3.349565	$\Sigma x_3 =$ 58.9086

Table – 2.9.1 B

x_1^2, M^{-2}	x_2^2, M^{-2}	x_3^2	(x_1x_2), M^{-2}	(x_1x_3), M^{-1}	(x_1y) min/M^2	(x_2y) min/M^2	(x_3y), min /M
2.4037	0.086907	27.65603	0.457054	8.15333	86.5965	16.466	293.7347
2.64387	0.091716	28.8272	0.492427	8.73015	93.27544	17.3728	307.9983
3.04572	0.099137	30.72215	0.549494	9.67322	104.26657	18.8112	331.1507
3.342132	0.104193	32.07659	0.590107	10.35394	112.24974	19.8195	347.7504
3.62811	0.108778	33.35337	0.628218	11.0004	119.86628	20.7552	363.43492
3.95242	0.113673	34.77013	0.670285	11.72289	128.4146	21.7776	380.8780
4.52694	0.121658	37.210	0.742119	12.07872	143.3579	23.5012	411.0071
4.98245	0.127459	39.09038	0.796907	13.95585	155.05103	24.7993	434.2983
5.615336	0.134865	41.63605	0.870237	15.29053	171.10609	26.5172	465.92108
6.09663	0.14006	43.5274	0.924074	16.2902	183.18762	27.7659	489.4772
$\Sigma x_1^2 =$ 40.237327	$\Sigma x_2^2 =$ 1.1284499	$\Sigma x_3^2 =$ 348.86934	$\Sigma x_1x_2 =$ 6.7209258	$\Sigma x_1x_3 =$ 118.14929	$\Sigma(x_1y) =$ 1297.37186	$\Sigma(x_2y) =$ 217.5861	$\Sigma(x_3y) =$ 3825.6508

To note that $\Sigma(x_2x_2) = \Sigma x_1 = 19.84117$.

On substituting the numerical values of all elements of the coefficient matrix and LHS matrix, equation (iii) becomes,

$$\begin{bmatrix} 645.12658 \\ 1297.37186 \\ 217.5861 \\ 3825.65083 \end{bmatrix} = \begin{bmatrix} 10 & 19.84117 & 3.349565 & 58.9086 \\ 19.84117 & 40.237317 & 6.7209258 & 118.14929 \\ 3.349565 & 6.7209258 & 1.1284499 & 19.84117 \\ 58.9086 & 118.14929 & 19.84117 & 348.86934 \end{bmatrix} \begin{bmatrix} a_0 \\ a_1 \\ a_2 \\ a_3 \end{bmatrix}$$

On solving,

$$a_0 = 0.51282$$

$$a_1 = a_3 = 8.10256$$

$$a_2 = 0.57435$$

Therefore, $r_{max} = (1/a_0) = 1.95$ M/min

$$K_{m1} = (a_2/a_0) = 1.12 \text{ mol /L}$$

$$K_{m2} = (a_1/a_0) = 15.8 \text{ mol /L}$$

$$K_{Si} = (a_1/a_3) = 1.0 \text{ mol /L}$$

The kinetic equation representing the trans-esterification process is, therefore,

$$\left(-dC_{S1}/dt\right)=\left(1.95\right)C_{S1}C_{S2}/\left[C_{S1}C_{S2}+1.12C_{S1}+15.8C_{S2}+15.8C_{S2}^{2}/1.0\right]$$

2.4 Kinetics of Heterogeneous Biochemical Processes

As discussed in Chapter-1, heterogeneous biochemical processes could be broadly classified into two categories, such as

(a) Biofilm systems (processes employing attached growth of microbes).
(b) Processes employing immobilized cells / immobilized enzyme.

Biofilm systems / reactors, as stated in Chapter -1, mostly employ support particles (silica granules, polymer beads, activated carbon particles), each particle being surrounded by a thin biofilm (a thin layer of microbial solution), each thus forming a particle- biofilm aggregate (see figure 1.3 of Chapter – 1). These particle – biofilm aggregates form a packed bed or fluidized bed inside a tall column (bioreactor). Due to sloughing phenomenon, the cell mass concentration in the biofilm (x_f) and average thickness (δ) of the biofilm remain more or less constant throughout the operation of the bioreactor. Microbes do grow and multiply in the biofilm and the thickness of the biofilm does tend to increase. However, when the thickness of the film crosses a particular value (around 0.3 mm), the film gets detached from the particle surface and falls into the surrounding liquid bulk, to be immediately replaced by a fresh biofilm on the solid surface. This is what is called the *sloughing phenomenon* (discussed in Section (1.1) of Chapter – 1). The size $\left(d_{pm}\right)$ and density $\left(\rho_{Sm}\right)$ of each particle – biofilm aggregate shall be thus,

$$d_{pm}=\left(d_{p}+2\delta\right) \tag{2.4.1}$$

$$\rho_{Sm}=\left(1-f\right)\rho_{S}+f\rho_{m} \tag{2.4.2}$$

where ρ_S, ρ_m = density of support particle and that of microbial solution, respectively

f = volume fraction of biofilm in particle – biofilm aggregate.

= (volume of biofilm) / (volume of particle – biofilm aggregate)

$$=\left[\left(\pi d_{Pm}^{3}/6\right)-\left(\pi d_{P}^{3}/6\right)\right]/\left(\pi d_{Pm}^{3}/6\right) \tag{2.4.3}$$

$$=1-\left(d_{P}/d_{Pm}\right)^{3} \tag{2.4.4}$$

d_P = diameter of support particle

δ = average thickness of biofilm

The substrate diffuses into the biofilm, where it is consumed by the microbial cells as food and partly converted into product(s). The product diffuses out into the liquid bulk. Due to the additional diffusional resistance that comes into play, the actual rate of bioconversion (also called the global rate of reaction), $\left(-r_S\right)$, shall be lower than the intrinsic rate, $\left(-r_S\right)\left(int\right)$. In other words,

$$\left(-r_S\right)=\eta\left(-r_S\right)\left(int\right) \tag{2.4.5}$$

where η is called the *effectiveness factor,* which takes into account the additional resistance to substrate transfer into the biofilm. The value of η usually ranges from 0.75 to 0.95.

In spite of this additional resistance to substrate transfer, the overall rate of bioconversion in attached growth systems like this shall be distinctly higher than that in suspended growth systems, mainly due to the substantially high cell mass concentration in the biofilm. The microbial cells are distributed within the very small volume of the biofilm and consequently, the magnitude of x_f shall be much larger than that of x in a tank reactor (in which suspended growth of microbes occurs).

The intrinsic rate of reaction $(-r_S)(int)$, is what is predicted by any of the kinetic equations discussed in the earlier sections. However, when applied to biofilm systems (in which attached growth of microbes occurs), a few modifications shall be necessary. Firstly, the cell mass concentration in the biofilm is no more a variable, since it remains constant at x_f throughout the process. Secondly, since x_f is defined as (mass of microbial cells)/(volume of biofilm), we must multiply it by $\left[f(1-\in)/\in\right]$ to re – express it as mass of microbial cells per unit of volume of solution/liquid. Here, $\in$ is the fractional liquid holdup (volume fraction of the liquid) in the system/ bioreactor. For example, in the case of a packed bed reactor, $\in$ shall be equal to the void fraction or voidage of the bed (also called the porosity of the bed), assuming all the voids are completely filled with the liquid/solution.

It is also necessary to keep in mind that the substrate has to first diffuse from the liquid bulk to the surface of the particle – biofilm aggregate (thereafter, it diffuses into the biofilm). The resistance to substrate transport from liquid bulk to biofilm surface (biofilm – liquid interface) is predicted by the liquid to particle mass transfer coefficient (k_{LP}). Accordingly, if C_S represents the substrate concentration in the liquid bulk, then that at the liquid – biofilm interface shall be C_{SP}, the magnitude of C_{SP} being lower than that of C_S. Therefore, it shall be more accurate to express equation (2.4.5) as

$$(-r_S)=\eta(-r_{SP})(int) \tag{2.4.6}$$

where, $(-r_{SP})(int)$ is the intrinsic rate of reaction expressed in terms of C_{SP}. No doubt, in many cases, the value of k_{LP} has been observed to be significantly large, and in such cases, $C_{SP} \approx C_S$.

Thus, if, for example, the intrinsic rate of bioconversion is predicted by Monod's kinetic equation (equations 2.2.8 and 2.2.13),

$$(-r_{SP})(int)=(\mu_m/Y)\left[x_f\, f(1-\in)/\in\right]C_{SP}/(K_S+C_{SP}) \tag{2.4.7}$$

$$=\mu_m(app)C_{SP}/(K_S+C_{SP}) \tag{2.4.8}$$

where $\mu_m(app)=(\mu_m/Y)\left[x_f\, f(1-\in)/\in\right]$. (2.4.9)

It can be seen that the above equation is analogous to Michaelis – Menten kinetic equation (equation 2.3.10), with r_{max} being replaced by $\mu_m(app)$ and K_m by K_S.

In the same way, if the bioconversion intrinsically follows Haldane – Andrews kinetic equation (equation 2.2.53), then,

$$(-r_{SP})(int) = (\mu_m / Y) C_{SP} \left[x_f f(1-\in)/\in \right] / \left[K_S + C_{SP} + \left(C_{SP}^2 / K_{Si} \right) \right] \quad (2.4.10)$$

$$= \mu_m (app) C_{SP} / \left[K_S + C_{SP} + \left(C_{SP}^2 / K_{Si} \right) \right] \quad (2.4.11)$$

However, as discussed in Chapter – 1, neither substrate inhibition nor product inhibition to microbial growth is likely to be significant in biofilms, since neither the substrate nor the product tends to accumulate in these films. Accordingly, in many applications, the term containing K_{Si} could be neglected and the above equation would reduce to the Monod – type kinetic equation given earlier (equation – 2.4.8).

In the case of immobilized cell systems also (popular methods of cell immobilization within solid particles/within solid matrix are discussed in Section 1.1 of Chapter – 1), the cell mass concentration remains constant at x_p. However, since x_p is expressed as mass of microbial cells per unit volume of support particle, it has to be multiplied by $\left[(1-\in)/\in\right]$, to re-express as mass of microbial cells per unit volume of liquid/solution. Thus, if the intrinsic rate follows Monod's kinetics, then,

$$(-r_{SP})(int) = \mu_m (app) C_{SP} / (K_S + C_{SP}) \quad (2.4.12)$$

where $\mu_m (app) = (\mu_m / Y) x_P (1-\in)/\in$ (2.4.12a)

In this case also, the substrate has to diffuse into the particle, where it meets the microbial cells and the bioconversion occurs. Accordingly, the actual rate of bioconversion (the global rate) shall be that predicted by equation (2.4.6).

As stated earlier, in most commercial bioreactors, enzymes are used in the immobilized form. The system thus becomes heterogeneous, the resistance to substrate transfer into the particle comes into play and the global rate / actual rate of bioconversion becomes that predicted by equation (2.4.6). The intrinsic rate of bioconversion could be predicted by Michaelis – Menten kinetic equation or by any of its modified forms (discussed in the earlier section). However, in the case of immobilized enzyme systems, the kinetic constant r_{max} is mostly expressed as moles of substrate converted per unit time per unit mass of catalyst, such as mol/(s. gm catalyst). In such cases, it would have to be multiplied by $[\rho_S (1-\in)/\in]$, to re-express it in mol /(L.s) or kmol / (m^3.s). Thus, for heterogeneous systems employing immobilized enzyme, equation (2.3.10) gets modified to

$$(-r_{SP})(int) = r_{max} (app) C_{SP} / (K_S + C_{SP}) \quad (2.4.13)$$

where $r_{max} (app) = r_{max} \rho_S (1-\in)/\in$. (2.4.14)

2.4.1 Computation of Effectiveness Factor

It is thus evident that the effectiveness factor (η) is an important parameter involved in the kinetic analysis of heterogeneous systems. It is the parameter that predicts the resistance to substrate transfer into the particle/ biofilm and is a complex function of the effective diffusivity (D_e) of substrate into the particle / biofilm, the characteristic dimension of the particle / biofilm (L^*) and the kinetic constants. Kinetic experiments / investigations have demonstrated that a reasonable estimate of η could be obtained from the following correlation:

$$\eta = [Coth(3\varnothing)/\varnothing] - 1/(3\varnothing^2) \quad (2.4.15)$$

where $\varnothing$ = Thiele – type modulus = $L^*(-r_{SP})(int)/\sqrt{2D_e I}$ (2.4.16)

D_e = effective diffusivity of substrate into biofilm / particle

L^* = characteristic dimension of the particle/biofilm = (V_P/A_P) (2.4.17)

V_P, A_P = volume and surface area of particle/biofilm respectively

$$I = \int_0^{C_{SP}} (-r_{SP})(int)\,dC_S \quad (2.4.18)$$

In the case of immobilized cell or immobilized enzyme systems, assuming the support particle to be spherical,

$$L^* = (V_P / A_P) = (\pi d_P^3 / 6)/(\pi d_P^2) = (d_P / 6) \quad (2.4.19)$$

In the case of the particle - biofilm aggregates,
L^* = (volume of the biofilm)/(surface area of biofilm)

$$= \left[(\pi d_{Pm}^3/6) - (\pi d_P^3/6)\right]/(\pi d_{Pm}^2)$$

$$= (d_{Pm}^3 - d_P^3)/(6 d_{Pm}^2) \quad (2.4.20)$$

The integral of equation (2.4.18) could be evaluated either analytically (if the kinetic equation is not too nonlinear) or numerically using any of the popular numerical techniques such as the Simpson's rule.

As for example, let us consider a bioprocess that intrinsically follows Monod's kinetic equation (equation 2.4.8). Then, we can evaluate the integral of equation (2.4.18) analytically, after substituting equation (2.4.8) (expressed in terms of C_S) into it. Thereafter, if we substitute the result in equation (2.4.16), we get the expression for Thiele – type modulus $(\varnothing)$ as

$$\varnothing = (1/\eta_d) \quad (2.4.21)$$

where $\eta_d = (\sqrt{2}/\varnothing')[(1+\beta)/\beta]\sqrt{\beta - \ln(1+\beta)}$ (2.4.22)

$$\beta = (C_{SP}/K_S) \quad (2.4.23)$$

$$\varnothing' = L^*\sqrt{\mu_m(app)/(K_S D_e)} \quad (2.4.24)$$

In the case of an enzyme - catalyzed process, that intrinsically follows Michaelis – Menten kinetics (equation 2.4.13) also, the expression for Thiele – Type Modulus ($\varnothing$) shall be same as that given above, except that

$$\beta = (C_{SP}/K_m) \quad (2.4.25)$$

$$\varnothing' = L^*\sqrt{r_{max}(app)/(K_m D_e)} \quad (2.4.26)$$

Once the value of $\varnothing$ is known, the value of effectiveness factor (η) could be computed from equation (2.4.15).

An alternate correlation for the computation of η has been proposed by Gottifreddi and Gonzo [16]. Their correlation is given by:

$$(1/\eta)^2 = (1/\eta_d)^2 + \exp\left[(6/5)(\varnothing')^2 /(1+\beta)^2 - (1/\eta_d)^2\right] \quad (2.4.27)$$

The above correlation has been developed principally for bioconversion processes whose intrinsic kinetics follow Monod – type kinetic equation (2.4.8) or Michaelis – Menten kinetic equation (2.4.13).

Narayanan [17] has reported that the above correlation and equation (2.4.15) predict comparable values of η, within 8–10 % deviation.

Semi-empirical correlations have also been reported in literature for the estimation of η. Such a correlation, basically developed for enzyme-catalyzed processes intrinsically following Michaelis – Menten kinetics, is presented below [18]:

$$\eta = 1 - \left[\tanh(\varnothing')/\varnothing'\right]\left[\eta_d / \tanh(\eta_d) - 1\right], \text{ for } \eta_d \le 1 \quad (2.4.28)$$

$$= \eta_d - \left[\tanh(\varnothing')/\varnothing'\right]\left[1/\tanh(\eta_d) - 1\right], \text{ for } \eta_d \ge 1 \quad (2.4.29)$$

Since semi-empirical in nature, application of this correlations to industrial bioreactor design must be done with caution.

In the case of bioprocesses that intrinsically follow any alternate type of kinetic equation, the effectiveness factor (η) could be computed from equations (2.4.15) to (2.4.18), the integral being evaluated either analytically or numerically.

For example, if the bioprocess intrinsically follows Tessier's kinetic equation , then

$$(-r_S)(int) = \mu_m(app)\left[1 - \exp(-C_S / K_S)\right] \quad (2.4.30)$$

Substituting the above expression for $(-r_S)(int)$ in equation (2.4.17), it can be integrated analytically. Therefore, from equation (2.4.16),

$$\varnothing = \left(\varnothing' / \sqrt{2}\right)\left[1 - \exp(-\beta)\right] / \sqrt{(\beta - 1) + \exp(-\beta)} \quad (2.4.31)$$

where $\varnothing'$ and β are those defined in equations (2.4.23) and (2.4.24) respectively. Equation (2.4.15) then predicts the value of .

Similarly, for bioprocesses whose intrinsic kinetics follow Contois' kinetic equation,

$$(-r_S)(int) = \mu_m(app) C_S / \left[K_C(app) + C_S\right] \quad (2.4.32)$$

where $K_C(app) = \left[K_C x_f f(1-\in)/\in\right]$ (2.4.33)

It can be seen that the above equation is exactly similar to equation - 2.4.8 (Monod's kinetic equation), except that K_S has been replaced by $K_C(app)$. Accordingly, η could be computed from equation (2.4.21) coupled with equation (2.4.15), or from Gottifreddi and Gonzo's correlation (equation – 2.4.27).

2.5 Computation of Liquid – to – Particle Mass Transfer Coefficient

As we discussed in the earlier section, in the case of heterogeneous systems, the kinetic equation representing the intrinsic kinetics of the bioprocess under consideration is to be expressed in terms of C_{SP}, which is the substrate concentration at the particle - liquid or biofilm – liquid interface. The magnitude of C_{SP} shall be lower than that of C_S (substrate concentration in the liquid bulk), though the difference, in many cases, shall not be significant. This is because the substrate has to first diffuse from the liquid bulk to the surface of the particle/biofilm and thereafter, it diffuses into the biofilm/particle. The resistance to the transport of the substrate from the liquid bulk to the particle/biofilm surface is dictated by the liquid – to – particle mass transfer coefficient, k_{LP}. Under steady state conditions,

$$k_{LP}(a)(C_S - C_{SP}) = \eta(-r_{SP})(int) \tag{2.5.1}$$

where k_{LP} = liquid - to - particle mass transfer coefficient, m/s

a = specific interfacial area for mass transfer, m^2/m^3

It shall not be too erroneous to assume that the interfacial area for mass transfer (a) is equal to the total surface area of the particles or the particle – biofilm aggregates. Accordingly,

$$a = s_P(1-\in)/\in \tag{2.5.2}$$

where s_P = specific surface area (surface area per unit volume) of each particle or particle - biofilm aggregate, m^2/m^3

$\in$ = fractional liquid holdup in the bioreactor

In the case of particle - biofilm aggregates,

$$s_P = (\pi d_{pm}^{\,2}) / \left[\pi(d_{pm}^{\,3})/6\right]$$

$$= (6/d_{pm}) \tag{2.5.3}$$

Similarly, in the case of immobilized cell / immobilized enzyme bioreactors,

$$s_P = (\pi d_p^{\,2}) / (\pi d_p^{\,3}/6)$$

$$= (6/d_p) \tag{2.5.4}$$

where d_p = diameter of support particle within which the cells / enzyme have been immobilized.

In the case of three phase (gas – liquid – solid) bioreactors,

$$a = s_P(1-\in)/\in_L \tag{2.5.5}$$

where $\in = (\in_g + \in_L)$. (2.5.5a)

$\in_g$, $\in_L$ = fractional gas holdup and fractional liquid holdup respectively in the bioreactor.

$\in$ = total fluid holdup in the bioreactor.

The mass transfer coefficient (k_{LP}) is to be estimated from any of the experimental correlations reported in literature. For example, Humphrey and Vanness [19] have proposed the following correlation for the estimation of liquid to particle mass transfer coefficient in two phase, agitated vessels (stirred tanks) :

$$Sh_T = 0.0032(Re_T)^{0.87}(Sc_L)^{0.5} \tag{2.5.6}$$

where Sh_T = Sherwood number based on tank diameter $= \left(k_{LP} D_t / D_{SL}\right)$ (2.5.7)

D_t = vessel / tank diameter

D_{SL} = liquid phase diffusivity of solute (substrate)

Re_T = Reynolds number based on tank diameter $= \left(nD_t^2 \rho_L / \mu_L\right)$ (2.5.8)

n = speed (rps) of impeller

Sc_L = liquid phase Schmidt number = $\mu_L / \left(D_{SL} \rho_L\right)$ (2.5.9)

The above correlation is for baffled vessels agitated by turbine impeller. In the case of propeller– agitated baffled vessels,

$$Sh_T = 0.13\left(Re_T\right)^{0.58}\left(Sc_L\right)^{0.5} \tag{2.5.10}$$

Barker and Treybal [20] have proposed an alternate correlation such as,

$$Sh_T = 0.02\left(Re_m\right)^{0.833}\left(Sc_L\right)^{0.5} \tag{2.5.11}$$

where Re_m = mixing Reynolds number (also called impeller Reynolds number)

$$= \left(nD_a^2 \rho_L / \mu_L\right) \tag{2.5.12}$$

D_a = impeller diameter

In the case of three phase (gas – liquid – solid) agitated vessels (such as stirred tank reactors conducting aerobic processes), Grisafi et.al [21] have proposed the following experimental correlation for the estimation of liquid – to – particle mass transfer coefficient (k_{LP}):

$$Sh_P = 2.0 + 0.30\left(Re_{Pm}\right)^{0.75}\left(Sc_L\right)^{1/3} \tag{2.5.13}$$

where Sh_P = Sherwood number based on particle diameter $= \left(k_{LP} d_P / D_{SL}\right)$ (2.5.14)

Re_{Pm} = modified particle Reynolds number = $\left[P/\left(V\rho_L\right)\right]^{1/3}\left(d_P\right)^{4/3}\rho_L / \mu_L$ (2.5.15)

$\left(P/V\right)$ = power input (power consumption of impeller) per unit volume of liquid

d_P = particle diameter

= d_{Pm}, for particle – biofilm aggregates (2.5.16)

The above correlation has been developed for stirred tanks fitted with four baffles, agitated by Rushton turbine or pitched turbine impeller with $\left(D_a / D_t\right)$ = (1/2) and d_P = 0.925 mm, 1.84 mm. The impeller being positioned at one – third tank diameter from the vessel bottom and the gas (air) being sparged at 3.0 to 10.0 L/min.

The value of power input per unit volume, $\left(P/V\right)$, in the case of a three phase stirred tank, agitated by a Rushton turbine impeller may be estimated from the *dimensional* correlation proposed by Patrick and Kennedy [22] :

$$P = \left(1.26 P_{gL}\right) - X\left(\rho_S / \rho_L\right)^{6.9} \tag{2.5.17}$$

where X = mass fraction of particles (particle – biofilm aggregates) in the reaction mixture

P_{gL} = power consumption for a gassed liquid = $\left[P_L^2 n D_a^5 / Q_g^{0.53}\right]^{0.43}$ (2.5.18)

Q_g = volume flow rate at which the gas (air) is sparged

P_L = power consumption for ungassed liquid

= $\left(K_T n^3 D_a^5 \rho_L\right)$, for $Re_m \geq 10{,}000$ (2.5.19)

ρ_S = particle density

= ρ_{Sm}, for particle – biofilm aggregates

K_T = turbulent constant

= 6.3, for Rushton turbine impeller operating in baffled tank

Rapala and Karcz [23] have proposed an alternate correlation for the estimation of (P/V) in three phase, baffled stirred tanks :

$$(P/V) = (45.8X + 0.926)\left[1.0 - (0.229 + 11.71X)(Q_g/V)\right]\rho_L \quad (2.5.20)$$

However, the above correlation is strictly valid to those cases in which the impeller is operating at its critical speed, which is given by

$$n(critical) = (48.8X + 4.296)\left[1.0 + (0.223 - 0.08X)(60Q_g/V)\right] \quad (2.5.21)$$

By critical speed, we mean the minimum speed of the impeller (rps) required to uniformly disperse the solid particles and the gas bubbles in the liquid.

A detailed survey on estimation of liquid – to – particle mass transfer coefficient (k_{LP}) in two phase and three phase agitated vessels (stirred tank reactors) has been presented by Pangarkar et.al [24].

For the estimation of liquid – to – particle mass transfer coefficient in two phase fluidized bed reactors (in liquid fluidized beds), Kaloini et.al [25] have proposed the following correlation:

For $0.1 < Re_P(\in_L)^n < 1.0$, $\quad Sh_P = 1.1(Re_P)(Sc_L)^{1/3}(\in_L)^{n-1}$ (2.5.22)

For $1.0 < Re_P(\in_L)^n < 100$, $\quad Sh_P = 0.81(Re_P)^{1/2}(Sc_L)^{1/3}(\in_L)^{0.5n-1}$ (2.5.23)

where Re_P = particle Reynolds number = $\left(d_P V_t \rho_L / \mu_L\right)$ (2.5.24)

$n = (5.1 + 2.7N)/(N+1)$ (2.5.25)

$N = 0.1(Re_P)^{0.9}$ (2.5.26)

V_t = terminal free settling velocity of each particle (particle-biofilm aggregate) in the liquid.

To note that d_p should be replaced by d_{Pm} in the case particle – biofilm aggregates. The terminal free settling velocity of the particle (V_t) could be estimated from the generalized law of settling, given below:

$$V_t = [(4/3)(\rho_S - \rho_L) g d_P / (\rho_L f_D)]^{1/2} \quad (2.5.27)$$

In the case of particle – biofilm aggregates, d_p and ρ_S are to be replaced by d_{Pm} and ρ_{Sm} respectively. The magnitude of the drag coefficient (f_D) depends on the particle Reynolds number (Re_P). In the case of spherical particles settling in the laminar zone $(Re_P \leq 0.1)$,

$$f_D = (24 / Re_P) \tag{2.5.28}$$

If the settling of the particle is occurring in the fully developed turbulent zone ($Re_P \geq 1000$), then f_D shall remain more or less constant and shall be independent of Re_P. For spherical particles settling in the fully developed turbulent zone,

$$f_D = 0.44 \tag{2.5.29}$$

If settling occurs in the transition zone ($0.1 < Re_P < 1000$), then the drag coefficient (f_D) shall vary randomly with Re_P and could be estimated from a polynomial fit of following form:

$$\ln(f_D) = a_0 + a_1\,[\ln(Re_P)] + a_2\,[\ln(Re_P)]^2 + a_3\,[\ln(Re_P)]^3 \tag{2.5.30}$$

The values of correlation coefficients (a_0, a_1, a_2, a_3) that depend on the magnitude of the particle Reynolds number (Re_P), are listed in Table (2.1).

Table 2.1

Values of correlation constants of equation (2.5.30)				
Particle Reynolds number (Re_p)	a_0	a_1	a_2	a_3
0.1 – 1.0	3.3185	– 0.8982	0.0501	0.014
1.0 – 100	3.320	– 0.929	0.0429	0.0014
100 – 1000	1.0721	0.110	– 0.103	0.0069

A reliable estimate of k_{LP} in three phase fluidized beds can be obtained from the correlation proposed by Arters and Fan [26] which is reproduced below:

$$Sh = 0.228\left(1 + 0.0826\,Re_g^{0.623}\right)(Ga)^{0.323}\left(\rho_{Sm} - \rho_L / \rho_L\right)^{0.3}(Sc)^{0.4} \tag{2.5.31}$$

where Sh = Sherwood number $= \left(k_{LP}\, d_{Pm} / D_{SL}\right)$ (2.5.32)

S_C = Schmidt number $= \left(\mu_L / \rho_L\, D_{SL}\right)$ (2.5.33)

Ga = Galileo number $= \left(d_{Pm}^3\, \rho_L^2\, g / \mu_L^2\right)$ (2.5.34)

Re_g = modified Reynolds number $= \left(d_{Pm}\, U_g\, \rho_L / \mu_L\right)$ (2.5.35)

The above correlation is valid for $Re_g < 98.0$, Re_P = 17.4 to 253.0, Ga = 22.3 to 355.0 and Sc = 1960 to 3550.

As stated earlier, in a large number of cases, mass transfer resistance in the liquid bulk has been observed to be negligibly small and the assumption, $C_{SP} \cong C_S$ has been observed to cause no serious error.

Example 2.10: Microbial fermentation of lactose present in cheese whey permeate to lactic acid using a culture of *Lactobacillus helveticus* is to be carried out in a moving bed biofilm reactor which consists of a baffled tank (diameter, D_t = 1.0 m) agitated with a 300 rpm Rushton impeller (diameter = D_t/3). The microbial culture forms a thin layer of 0.3 mm thickness around 1.5 mm silica granules and these particle – biofilm aggregates are kept in suspension in the substrate solution (cheese whey permeate diluted to a lactose concentration of 9.0 g/L) by the Rushton impeller. If the feed solution is being admitted at 0.35 m^3/hr, determine the volume of the reactor required to affect 70% conversion of lactose. The bioconversion reportedly follows Monod – type kinetics with μ_m = 0.7 hr^{-1}, K_S = 0.22 g/L and overall yield coefficient for cell mass production = 0.65. Take cell mass concentration in the biofilm = 100 g/L and volume fraction of catalyst particles (particle – biofilm aggregates) in the slurry = 0.02. Liquid phase diffusivity of substrate in solution = 1.0 × 10^{-9} m^2/s and effective diffusivity of substrate into the biofilm = 3.0 × 10^{-10} m^2/s. The performance of the bioreactor may be assumed to be equivalent to that of an ideal CSTR (completely backmixed).

Solution:

We have discussed in Chapter – 1 on cheese whey permeate (de-proteinised cheese whey) as an economical raw material for lactic acid synthesis (lactic acid being the monomer for the manufacture of the bioplastic, PLLA). When approximated to an ideal CSTR, the performance equation of the moving bed biofilm reactor shall be,

$$\tau = \left(C_{S0} - C_{Se}\right) / \eta\left(-r_{SP}\right)_e \qquad \text{(i)}$$

where $\left(-r_{SP}\right)_e$ = intrinsic rate of bioconversion at $C_S = \left(C_{SP}\right)_e$

η = effectiveness factor at $C_S = \left(C_{SP}\right)_e$

Based on Monod – type kinetics, from equation (2.4.8) of section (2.4),

$$\left(-r_{SP}\right)_e = \mu_m\left(app\right) C_{SP} / \left(K_S + C_{SP}\right) \qquad \text{(ii)}$$

where $\mu_m\left(app\right) = \left(\mu_m / Y\right) x_f\, f\left(1 - \epsilon_L\right) / \epsilon_L$ (iii)

To note that in the above equation, $C_{SP} = \left(C_{SP}\right)_e$, we have avoided the suffix "*e*" for sake of convenience. This notation has been followed throughout this Example. Since d_P = 1.5 mm and d_{Pm} = 1.5 + 2(0.3) = 2.1 mm, from equation (2.4.4),

$$f = 1 - (1.5/2.1)^3 = 0.36443$$

Given that x_f = 100 g/L, μ_m = 0.7 hr^{-1}, Y = 0.65 and $\left(1 - \epsilon_L\right)$ = volume fraction of particle – biofilm aggregates (solid holdup) in the slurry/bioreactor = 0.02. Therefore, from equation (iii),

$$\mu_m\left(app\right) = (0.7/0.65)\,(100)\,(0.36443)\,(0.02)\,/\,(0.98) = 0.8 \text{ g /(L.hr)}$$

To compute the space time (τ) and thereby the required volume of the bioreactor, we need to first compute the value of C_{SP}, namely the steady state concentration of substrate at the surface of each particle – biofilm aggregate. For this, we have to solve equation (2.5.1) of Section (2.5) by trial:

$$\left(k_{LP}a\right)\left(C_{Se} - C_{SP}\right) = \eta\,\mu_m\left(app\right) C_{SP} / \left(K_S + C_{SP}\right) \qquad \text{(iv)}$$

The liquid – to – particle mass transfer coefficient (k_{LP}) could be computed from the correlation proposed by Barker and Treybal (equation – 2.5.11):

$$Sh_T = 0.02\left(Re_m\right)^{0.833}\left(Sc_L\right)^{0.5} \quad \text{(v)}$$

where $Re_m = \left(n D_a^2 \rho_L / \mu_L\right)$. (vi)

Since n = speed of the impeller = (300/60) = 5.0 *rps*, $D_a = (D_t/3) = (1/3)\ m$ and taking ρ_L = 1000 kg/m³ and μ_L = 1.0 cp = 0.001 kg/(m.s),

$$Re_m = 5.555 \times 10^5$$

The impeller is thus operating in fully developed turbulent zone. Now,

$Sc_L = \left[\mu_L / \left(\rho_L D_{SL}\right)\right]$

$= (0.001)/[(1000)(1.0 \times 10^{-9})] = 1000.0$

Therefore, from equation (v),

Sh_T = Sherwood number = 38582.3

Since $Sh_T = \left(k_{LP} D_t / D_{SL}\right)$,

$k_{LP} = (38582.3)(1.0 \times 10^{-9})/(1.0)$

$= 38.5823 \times 10^{-6}\ m^2/s$

The specific interfacial area (a) for mass transfer is obtained from equation (2.5.2):

$$a = s_P\left(1 - \in\right)/\in_L \quad \text{(vii)}$$

where S_P = specific surface (surface area per unit volume) of particle – biofilm aggregates (see equation – 2.5.3) = $\left(6/d_{Pm}\right)$ = (6/0.0021) = 2857.1428 m²/m³.

Therefore, a = (2857.1428) (0.02) / (0.98)

= 58. 309 m² / m³

Now, equation (iv), on rearrangement, becomes

$$C_{SP}^{\ 2} + \left[K_S - C_{Se} + \eta\mu_m(app)/(k_{LP}a)\right]C_{SP} - \left(K_S C_{Se}\right) = 0$$

Substituting the known values of $\mu_m(app)$, $\left(k_{LP}a\right)$, K_S = (0.22 g/L) and $C_{Se} = \left(0.3\ C_{S0}\right) = 2.7 g/L$,, we get,

$$C_{SP}^{\ 2} + (0.0987789\ \eta - 2.48)\ C_{SP} - 0.594 = 0 \quad \text{(viii)}$$

The above equation can be solved for C_{SP} only by trail, since η is a function of C_{SP}. Thus, let

$$C_{SP} = \left(C_{Se} - 0.1\right) = (2.7 - 0.1) = 2.6\ g/L$$

The value of η can be computed from equation (2.4.15), coupled with equations (2.4.20) to (2.4.24). Thus, from equation (2.4.23),

$$\beta = \left(C_{SP}/K_S\right) = (2.6 / 0.22) = 11.81818$$

From equation (2.4.20),

$$L^{*} = \left(d_{Pm}^{\ 3} - d_{P}^{\ 3}\right)/\left(6d_{Pm}^{\ 2}\right) = 0.22245 \text{ mm}$$

Now, $\varnothing'$ (from equation 2.4.24) = $L^{*}\sqrt{\mu_m(app)/(K_S D_e)}$

Since $\mu_m(app)$ = (0.8/3600) g/(L.s), K_S = 0.22 g /L, and D_e = 3.0 × 10^{-10} m²/s,

$$\varnothing' = 0.4084$$

Then, from equation (2.4.22), η_d = 11.4313

And from equation (2.4.21),

$$\varnothing = \left(1/\eta_d\right) = 0.087479$$

Finally, from equation (2.4.15),

$$\eta = 0.99545$$

The value of η is very close to 1.0, showing that the resistance to substrate transfer from liquid bulk to biofilm is not substantial.

Equation (viii) now becomes,

$$C_{SP}^{\ 2} - (2.38167)C_{SP} - 0.594 = 0$$

On solving,

$$C_{SP} = 2.609 \text{ g /L}$$

This value of C_{SP} does not differ significantly from that assumed at the outset. Hence, we could safely take

$$C_{SP} = 2.6 \text{ g/L}$$

Therefore, from equation (ii),

$$\left(-r_{SP}\right)_e = (0.8)(2.6)/(0.22+2.6)$$

$$= 0.7376 \text{ g /(L.hr)}$$

And from equation (i),

$$= (9.0 - 2.7) / [(0.99545)\ (0.7376)]$$

$$= 8.58 \text{ hr}$$

Since $\tau = \left(V \in_L\right)/Q_0$,

$$V = (8.58)\ (0.35)/(0.98) = \mathbf{3.064\ m^3}$$

This is the required volume of the bioreactor. Since diameter (D_t) of the tank = 1.0 m and $V = \left(\pi D_t^{\ 2}/4\right)H$,

$$H = (4V)/\left(\pi D_t^{\ 2}\right) = \mathbf{3.9\ m}$$

This the required height of the tank (bioreactor).

Nomenclature

a = specific interfacial area for mass transfer, m^2/m^3

b = parameter defined in equation (2.2.12), g/L

C_g = concentration of dissolved oxygen, g/L

C_{hp} = concentration of hydrogen peroxide in solution, g/L

C_I = concentration of inhibitor, g/L or mol/L

C_P = product concentration, g/L

C_{Se} = concentration of substrate in the exit stream (product stream), g/L or mol/L

C_{S0} = substrate concentration in feed; initial concentration of substrate, g/L or mol/L

C_{SP} = substrate concentration at particle surface or at the surface of particle-biofilm aggregate, g/L or mol/L

d_P = diameter of support particle, m

d_{Pm} = diameter (size) of particle – biofilm aggregate, m

D_g = dilution rate, s^{-1}

D_a = impeller (agitator) diameter, m

D_e = effective diffusivity of substrate into particle / biofilm, m^2/s

D_{SL} = liquid phase diffusivity of substrate, m^2/s

D_t = tank diameter, m

f = volume fraction of biofilm in particle – biofilm aggregate, m^3/m^3

f_D = drag coefficient, dimensionless

k = second order decay coefficient, L/(g.s)

k_d = endogeneous decay coefficient, g/L

k_e = cell maintenance coefficient, s^{-1}

k_{LP} = liquid – to – particle mass transfer coefficient, m/s

K_C = Contois' kinetic constant, g/g

K_g = oxygen Utilization Coefficient, g/L

K_i = competitive inhibition coefficient, g/L

K_i' = uncompetitive inhibition coefficient, g/L

K_{LP} = Liquid Phase Oxygen (LPO) Utilization coefficient, g/L

K_m = Michaelis – Menten kinetic constant, g/L or mol/L

K_P = product inhibition, coefficient, g/L

K_S = saturation constant, , g/L

K_{Si} = substrate inhibition, coefficient, g/L

L^* = characteristic dimension of particle / particle – biofilm aggregate, m

n = speed of impeller, rps

P = impeller power consumption, W

p_{gL} = impeller power consumption in gassed liquid, W

P_L = impeller power consumption in ungassed liquid, W

Q_g = volumetric flow rate of gas (air), m^3/s

Q_0 = volume flow rate of feed solution, m^3/s

r_{max} = maximum rate of enzyme-catalyzed reaction (Michaelis – Menten kinetic constant), g /(L.s) or mol /(L.s)

r_0 = initial rate of reaction, g /(L.s) or mol /(L.s)

$(-r_S)$ = rate of reaction (Rate of substrate consumption), g /(L.s) or mol /(L.s)

$(-r_S)(int)$ = intrinsic rate of reaction, g /(L.s) or mol /(L.s)

$(-r_{SP})$ = rate of reaction at $C_S = C_{SP}$, g /(L.s) or mol /(L.s)

Re_m = mixing Reynold's number (impeller Reynold's number), dimensionless

Re_P = particle Reynold's number, dimensionless

Re_{Pm} = modified Reynold's number (equation 2.5.13), dimensionless

Re_T = modified Reynold's number (equation 2.5.8), dimensionless

s_P = specific surface of particle / particle – biofilm aggregate, m^2/m^3

Sc_L = liquid phase Schmidt number, dimensionless

Sh_P = modified Sherwood number (equation 2.5.14), dimensionless

Sh_T = modified Sherwood number (equation 2.5.7), dimensionless

V = volume of reactor / reaction mixture, m^3

V_t = terminal free settling velocity of particle / particle – biofilm aggregate, m/s

x = cell mass concentration, g/L

x_e = cell mass concentration in exit stream, g/L

x_f = cell mass concentration in biofilm, g/L

x_0 = cell mass concentration in feed solution/ initial cell mass concentration, g/L

x_S = cell mass concentration in stationary phase (maximum cell mass concentration), g/L

Y = overall yield coefficient for cell mass production, g/g

Y_e = true yield coefficient for cell mass production, g/g

Y_P = yield coefficient for product formation, g/g

Y_S = selective yield coefficient for cell mass production, g/g

Y_{SP} = selective yield coefficient for product formation, g/g

Greek Letters

α = fractional substrate conversion, dimensionless

α_0 = parameter defined in equation (2.2.99), L/g

α_1, α_2 = coefficients in Leudeking – Piret equation (equation 2.2.107)

β = parameter defined in equation (2.4.23), dimensionless

δ = biofilm thickness, m

$\in$ = total fractional fluid holdup in bioreactor, total voidage (void fraction) of the bed, dimensionless

$\in_g$ = fractional gas holdup, dimensionless

$\in_L$ = fractional liquid holdup, dimensionless

η = effectiveness factor, dimensionless

η_d = parameter defined in equation (2.4.22), dimensionless

μ = specific growth rate, s^{-1}

μ_L = liquid viscosity, kg/(m.s)

μ_m = maximum specific growth rate, s^{-1}

ρ_L = liquid density, kg / m^3

ρ_m = density of microbial solution, kg / m^3

ρ_S = density of support particle, kg / m^3

ρ_{Sm} = density of particle – biofilm aggregate, kg / m^3

τ = space time, s

$\varnothing$ = Thiele – type modulus, dimensionless

$\varnothing'$ = parameter defined in equations (2.4.24) and (2.4.26), dimensionless

Exercises

2.A Comment on the following statements giving reasons. Use sketches /examples to elucidate your answers wherever possible:

(a) A large size of inoculum minimizes the lag phase associated with microbial growth.

(b) Monod's kinetic model describes microbial fermentation processes more accurately than logistic equation, since the former accounts for product inhibition to microbial activity, while the latter does not.

(c) The rate of an enzyme – catalyzed reaction increases continuously with increase in substrate concentration.

(d) Diauxic growth of microbes occurs due to nutrient deficiency in the substrate.

(e) Sloughing is a phenomenon that is desirable in biofilm reactors, since this helps in maintaining the biofilm thickness and the cell mass concentration in biofilm as high as possible.

(f) A low substrate concentrations, the effectiveness factor shall be independent of substrate concentration at the catalyst surface, in the case of enzyme – catalysed reactions following Michaelis – Menton kinetics.

(g) In the case of an ideal CSTR conducting a microbial fermentation process (that follows the Haldane – Andrews kinetic model) and receiving a sterile feed, the dilution rate shall be equal to the specific growth rate of microbes.

2.B Explain briefly:

(a) Lineweaver – Burk plot

(b) Leudeking – Piret equation

(c) Endogenous decay of microbes

(d) Competitive and uncompetitive inhibition to enzyme activity

(e) Effectiveness factor and its estimation for biofilm systems.

2.C Chemostat studies using *K. aerogenes* in a synthetic medium are given below. The feed to the fermenter was sterile and the volume of the reaction mixture in the fermenter was kept constant at 1.5 L.

Feed rate, L/hr	Steady state substrate concentration, mg/L
0.1	0.7659
0.2	1.6744
0.3	2.7692
0.4	4.1143
0.5	5.8064
0.6	8.00
0.7	10.9565

Check whether the process follows Monod's kinetics and if so, evaluate the kinetic constants.

2.D Aerobic treatment of wastewater from a food processing unit by activated sludge process is studied using a 5.0 L chemostat and the data collected are given below:

Flow rate, L/d	BOD in feed, mg/L	Steady state BOD in effluent, mg/L	Steady state concentration of VSS in effluent, mg/L
5.0	200.0	27.32	97.744
3.85	200.0	19.124	100.680
2.78	200.0	12.920	101.315
2.50	200.0	11.475	101.00
2.27	200.0	10.338	100.5135
1.666	200.0	7.545	97.860
1.35	200.0	6.180	95.1481

Determine whether the above data are in consonance with Monod's kinetic model that incorporates the effect of endogenous decay of microbes as well and if so, evaluate the kinetic parameters. Assume that the fresh feed to the chemostat was essentially sterile.

2.E Prove that the integrated form of Aiba et. al's correlation (equation 2.2.83 of Section – 2.2) that considers product inhibition to microbial activity as well, is as follows:

$$\left(C_{S0}-C_S\right)/t=\left(K_P\mu_m/Y_P\right)+\left(pK_S/b\right)\left(1/t\right)\ln\left[(1-b/C_{S0})/(1-b/C_S)\right]$$

$$+\left(p-K_S-b\right)\left(1/t\right)\ln\left[\left(C_{SO}-b\right)/\left(C_S-b\right)\right]$$

where, $p=C_{S0}+\left(K_P/Y_P\right)$, $b=C_{S0}+\left(x_o/Y\right)$.

2.F Anaerobic microbial fermentation of a sample of glucose solution (glucose concentration = 70 g/L) using yeast cells to produce ethanol was studied in a laboratory batch reactor and the experimental data collected are presented below:

Time (hr)	Cell mass concentration (g/L)
0.0	0.001
2.4876	0.00275
5.8983	0.011
10.2247	0.0635
12.5033	0.1585
14.0475	0.2920
14.9814	0.420
15.7733	0.5687
16.1580	0.6575
16.6886	0.8010

Product inhibition to microbial activity is anticipated during ethanol synthesis from glucose solution. Determine the above data fit into the kinetic equation proposed by Aiba et.al (that accounts for product inhibition to microbial activity as well) and if so, evaluate the kinetic constants. Use integral analysis and it maybe assumed that the overall yield coefficient for cell mass production and that for product formation are constants and are equal to 0.25 and 0.51 respectively.

2.G Synthesis of Xanthan gum through microbial fermentation of a starchy substrate using a culture of *Xanthomonas campestris* was studied in a laboratory batch reactor. Samples of reaction mixture were withdrawn at regular intervals and analyzed to determine the cell mass concentration, substrate concentration and product concentration. The results are tabulated below:

Time (hr)	Cell mass concentration, g/L	Substrate concentration, g/L	Product concentration, g/L
0.0	0.01	20.0	0.0
5.0	0.0207	19.95768	0.0310
10.0	0.0419	19.8724	0.09315
15.0	0.0811	19.7094	0.2111
20.0	0.145367	19.4232	0.41477
25.0	0.2323	18.9830	0.71916

Time (hr)	Cell mass concentration, g/L	Substrate concentration, g/L	Product concentration, g/L
30.0	0.32376	18.4044	1.1025
35.0	0.3977	17.7413	1.5191
40.0	0.44585	17.0430	1.9356
45.0	0.47287	16.3348	2.3422
50.0	0.4868	15.6244	2.7402
55.0	0.4937	14.9140	3.1330
60.0	0.4970	14.2037	3.5231
65.0	0.49857	13.4935	3.9118
70.0	0.4993	12.7834	4.30
75.0	0.49968	12.0734	4.6877
80.0	0.4998	11.3634	5.0754
100.0	0.50	8.5233	6.6255

The lag phase during microbial growth was found to be negligible.

(a) Determine whether the above data satisfy Ollis' kinetic model for cell growth, substrate utilisation and product formation and if so, evaluate the kinetic parameters.

(b) Based on the Ollis' kinetic model, what would be the product yield at the end of 10 hr, if the process is conducted in an industrial batch reactor with a feed solution containing 60 g of substrate per litre and an initial cell mass concentration of 0.05 g/L? What would be the cell mass yield at that instant?

2.H Experimental data on hydrolysis of methyl hydrocinnamate using the enzyme *chymotrypsin* are given below:

Initial concentration of ester, mM	Initial rate of reaction, mM/min
1.0	0.003184
0.8	0.00265
0.7	0.002374
0.5	0.001773
0.4	0.001450
0.3	0.001114

Determine whether process follows Michaelis – Menten kinetics and if so, evaluate the kinetic constants, using

(a) Lineweaver – Burk equation

(b) Hanes – Woolf equation

(c) Eadie – Hofstee equation

2.I Glutamic acid, produced during the enzymatic hydrolysis of L- glutamyl- L – tyrosine to carbobenzoxy-L-glutamic acid and L-tyrosine using the enzyme *cathepsin*, has been reported to inhibit the enzyme activity and thereby the rate of hydrolysis. From the following experimental data, determine the nature of inhibition (competitive, uncompetitive or mixed) and the kinetic equation for the reaction:

Initial substrate concentration, mM	**Initial rate of reaction, mM/min**	
	Inhibitor concentration = nil	**Inhibitor concentration = 30.3 mM**
4.70	0.0434	0.0133
5.88	0.0505	0.0162
7.143	0.05714	0.01922
10.80	0.0713	0.0266
30.30	0.1111	0.0581

2.J In a typical wastewater treatment process, the p-cresol present is removed by oxidation using hydrogen peroxide, with peroxidase enzyme being used as the catalyst. The reaction may be represented as

$$2S_1 + S_2 \longrightarrow 2P$$

Where S_1 is p-cresol and S_2 is hydrogen peroxide. The following kinetic equation has been proposed for the reaction:

$$\left(dC_P / dt\right) = \left(r_{max} C_{S1} C_{S2}\right) / \left[K_{m1} C_{S1} + K_{m2} C_{S2} + C_{S1} C_{S2}\right].$$

Evaluate the kinetic constants (r_{max}, K_{m1}, K_{m2}) from the following experimental data collected on a sample of wastewater in the laboratory:

$(C_{S2})_0$, **μM**	**Initial rate of reaction, μM/min**				
	$(C_{S2})_0$ **= 300 μM**	**700 μM**	**1400 μM**	**2800 μM**	**4100 μM**
100	85.0	102.5	110.0	115.0	116.0
300	144.0	200.0	232.0	254.0	261.0
500	166.0	245.5	298.0	335.0	347.0
800	183.0	282.5	355.0	407.0	427.0
1800	201.0	328.5	430.0	510.0	541.0

2.K A two substrate microbial fermentation process is being conducted in a 1.5 m^3 CSTR using a mixed culture of two microbes which exhibit mutualistic relationship with each other. The scheme may be represented as

$$S_1 \xrightarrow[\text{Microbe - 1}]{} S_2$$

$$S_2 \xrightarrow[\text{Microbe - 2}]{} S_1$$

The feed solution containing 0.01g of S_1 per litre and 0.03 g of S_2 per litre (but is essentially sterile) is admitted at 250 L/hr. If the experimental values of kinetic constants are those listed below, estimate the fractional conversion of S_1 and that of S_2 attained and the steady state concentrations of microbe - 1 and microbe - 2 in the exit stream. Assume that the reactor performs close to ideal and the density of the reaction mixture changes little due to the bioconversion process.

Parameter	Microbe - 1	Microbe - 2
μ_m, hr $^{-1}$	0.30	0.20
K_S, g /L	0.001	0.002
K_{Si}, g /L	0.023	0.20
Y_P, g /g	2.0	10.0
Y_S, g /g	0.25	0.10

Note: It maybe noted that the performance equation of an ideal CSTR could yield a quadratic equation in C_{Se} and it is not impossible that it provides two positive roots (two positive values of C_{Se}). In such cases, that value of C_{Se} is to be chosen which satisfies the stability criterion of reactor operation. For example, in the case of the bioreactor considered above, a simple stability analysis could predict that the system shall be stable only if,

$$\left(C_{Se}\right)_1 < \sqrt{K_{S1}\left(K_{Si}\right)_1}$$

$$\left(C_{Se}\right)_2 > \sqrt{K_{S2}\left(K_{Si}\right)_2}$$

In other words, one of the microbes must be substrate - limited, while other the substrate - inhibited.

References

1. Monod, J., The Growth of Bacterial Cultures, Ann. Rev. Microbiol., **3**, 371 – 94, 1949.
2. Andrews, J.F., A Mathematical Model for the Continuous Culture of Microorganisms using Inhibitory Substrates, Biotechnol. Bioeng., **10**, 707 – 23, 1968.
3. Aiba, S., Shoda, M. and Nagatani, M., Kinetics of Product Inhibition in Alcohol Fermentation, Biotechnol. Bioeng., **10**, 845 – 64, 1968.
4. Anjana, D.N., and Kumar, S., Kinetic Modelling of Lactic Acid Production from Molasses using *Enterococcus faecalis RKY1*, Biochemical Eng. J., **38**(3), 277 – 84, 2008.
5. Tessier, G., Croissance des Populations Bacteriennes et QuantiteD'aliment Disponible, Rev. Sci. Paris, **80**, 1942.

6. Contois, D.E., Kinetics of Bacterial Growth : Relationship between Population Density and Specific Growth Rate of Continuous Cultures, Microbiology, **21**, 40 – 50, 1959.
7. Zabot G.L., Mecca, J., Mesomo, M., Silva, M.F., Pra, V.D., Oliveira, D. et al., Hybrid Modeling of Xanthan Gum Bioproduction in Batch Bioreactor, Bioproc. Biosyst. Eng., 34(8), 975–86, 2011.
8. Ollis, D.F., A Simple Batch Fermentation Model : Theme and Variations, Annals N.Y. Acad. Sci., **413**, 144 – 150, 1983.
9. Leudeking, R. and Piret, E.L., A Kinetic Study of Lactic Acid Fermentation, J. Biochem. Microbiol. Technol. Eng., **1**, 393 – 98, 1959.
10. Michaelis, L. and Menten, M.L., Die Kinetik der Invertinwirkung, Biochem. Zeitschr., **49**, 333 – 69, 1913.
11. Briggs, G.E. and Haldane, J.B.S., A Note on the Kinetics of Enzyme Action, Biochem. J., **19**, 338 – 39, 1925.
12. Du W., Xu Y.Y., Liu D.H. (2005). Study on Kinetics of Enzymatic Interesterification of Triglycerides for Biodiesel Production with Methyl Acetate as Acyl Acceptor, J. Molecular Catalysis, **32**, 241 – 245, 2005.
13. Narayanan, C.M., De, T. and Sikder, J., Studies on Synthesis of Biodiesel by Transesterification of Neem Oil with Immobilized Lipase in Cylindrical and Diverging – Converging Fluidized Bed Bioreactors, Int. J. Chem. Environ. Eng., **6**(5), 322 - 336, 2015.
14. De, T., Sikder, J. and Narayanan, C.M., Biodiesel Synthesis Using Immobilized Lipase Enzyme in Semifluidised Bed Bioreactors – Bioreactor Design and Performance Analysis, AIChE – Environ. Progr. Sustainable Energy, **36**(5), 1537-45, 2017. DOI : 10.1002/ep.12602.
15. De, T., Sikder, J. and Narayanan, C.M., Biodiesel Synthesis and Bioreactor Design – An Overview, Ind. J. Chem. Technol., **24**, 575-92, November, 2017.
16. Gottifredi, J.C. and Gonzo, E.E., An Approximate Expression for Estimation of Effectiveness Factor, Chem. Eng. J., **109**, 83-85, 2005.
17. Narayanan, C.M., Biotechnology and Bioprocess Engineering, Galgotia Publishers, New Delhi, 2011.
18. Bailey, J.E. and Ollis, D.F., Biochemical Engineering Fundamentals, 2nd ed., McGraw Hill, Singapore, 1986.
19. Humphrey, D.W. and Van Ness, H.C., Mass Transfer in Continuous Flow Mixing Vessel, AIChE J., **3**(2), 293-86, 1957.
20. Barker, J.J. and Treybal, R.E., Mass Transfer Coefficients for Solids Suspended in Agitated Liquids, AIChE J., **6**(2), 289-95, 1960.
21. Grisafi, F., Brucato, A. and Rizzuti, L., Solid-liquid Mass Transfer Coefficients in Gas – Liquid – Solid Agitated Vessels, Canadian J. Chem. Eng., **76**, 446-55, 1998.

22. Patrick. A.J. and Kennedy, M.J., Power Consumption of Agitated Gas – Liquid – Solid Systems with Particular Relevance to Solid Substrate Submerged Culture Fermentations, Biotech Letters, **17**(5), 487-92, 1995.

23. Rapala, A.K. and Karcz, J., Solid Suspension and Gas Dispersion in a Gas – Liquid – Solid Agitated Vessel, Chem. Papers, **64**(2), 154-62, 2010.

24. Pangarkar, V.G., Yawalkar, A.A., Sharma, M.M. and Beenackers, A.A.C.M., Particle-Liquid Mass Transfer Coefficient in Two and Three Phase Stirred Tank Reactors, Ind. Eng. Chem. Res., **41**, 4141-67, 2002.

25. Koloini T., and others, Mass Transfer Coefficients in Fluidized Beds, Chem. Eng. Commun., **5**, 233- 238, 1980.

26. Arters, D.C. and Fan, L.S., Solid–Liquid Mass Transfer in a Gas–Liquid–Solid Fluidized Bed, Chem. Eng. Sci., **41** (1), 107-115, 1986

Chapter - 3

Design and Analysis of Industrial Bioreactors

Performance analysis and computer aided design of industrial bioreactors is relatively more complex (as compared to design and analysis of chemical reactors). However, that itself makes the task most challenging and most exciting. The complexity associated with the performance analysis of industrial bioreactors stems chiefly from the non-linear nature of kinetic equations representing biochemical processes (Chapter – 2).

3.1 Concept of Ideal Reactors

Let us start with the concept of ideal reactors. We have partly discussed on this in Chapter – 2. Ideality is, no doubt, an Utopian concept. However, the performance of some of the industrial reactors (bioreactors) could often approach close to ideality and in such cases, it shall not be too erroneous to perform their design based on the ideal reactor approach. Such situations, nevertheless, arise mainly in small capacity installations.

Stirred tank reactors, as already stated in Chapter – 2, could be operated either continuously (as flow reactors) or batchwise. When operated as a batch reactor, its operation is termed as ideal, if the composition of the reaction mixture at any time remains the same at every point inside the reactor (the composition does not vary from point to point within the reactor). In other words, the composition of the reaction mixture remains uniform at any instant and any sample of reaction mixture drawn randomly from any point in the reaction mixture, shall analyze for the same substrate (unreacted) concentration and product concentration. For such an ideal batch reactor (ideal batch stirred tank), the performance equation shall be

$$t = \int_{C_S}^{C_{S0}} dC_S / (-r_S) \tag{3.1.1}$$

where, t = batch time.

After substituting the appropriate rate equation (kinetic equation), the integral on the right hand side of the above equation could be evaluated either analytically or numerically.

It is, therefore, apparent that the performance equation of an ideal batch reactor (ideal batch stirred tank) is nothing but the integrated form of the kinetic equation (rate equation). For example, equations (2.2.16) of Chapter-2 forms the performance equation of an ideal batch reactor conducting a microbial fermentation process that follows Monod's kinetic equation and equation (2.2.58) represents the performance equation, when the microbial process follows Haldane – Andrews kinetics. In the same way, equation (2.3.23) stands for the performance equation of ideal batch reactor conducting an enzyme – catalyzed reaction that follows

Michaelis-Menten scheme and equation (2.3.44) that when substrate inhibition to enzyme activity is anticipated.

However, equation (3.1.1) forms the performance equation for an ideal batch stirred tank reactor handling a homogeneous system. In other words, all the constituents of the reaction mixture (reactants, catalyst, products) are in the same phase. It is also restricted to constant density systems (for those systems in which the density of the reaction mixture changes little due to the biochemical reaction). Microbial fermentation of aqueous solutions using suspended growth of microbes is an excellent example in this connection. Even in the case of the aerobic treatment of wastewater in agitated vessels (when operated batch wise), the performance equation of the bioreactor is often approximated to equation (3.1.1), with the assumption that the air that is sparged into the reactor moves up in the form of dispersed, tiny bubbles (and these do not disturb the overall kinetics of the bioconversion process that is taking place in the liquid phase).

When the system is heterogeneous in nature as in the case of batch-operated moving bed biofilm reactors, equation (3.1.1) would have to be modified to account for the heterogeneous nature of the system handled. We have already discussed in Chapter-1 (and also in Chapter-2) that moving bed biofilm reactors are, in fact, slurry reactors which consist of mechanically agitated vessels containing the substrate solution, in which the particle-biofilm aggregates remain in suspension. In the case of such reactors (assuming ideal performance), equation (3.1.1) would have to be modified to

$$t = \int_{C_S}^{C_{S0}} dC_S / \left[\eta(-r_{SP})(int) \right] \tag{3.1.2}$$

where η = effectiveness factor

$(-r_{SP})(int)$ = intrinsic rate of bioconversion (at $C_S = C_{SP}$)

For example, if the bioconversion intrinsically follows Monod's kinetic equation, then from equation (2.4.8) of Chapter – 2,

$$(-r_{SP})(int) = \mu_m(app) C_{SP} / (K_S + C_{SP}) \tag{3.1.3}$$

where $\mu_m(app) = (\mu_m / Y) x_f \, f (1 - \in_L) / \in_L$ (3.1.4)

$\in_L$ = fractional liquid holdup in the bioreactor

In the case of three phase operation (for example, an aerobic process is being conducted in the bioreactor and air is being sparged into the reactor vessel at a specified flow rate),

$$\mu_m(app) = (\mu_m / Y) x_f \, f (1 - \in) / \in_L \tag{3.1.5}$$

where $\in$ = total fluid holdup in the bioreactor

$= (\in_L + \in_g)$ (3.1.6)

$\in_g$ = fractional gas holdup in the bioreactor

The value of η is to be computed from reported correlations such as equation (2.4.15) or (2.4.27) of Chapter – 2. Incidentally, since η is a nonlinear function in C_{SP}, the integral of equation (3.1.2) can be evaluated only numerically (by using any of the well-established numerical methods such as the Simpson's rule).

Stirred tank bioreactors are also operated continuously as flow reactors (as CSTRs – as continuous stirred tank reactors). This is preferred in large capacity installations. Once again, considering a homogeneous, constant density system (see figure – 2.1 of Chapter – 2) and assuming ideal behavior, the performance equation of the bioreactor shall be that given in equation (2.1.1) of Chapter – 2 (reproduced below) :

$$\tau = (V/Q_0) = (C_{S0} - C_{Se})/(-r_{Se}) \tag{3.1.7}$$

As discussed in Chapter-2, an ideal CSTR is a completely backmixed reactor (in which the degree of backmixing is by and large cent percent) and consequently, the concentration of any constituent of reaction mixture inside the reactor shall be same as that in the exit stream (product stream).

If the CSTR is handling a heterogeneous system (such as in the case of a continuous flow, moving bed biofilm reactor) and if ideal performance has been assumed, then the performance equation (equation – 3.1.7) gets modified to

$$\tau = (V \in_L)/Q_0 = (C_{S0} - C_{Se})/\left[\eta_e(-r_{Se})(int)\right] \tag{3.1.8}$$

where η_e = effectiveness factor at $C_S = C_{Se}$

$(-r_{Se})(int)$ = intrinsic rate of bioconversion at $C_S = C_{Se.}$

It has been, nevertheless, assumed that $C_{SP} \approx C_S$. In a completely backmixed reactor, this is a reasonable approximation, since due to the intense mixing between fluid elements, the resistance to substrate transport in the fluid bulk (from fluid bulk to particle surface or biofilm surface) shall be negligibly small.

Here also, the expression for the intrinsic rate, $(-r_{Se})(int)$, is to be substituted from equation (2.4.8) or the like, after replacing C_{SP} by C_{Se} and the value of η_e estimated from equation (2.4.15) or (2.4.27), after substituting $C_{SP} = C_S = C_{Se}$.

Incidentally, due to the nonlinear dependence of η_e on substrate concentration (C_{Se}), equation (3.1.8) can be solved for C_{Se} only by trial (through a trial and error procedure).

An alternative to tank reactors are the *tubular reactors*. These reactors could consist of a single tube (or tube coil) or a bundle of tubes, depending on the capacity (amount of feed solution to be handled per hour) of the reactor. An *ideal* tubular reactor is one, in which the fluid executes essentially plug flow. In other words, each fluid element passes through the tube (flow channel) in the form of a plug, unconcerned of the fluid element moving ahead of it or behind it. In such a situation, both the velocity profile as well as the concentration profile at any cross-section of the tubular reactor shall be more or less flat (neither the velocity nor the concentration shall vary in the radial direction). The concentration shall vary in the axial direction as more and more substrate undergoes bioconversion, but the fluid (the substrate solution or the reaction mixtures) flows through the tube at a specific velocity (plug flow velocity), U_0. Such a reactor is called the *plug flow reactor* (PFR).

True plug flow is an ideal concept, that can hardly be attained in industrial practices. However, the performance of the bioreactor could be approximated to plug flow in selected cases. For example, the flow through packed beds is often close to plug flow and consequently, the performance of packed bed bioreactors could often be approximated to that of a PFR.

When the fluid executes fully developed turbulent flow, the velocity profile across any cross- section of the flow channel shall be essentially flat and when this occurs in narrow flow channels or low diameter conduits, plug flow could be assumed with allowable error.

If we consider a differential element of volume dV (of length dz) in an ideal tubular reactor (see figure- 3.1), then a steady state mass balance across this element yields (considering a constant density system),

$$Q_0 C_S - Q_0 (C_S + dC_S) = (-r_S) dV \tag{3.1.9}$$

or,

$$(1/Q_0) dV = -dC_S / (-r_S) \tag{3.1.10}$$

On integration,

$$(V/Q_0) = \tau = \int_{C_{Se}}^{C_{S0}} dC_S / (-r_S) \tag{3.1.11}$$

It can be seen that the above equation is exactly analogous to equation (3.1.1), the only difference being the batch time (t) having been replaced by the space time (τ). In other words, for constant density systems, the performance equation of a plug flow reactor (ideal tubular reactor) shall be the same as that for an ideal batch reactor, except for the batch time (t) being replaced by the space time (τ).

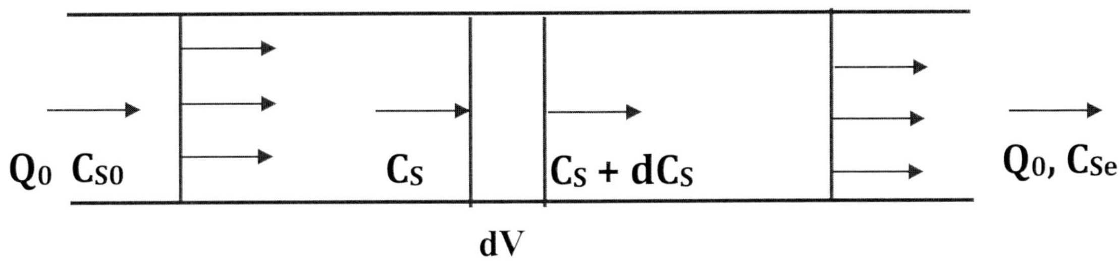

Figure 3.1: Schematic of a Plug Flow Reactor

It must be, however, kept in mind that this analogy is applicable only for constant density systems. No doubt, as stated earlier, since bioconversion processes are mostly conducted in dilute solutions, the assumption of constant density system shall be more or less valid in many cases.

In the case of heterogeneous systems and when the plug flow approximation is still applied, equation (3.1.11) gets modified to

$$\tau = \int_{C_{Se}}^{C_{S0}} dC_S / [\eta(-r_{SP})(int)] \tag{3.1.12}$$

Both stirred tank reactors and plug flow (tubular) reactors have their own advantages and disadvantages. It must nevertheless be kept in mind that the ideal CSTR (with 100% backmixing) and the PFR (with zero backmixing) form the two extremes of ideal behavior. A real-life industrial reactor's performance shall fall in between these two extremes.

In the case of many conventional chemical reactions (that follow first order, second order or fractional order kinetics), the PFR reportedly provides much higher fractional conversion of the reactant than the completely backmixed CSTR. Also, two or more ideal CSTRs in series (of total volume, V) provide higher fractional conversion of the reactant than a single CSTR of volume, V.

However, when the kinetics of the process is too nonlinear and the kinetic equation is composed of two to four kinetic constants (as is the case with most bioconversion processes discussed in Chapter – 2), the above observations could become far from true.

For many microbial fermentation processes involving suspended growth of microbes, the tank reactors (CSTRs) are being preferred, since they provide more effective (more intimate) contacting between microbial cells and the substrate solution. As illustrated in Example (3.1) given below, for a microbial fermentation process following Monod-type kinetics, a single stirred tank reactor (assuming ideal behavior) provides much better performance than a PFR. The PFR is seen to demand much larger reactor volume to achieve the specified degree of bioconversion. Also, when two stirred tanks (of equal volume) are used in series, the combined volume of the two, required to achieve the desired degree of fractional substrate conversion, is seen to be slightly higher than that of the single CSTR. A series cascade of CSTRs thus appears to be offering no additional benefit. In the case of a CSTR – PFR combination, the combined volume of the system required to achieve the desired bioconversion is lower than that of a single PFR, but higher than that of a single CSTR (see Example 3.1). Incidentally, an interesting observation is that in such a combination, only one-third of desired bioconversion occurs in the CSTR, the rest two-third is affected by the PFR, both being of the same volume.

Example 3.1

Synthesis of an organic acid through microbial fermentation of a starchy substrate (with suspended growth of microbes) is to be performed. The following options are available:

(a) A single ideal CSTR
(b) A single plug flow reactor (PFR)
(c) An ideal CSTR with partial cell recycle, recycle ratio = 0.75 and cell concentration coefficient = 1.8.
(d) Two ideal CSTRs of equal volume in series.
(e) An ideal CSTR followed by a PFR, each of same space time (τ).

The feed solution with a substrate concentration of 30 g/L and a cell mass concentration of 0.1 g/L is being admitted at 180 L/hr. The bioconversion has been observed to follow Monod – type kinetics, with μ_m = 1.6 hr^{-1}, K_S = 10.795 g/L. Determine which among the above options would demand the least reactor volume, if the bioconversion desired to 75 %. Assume a constant density system and take the overall yield coefficient for cell mass production = 0.48.

Solution:

(a) For an ideal CSTR, the performance equation is,

$$\tau = (C_{S0} - C_{Se})(K_S - C_{Se}) / [\mu_m C_{Se}(b - C_{Se})] \qquad \text{(i)}$$

where, $C_{S0} = 30$ g/L, $K_S = 10.795$ g/L, $\mu_m = 1.6$ hr^{-1},
$b = (x_0 / Y) + C_{S0} = (0.21/0.48) + 30 = 30.2083$ g/L. Also, since $\alpha = 75$ %,
$C_{Se} = (0.25 C_{S0}) = 7.5$ g/L.

Therefore, $\tau = 1.5106$ hr

and $V = (1.5106)(180) =$ **271.908 L**

(b) For a PFR, the performance equation is that given in equation (2.2.16), with batch time (t) replaced with space time (τ). Thus,

$$(\tau \mu_m) = \ln [(b - C_{Se})/(b - C_{S0})] + (K_S / b)\ln[(1 - b/C_{Se})/(1 - b/C_{S0})] \qquad \text{(ii)}$$

Substituting the known values of parameters,

$$\tau = 4.2895 \text{ hr}$$

and $V = (4.2895)(180.0) =$ **772.11 L**.

The volume requirement of the PFR is thus much larger than that of CSTR.

(c) The operation of an ideal CSTR with partial cell recycle is sketched in figure (3.2). In the present case,

$$\text{R} = \text{recycle ratio} = 0.75$$

$$\text{g} = \text{cell concentration coefficient} = 1.8$$

Now, a cell balance around the cell separator yields,

$$(1 + R)Q_0 x_e = Q_0 (W x_e) + (R Q_0)(g x_e)$$

or $\quad W = 1 - R(g - 1)$

$\quad = 1 - 0.75(1.8 - 1) = 0.40$

An overall cell balance (for the entire system) yields,

$$Q_0 (W x_e) - (Q_0 x_0) = (dx/dt)_e V$$

or $\quad V = Q_0 (W x_e - x_0) / (dx/dt)_e$

Since $\quad (dx/dt)_e = (\mu_m C_{Se} x_e)/(K_S + C_{Se})$,

$$V = Q_0 (W x_e - x_0)(K_S + C_{Se})/(\mu_m C_{Se} x_e) \qquad \text{(iii)}$$

Now, $\quad x_e = x_0 + Y(C_{S0} - C_{Se})$

$\quad = 0.1 + 0.48(30 - 7.5) = 10.9$ g/L

Therefore, substituting the known values of all parameters in equation (iii), we get

$$V = \mathbf{107.253\ L}$$

The volume requirement of the bioreactor is thus much lower than that of CSTR without cell recycle (there is more than 50 % reduction in volume). The benefit of cell recycle is, therefore, obvious.

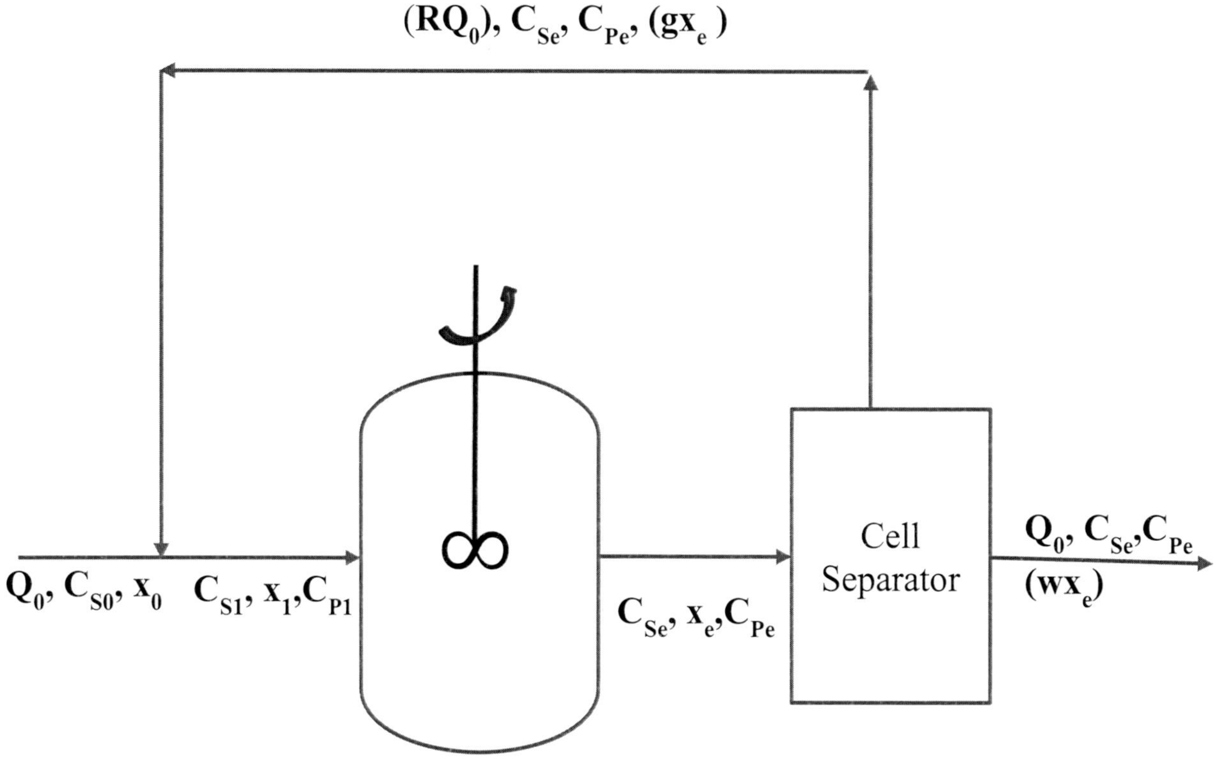

Figure 3.2: CSTR with recycle

(d) For a series cascade of CSTRs, we have to write the performance equation separately for each reactor. To note that since both reactors are of same volume, $\tau_1 = \tau_2 = \tau$. Thus, for CSTR – 1, the performance equation is [it is same as that given in equation (i) of part (a)],

$$(\tau \mu_m) = (C_{S0} - C_{S1})(K_S + C_{S1}) / [C_{S1}(b - C_{S1})]$$

$$= (30 - C_{S1})(10.795 + C_{S1}) / [C_{S1}(30.2083 - C_{S1})] \quad \text{(iv)}$$

where C_{S1} = substrate concentration in the exit stream from CSTR – 1.
Similarly, for CSTR – 2,

$$(\tau \mu_m) = (C_{S1} - C_{Se})(K_S + C_{Se}) / [C_{Se}(b - C_{Se})]$$

$$= (C_{S1} - 7.5)(10.795 + 7.5) / [7.5(30.2083 - 7.5)]$$

$$= 0.11(C_{S1} - 7.5) \quad \text{(v)}$$

To note that, for CSTR – 1,

$$(x_1 - x_0) = Y(C_{S0} - C_{S1})$$

or $$(x_1 / Y) = (C_{S0} + x_0 / Y) - C_{S1}$$
$$= (b - C_{S1}) \quad \text{(vi)}$$

In the case of CSTR – 2 also,

$$(x_2 - x_1) = Y(C_{S1} - C_{Se})$$

or $$(x_2 / Y) = (x_1 / Y) + C_{S1} - C_{Se}$$

Since $C_{S1} = b - (x_1 / Y)$,

$$(x_2 / Y) = (x_1 / Y) + b - (x_1 / Y) - C_{Se}$$
$$= (b - C_{Se}) \quad \text{(vii)}$$

Solving equations (iv) and (v) simultaneously by trial,

$$C_{S1} = 21.0 \text{ g/L}$$
$$\tau = 0.925 \text{ hr}$$

The total volume of the reactors is,

$$V = 2\ (0.925)\ Q_0$$
$$= 2\ (0.925)\ (180.0) = \mathbf{333.0\ L}$$

The total volume requirement of both reactors together is larger than that of a single CSTR. It should be inferred therefore, that the use of a series cascade does not fetch any improved results, in the present case.

(e) Let us now consider the CSTR – PFR combination. For the ideal CSTR, the performance equation is same as equation (i) of part (a):

$$(\tau \mu_m) = (30 - C_{S1})(10.795 + C_{S1}) / [C_{S1}(30.2083 - C_{S1})] \quad \text{(viii)}$$

For the PFR that follows, the performance equation is that given in equation (ii) of part (b):

$$(\tau \mu_m) = \ln [(b - C_{Se}) / (b - C_{S1})] + (K_S / b) \ln[(b / C_{Se} - 1) / (b / C_{S1} - 1)]$$

Substituting the known values of $b = 30.2083$ g/L, $C_{Se} = 7.5$ g /L, $K_S = 10.795$ g/L, we get

$$(\tau \mu_m) = \ln[22.7083 / (30.2083 - C_{S1})](0.35735) \ln[3.0277 C_{S1} / (30.2083 - C_{S1})] \quad \text{(ix)}$$

To note that since the CSTR and PFR are of same volume, $\tau_m = \tau_P = \tau$. Solving equations (viii) and (ix) simultaneously by trial,

$$C_{S1} = 20.5 \text{ g/ L}$$
$$\tau = 0.937 \text{ hr}$$

The total volume requirement of the system (V) is, therefore,

$$V = (2)\ (0.937)\ Q_0$$
$$= (1.874)\ (180) = \mathbf{337.32\ L}$$

Fractional substrate conversion attained in CSTR is,

$$= (30.0 - 20.5) / (30.0) = 31.66\%$$

Conversion attained in PFR = (20.5 – 7.5) / (20.5) = 63.41%

Among the two reactors in the combination therefore, the PFR offers the major share of substrate conversion. The results obtained are summarized below :

Bioreactor	**Volume requirement, liters (α = 75%)**
1. Single CSTR	271.9
2. Single PFR	772.1
3. CSTR with partial cell recycle	107.25
4. Two CSTRs in series	333.0
5. CSTR – PFR combination	337.32

Microbial fermentation systems (even when handling suspended growth of microbes) can not be treated as strictly homogeneous. The microbial cells do form a discrete phase and their characteristics are entirely different from those of the substrate solution. There are two independent parameters affecting the rate of reaction, such as the substrate concentration (C_S) and the cell mass concentration (x), the former decreasing with time, while the latter increasing. Cell death or cell decay does occur, but the net value of x increases with time. Neither very high value of C_S nor very high value of x are preferable here, unlike chemical systems. Also, the entire substrate is not converted into product(s), a part is used as food by the microbes for their growth and multiplication and also for their maintenance. All these factors affect the performance of the bioreactor.

In the case of enzyme-catalyzed reactions that employ free enzyme (enzyme solution added directly to the reaction mixture, the system thus being homogeneous), the observations are different. Example (3.2), given below, illustrates the case of an enzyme-catalyzed reaction that follows Michaelis-Menten kinetics (with substrate inhibition to enzyme activity included). It is seen that the PFR provides the highest performance (highest fractional substrate conversion, α), followed by CSTR – PFR combination, the single CSTR providing the lowest substrate conversion, the reactor volume being the same in all the three cases. This order is more or less the same as that observed in conventional chemical systems.

However, when a series cascade of two ideal CSTRs is used (whose total volume is same as the single CSTR considered earlier), the fractional substrate conversion attained is observed to be more or less the same (see part (d) of Example – 3.2). At the same time, if substrate inhibition to enzyme activity is considered negligible, then such a series cascade is seen to provide higher value of substrate conversion (not a substantial increase, nevertheless).

Example 3.2 Hydrolysis of sugar cane molasses using *invertase* enzyme to produce glucose is being attempted. The following options are to be experimented with:

(a) A single CSTR (may be assumed ideal) of volume 1.5 m^3,

(b) A PFR of volume 1.5 m^3,

(c) An ideal CSTR followed by a PFR, each of volume of 0.75 m^3,

(d) Two ideal CSTRs in series each of volume 0.75 m^3,

Determine which option would provide the highest fractional conversion of sucrose. Substrate inhibition to enzyme activity has been reported in the process and the bioconversion follows modified Michaelis – Menten kinetic scheme with r_{max} = 0.35 g/(L.min), K_m = 50 g/L and K_{S1} (substrate inhibition coefficient) = 700 g/L. Molasses diluted to a sucrose concentration of 250 g/L is being admitted at the rate of 200 L/hr. Assume constant density system.

(e) If substrate inhibition to enzyme activity is assumed negligible, what would be the answers to part (a) and part (d)?

Solution:

(a) For the ideal CSTR, the performance equation is,

$$\tau=\left(C_{S0}-C_{Se}\right)\left(K_m+C_{Se}+C_{Se}^{\ 2}/K_{Si}\right)/(r_{max}\,C_{Se})$$

or, $$\left(\tau r_{max}\right)=\left(C_{S0}-C_{Se}\right)\left(1+K_m/C_{Se}+C_{Se}/K_{Si}\right) \qquad \text{(i)}$$

where, $\tau=\left(V/Q_0\right)$ = (1500/200) = 7.5 hr = 450 min, r_{max} = 0.35 g /(L.min), K_m = 50 g/L, K_{Si} = 700 g/L, C_{S0} = 250 g/L.

Accordingly, equation (i) reduces to

$$(157.50) = (250 - C_{Se})\ [1 + 50/C_{Se} + C_{Se}/700] \qquad \text{(ii)}$$

Solving by trial, C_{Se} = 148.5 g/L

The fractional substrate conversion (α) attained is,

α = (250 – 148.5)/(250.0) = **40.6 %**

(b) The performance equation of the PFR is that given in equation (2.3.40) of Chapter – 2, with the batch time (t) having been replaced by the space time (τ). If we rewrite this equation in terms of the fractional substrate conversion (α), then

$$\left(\tau r_{max}\right)=\left(C_{S0}\,\alpha\right)-K_m\ln\left(1-\alpha\right)+\left(C_{S0}^{\ 2}/K_{Si}\right)\left(\alpha-\alpha^2/2\right) \qquad \text{(iii)}$$

Substituting the known values of the kinetic constants, τ and C_{S0}, we get

$$(157.50) = \left(250\,\alpha\right)-50\ln\left(1-\alpha\right)+\left(89.2857\right)\left(\alpha-\alpha^2/2\right) \qquad \text{(iv)}$$

Solving by trial, α = 0.72 = **72 %**

(c) In the case of CSTR – PFR combination the performance equation of the CSTR shall be same as equation (ii) of part (a), except that

$\tau_m = \tau$ = (750/200) = 3.75 hr = 225 min

Thus, for the CSTR,

$$(78.75) = (250 - C_{S1})\ [1+ 50/C_{S1} + C_{S1}/700] \qquad \text{(v)}$$

Solving by trial, C_{S1} = 199 g/L

For the PFR that follows, the performance equation once again shall be same as equation (iv) of part (b), except for the change in the value of $\tau(\tau_p = \tau = 225$ min) and C_{S0} is to be replaced by C_{S1} (= 199 g /L). Thus, for the PFR,

$$(78.75) = (199\ \alpha) - 50\ \ln(1-\alpha) + (56.5728)\ (\alpha - \alpha^2/2) \tag{vi}$$

Solving by trial, α = 0.445

and C_{Se} = 199 (1 – 0.445) = 110.445 g/L

The overall fractional substrate conversion attained is,

α = (250 – 110.445)/(250) = **55.82 %**

(d) In the case of the series cascade of CSTRs, the performance equation for CSTR – 1 shall be same as equation (v) of part (c). Accordingly,

$$C_{S1} = 199 \text{ g/L}$$

This is the substrate concentration in the exit stream from CSTR – 1. For CSTR -2 similarly,

$$(78.75) = (199 - C_{Se})[1 + 50/C_{Se} + C_{Se}/700] \tag{vii}$$

Solving by trial, C_{Se} = 148.5 g/L

The fractional substrate conversion attained shall be,

α = (250 – 148.5) / (250) = **40.6 %**

Interestingly, this is same as that attained in a single CSTR, discussed in part (a).

(e) If substrate inhibition to enzyme activity is negligible, the performance equation for a single CSTR, namely equation (ii) of part (a), reduces to

$$(157.5) = (250 - C_{Se})[1 + 50/C_{Se}]$$

or $C_{Se}^2 - 42.5C_{Se} - 12500 = 0$

On solving, C_{Se} = 135 g/L

and α = (250 – 135)/(250) = **46 %**

This is larger than that attained when substrate inhibition to enzyme activity is present (which is understandable, no doubt).

In the case of CSTR – cascade, the performance equation for CSTR – 1 shall be,

$$(78.75) = (250 - C_{S1})[1 + 50/C_{S1}]$$

or, $C_{S1}^2 - 121.25C_{S1} - 12500 = 0$

On solving, C_{S1} = 187.8 g /L

Similarly for CSTR – 2,

$$(78.75) = (187.8 - C_{Se})[1 + 50/C_{Se}]$$

or $C_{Se}^{2} - 59.05C_{Se} - 9390 = 0$

On solving, $C_{Se} = 130.825$ g /L

and $\alpha = (250 - 130.825)/(250) =$ **47.67 %**

This is larger than that attained in the single CSTR of same volume (= 1.5 m^3). Thus, a series cascade does offer advantage, in this case.

The results are summarized below:

Bioreactor (V = 1.5 m^3)	**Fractional substrate conversion (α), attained**
1. Single CSTR	40.6 %
2. Single PFR	72.0 %
3. Two CSTRs in series	40.6 %
4. CSTR – PFR combination	55.82 %
5. Single CSTR (with no substrate inhibition)	46.0 %
6. Two CSTRs in series (with no substrate inhibition)	47.67 %

Let this be kept in mind that the results reported in Examples (3.1) and (3.2) are based on ideal reactor performances. In actual industrial practices, the bioconversion obtained shall be lower than that predicted through these calculations. The inferences could also differ. For example, in the case of a series – cascade of lower size CSTRs, the performance of each CSTR in the cascade could approach close to ideal behavior (close to cent percent back mixing) due to the low volume of each. However, in the case of a single, large volume CSTR, the deviation from ideality shall be more pronounced (as thorough back mixing is difficult to be achieved in such large tanks). Accordingly, in actual industrial practices, the series cascade could perform better than a single CSTR of same (total) volume.

Series cascades, however, demand larger installation cost, larger maintenance cost, larger floor space. The alignment of reactors could often become cumbersome when the number of reactor tanks in the cascade is large.

Deviation from ideality could be due to presence of dead zones in the reactor (this is common in tank reactors). Within these zones, there shall hardly be any intermixing of fluid elements and this brings down the reactor performance. Presence of bypass streams is another possibility. An excellent example is the phenomenon of *channeling* that occurs in packed bed reactors. Here, the substrate solution tends to predominantly pass through that portion of the packed bed which is relatively loosely packed (has relatively larger voidage and thus offers lesser resistance to substrate flow), thereby bypassing the remaining portion of the bed (that is more tightly packed and hence has lower voidage). All the particles are, therefore, not uniformly wet be the substrate solution and this adversely affects the performance of the reactor. No doubt, channeling occurs only in beds that are non-uniformly packed. If the voidage of the bed is uniform throughout, then the effect of channeling would be negligible. However, this is difficult to accomplish in large capacity, industrial bioreactors.

3.2 Design of Industrial Bioreactors – Basic Concepts

As stated earlier, the actual performance of an industrial bioreactor falls in between the two ideal extremes such as 100 % backmixing (provided by ideal CSTR) and zero percent backmixing (true plug flow). The industrial bioreactor operates with a given degree of backmixing.

Mathematical analysis and simulation of the performance of industrial bioreactors (leading to their computer aided design) could be carried out using several approaches. Some of the popularly employed procedures are those based on,

(a) tanks – in – series model,
(b) PFDR (Plug Flow Dispersion Reactor) model,
(c) ideal reactor combinations.

Under tanks – in – series model, the industrial bioreactor under consideration is assumed equivalent to n number ideal CSTRs in series, the value of n being determined through RTD (residence time distribution) studies that utilize tracer technique [1] or by trial. This is based on the observation that the performance of a series cascade of ideal CSTRs of total volume v falls in between that of a single ideal CSTR of volume v and a PFR of volume v. In other words, through the bioreactor under consideration could not be treated as an ideal CSTR, it could be assumed to be equivalent to a number of small volume ideal CSTRs in series.

This approach has been successful with a good number of chemical reactors, but has not been sufficiently successful with bioreactors. An interesting successful case study in this connection is that reported by Narayanan and Vikas Narayan [2]. They have successfully demonstrated that moving bed biofilm reactors (which are heterogeneous bioreactors), both two phase and three phase systems, could be reliably modeled as equivalent to a number of ideal CSTRs in series. This is illustrated through a numerical example (Example 3.4) in the next section.

The PFDR concept is, by far, one among the most versatile and most successful approaches in bioreactor modeling, simulation and design. In this case, the bioreactor is assumed equivalent to a plug flow dispersion reactor (PFDR), which, in essence, is a plug flow reactor but with a specific degree of axial dispersion. At any particular cross-section of the reactor, neither the velocity nor the concentration has been assumed to vary in the radial direction (equivalent to plug flow), put there exists a given degree of intermixing among fluid elements in the axial or longitudinal direction. The degree of backmixing in the reactor is thus neither zero percent nor 100 %. The extent of axial dispersion is predicted by an appropriately defined axial dispersion coefficient (D_L), the value of which could be determined experimentally through RTD studies (tracer experiments) or evaluated from available experimental correlations. The performance equation of a PFDR can be thus written as

$$-U_L\left(dC_S / dz\right)+D_L\left(d^2C_S / dz^2\right)=\left(-r_S\right) \tag{3.2.1}$$

where U_L = velocity of feed solution through the reactor

$$= Q_0 / \left(\pi D^2 / 4\right) \tag{3.2.2}$$

D = diameter of reactor column

D_L = axial dispersion coefficient, m^2/s

Note that when $D_L = 0$, the above equation reduces to the performance equation of a PFR (discussed earlier). When D_L is very large ($D_L = \infty$), the performance of the PFDR approaches that of an ideal CSTR. After substituting the kinetic equation representing the bioconversion process, equation (3.2.1) is to be solved numerically (based on appropriate boundary conditions) to obtain the performance characteristics of the bioreactor. Numerical techniques such as line successive over relaxation method (SOR) or fourth order Runge – Kutta method could be utilized here for solving the above differential equation.

It must be kept in mind that the axial dispersion coefficient (D_L) is a flow parameter and its magnitude depends on the fluid flow rate, fluid properties (viscosity, density). It could also be a function of the system characteristics (liquid holdup, gas holdup).

Experimental correlations are available for the estimation of D_L, though many of them are enormously empirical in nature. For example, Levenspiel [3] has presented elaborate graphical data for the estimation of axial dispersion in turbulent flow through straight, cylindrical pipes or tubes. His graphical data could be fitted into a correlation of following form [4]:

$$\ln(N_d) = 5.2886 - 0.8874\ln(Re) + 0.025\left[\ln(Re)\right]^2 \tag{3.2.3}$$

where, N_d = dispersion number = $D_L/(U_L D)$ (3.2.4)

$$= (\pi D)D_L/(4Q_0) \tag{3.2.5}$$

Re = Reynolds numb $= (DG_t/\mu_f)$ (3.2.6)

G_t = mass velocity of fluid through the pipe, kg / (m^2.s)

$$= \dot{m}/(\pi D^2/4) \tag{3.2.7}$$

$\dot{m}$ = mass flow rate of fluid, kg /s

The above correlation is valid for $10^4 \le Re \le 10^5$. At high values of $Re(Re > 10^5)$, the value of N_d reportedly remains more or less constant at 0.2. Thus, at $Re > 10^5$,

$$N_d = D_L/(U_L D) = 0.2$$

$$D_L = (0.2)(U_L D) \tag{3.2.8}$$

$$= (0.8Q_0)/(\pi D) \tag{3.2.9}$$

Studies on axial dispersion in Newtonian flow through packed beds have been reported by Bischoff and Levenspiel [5] and also by Wilhellm [6]. It is reported that at very low values of Reynolds number ($Re < 10$) and also at high values of $Re(Re \ge 1000)$, the dispersion number (N_d) remains more or less constant. Thus, at $Re < 10$,

$$N_d = D_{LP}/\left[d_P(U_L/\in_L)\right] = 1.8$$

or,

$$D_{LP} = (1.8)d_P(U_L/\in_L) \tag{3.2.10}$$

where, $\in_L$ = voidage of the bed (= fractional liquid holdup in the bed)

Here, Re is the modified Reynolds number and is defined as,

$$Re = \left(d_P U_L \rho_L / \mu_L\right) \tag{3.2.11}$$

To note that U_L represents the superficial velocity of fluid through the packed bed, while $\left(U_L / \in_L\right)$ represents the actual velocity of flow of fluid through the voids of the bed.

Similarly, at $Re \geq 1000$ also, N_d remains more less constant at 0.5. Therefore, at $Re \geq 1000$,

$$D_{LP} = (0.5) d_P \left(U_L / \in_L\right) \tag{3.2.12}$$

Carberry [7] has presented a comprehensive discussion on axial and radial dispersion in packed beds.

In the case of heterogeneous systems (biofilm reactors, immobilized enzyme or immobilized cell bioreactors), equation (3.2.1) is to be modified to

$$-U_L \left(dC_S / dz\right) + D_L \left(d^2 C_S / dz^2\right) = \eta(-r_S)(int) \tag{3.2.13}$$

where η = effectiveness factor

$(-r_S)(int)$ = intrinsic rate of bioconversion

The effectiveness factor (η), as discussed earlier, is to be estimated from equation (2.4.15) or (2.4.27) of Chapter – 2. The differential equation (3.2.13) is thus more nonlinear, since the expression for effectiveness factor (η) is also a complex, nonlinear function in C_S. However, solution of equation (3.2.13) using numerical techniques such as fourth order Runge – Kutta method or line SOR method has been fairly successful. These techniques, though iterative in nature, have been observed to provide convergence within a limited number of iterations and they are also seen to provide reliable prediction of bioreactor performance. This is highlighted more elaborately in the subsequent sections.

A further improvement or modification to the performance equation of PFDR is the incorporation of liquid - to - particle mass transfer coefficient, k_{LP}. As discussed in Chapter – 2 (Section 2.5), the mass transfer resistance to substrate transport in the liquid bulk is usually negligible and with allowable error, we could assume that substrate concentration in the liquid bulk (C_S) is more or less equal to that at the particle surface (C_{SP}). If such an assumption is to be avoided, then the performance equation of PFDR would have to be modified and made more rigorous as

$$-U_L \left(dC_S / dz\right) + D_L \left(d^2 C_S / dz^2\right) = \left(k_{LP}\, a\right)\left(C_S - C_{SP}\right) \tag{3.2.14}$$

$$\left(k_{LP}\, a\right)\left(C_S - C_{SP}\right) = \eta\left(at C_S = C_{SP}\right)\left(-r_{SP}\right)(int) \tag{3.2.15}$$

where k_{LP} = liquid - to - particle mass transfer coefficient, m/s

a = specific interfacial area for mass transfer, m^2/m^3

$\left(-r_{SP}\right)\left(int\right)=$ intrinsic rate of bioconversion at $C_S = C_{SP}$

The above two equations are to be solved simultaneously (using appropriate boundary conditions) to obtain the performance characteristics of the bioreactor. It may be noted that equation (3.2.15) would have to be solved by trial (using a trial and error procedure) to obtain the value of C_{SP} at any specified value of C_S. For example, let us assume that the bioconversion intrinsically follows Monod's kinetics such that

$$\left(-r_{SP}\right)\left(int\right)=\mu_m\left(app\right)C_{SP}/(K_S+C_{SP}) \tag{3.2.16}$$

where $\mu_m\left(app\right)$ is that defined in equation (2.4.9) of Chapter – 2. We are thus considering a biofilm reactor. Substituting this in equation (3.2.15) and re-arranging, we get a quadratic equation in C_{SP} as,

$$C_{SP}^{\ 2}+\left[K_S-C_S+\eta\mu_m\left(app\right)/\left(k_{LP}\,a\right)\right]C_{SP}-\left(K_S C_S\right)=0 \tag{3.2.17}$$

The above equation is to be now solved for C_{SP} by trial, as outlined below :

Step 1: Compute the liquid - to - particle mass transfer coefficient (k_{LP}) and the specific interfacial area for mass transfer (a) from available correlations.

Step 2: Assuming η = 1.0, solve the quadratic equation (3.2.17) for C_{SP}.

Step 3: Compute the value of effectiveness factor (η) at $C_S = C_{SP}$ from equation (2.4.27) or (2.4.15).

Step 4: Substituting the above - computed value of η in equation (3.2.17), solve it for C_{SP}.

Step 5: Check whether the above-computed value of C_{SP} agrees closely with that computed earlier. If not, repeat the computations starting from Step - 3.

In most cases, the value of C_{SP} would get finalized within 2 to 3 trials. Also, in the case of many bioreactors, the value of C_{SP} shall be very close to C_S. This procedure is illustrated in Example (3.6) of Section (3.4).

As stated earlier, the performance equations, such as equations (3.2.14) and (3.2.15), are to be solved simultaneously using any of the popular numerical algorithms such as the fourth order Runge - Kutta method. As for an example, the procedure, when the intrinsic rate of bioconversion follows equation (3.2.16), is summarized below. Let us consider a column reactor such as a packed bed reactor or a fluidized bed reactor. In either case, computations are started from the top of the bed where $C_S = C_{Se}$ and proceeded downwards. For example, let it be that we have to compute the height of the packed bed (L_P) required to achieve a specific fractional substrate conversion (α).

Equations (3.2.14) and (3.2.15) can be clubbed together to yield,

$$-U_L\left(dC_S/dz\right)+D_L\left(d^2C_S/dz^2\right)=\eta\mu_m\left(app\right)C_{SP}/\left(K_S+C_{SP}\right) \tag{3.2.18}$$

Now, let

$$W=\left(dC_S/dz\right) \tag{3.2.19}$$

Then, from equation (3.2.18),

$$(dW/dz) = F(z) = (U_L/D_L)W(z) + \eta(z)\mu_m(app)C_{SP}/\left[D_L(K_S + C_{SP})\right] \quad (3.2.20)$$

To note that W, η and C_{SP} are all functions of C_S or $C_S(z)$ and thereby are functions of axial distance (z). The procedure for the solution of equation (3.2.18) using fourth order Runge – Kutta method is then as follows:

Step 1: Select the step size, Δz. For example, let $\Delta z = 1.0$ mm = 0.001 m.

Step 2: Since α has been specified, compute C_{Se} as

$$C_{Se} = C_{SO}(1-\alpha) \quad (3.2.21)$$

Step 3: Put $i = 1$

$$C_S(1) = C_{Se}$$

$$W(1) = 0.0$$

Step 4: Put $j = 1$.

$$C_S(j) = C_S(i)$$

$$W(j) = W(i)$$

Step 5: Compute $C_{SP}(j)$ by solving equation (3.2.17) by trial. Thus,

(i) Put $\eta(j) = 1.0$

(ii) Solve equation (3.2.17) for $C_{SP}(j)$ as

$$C_{SP}(j) = -B + \sqrt{B^2 - 4AC}\,/\,2A \quad (3.2.22)$$

where $A = 1.0$ (3.2.23)

$$B = \left[K_S - C_S(j) + \eta(j)\mu_m(app)/(k_{LP}a)\right] \quad (3.2.24)$$

$$C = \left[-K_S C_S(j)\right] \quad (3.2.25)$$

(iii) Put $X = C_{SP}(j)$ (3.2.26)

(iv) Compute the characteristic dimension (L^*) from equation (2.4.20) and ϕ' from equation (2.4.24). Now, compute $\beta(j)$ as

$$\beta(j) = C_{SP}(j)/K_S \quad (3.2.27)$$

(v) Compute $\eta_d(j)$ from equation (2.4.22) and $\eta(j)$ from equation (2.4.27).

(vi) Re-compute $C_{SP}(j)$ as,

$$C_{SP}(j) = -B + \sqrt{B^2 - 4AC}\,/\,2A \quad (3.2.28)$$

where A, B, C are those defined in Step (ii).

(vii) Compute the % deviation (σ) as

$$\sigma = \left[X - C_{SP}(j)\right] / C_{SP}(j) \tag{3.2.29}$$

(viii) If the deviation is appreciable, repeat the computations starting from Step – (iii). Otherwise, proceed to Step – 6.

Step 6: Print $C_{SP}(j)$, $\eta(j)$.

Step 7: Compute $F(j)$ from equation (3.2.20). To note that $F(j)$ is nothing but the value of $F(z)$ at $C_{SP}(z) = C_{SP}(j)$, $W(z) = W(j)$ and $\eta(z) = \eta(j)$.

Step 8: $m(j) = (\Delta z)F(j)$ (3.2.30)

If $j = 4$, GO TO Step -10.

Step 9: $j = j + 1$ (3.2.31)

If $j = 2$, $C_S(j) = C_S(i) + (\Delta z / 2)W(i)$ (3.2.32)

If $j = 3$, $C_S(j) = C_S(i) + (\Delta z / 2)W(i) + (\Delta z / 4)m(j-2)$ (3.2.33)

If $j = 4$, $C_S(j) = C_S(i) + (\Delta z)W(i) + (\Delta z / 2)m(j-2)$ (3.2.34)

If $j < 4$, $W(j) = W(i) + (1/2)m(j-1)$ (3.2.35)

If $j = 4$, $W(j) = W(i) + m(j-1)$ (3.2.36)

GO TO Step - 5 (Repeat the computations starting from Step - 5).

Step 10: Compute $C_S(i+1)$ and $W(i+1)$ as

$$C_S(i+1) = C_S(i) + (\Delta z)W(i) + (\Delta z / 6)\left[m(1) + m(2) + m(3)\right] \tag{3.2.37}$$

$$W(i+1) = W(i) + (1/6)\left[m(1) + 2m(2) + 2m(3) + m(4)\right] \tag{3.2.38}$$

Step 11: If $C_S(i+1)$ is very close to C_{S0}, proceed to Step – 12. Otherwise, put $i = i + 1$ and repeat the computations starting from Step - 4.

Step 12: Compute the required height (L_P) of the packed bed as,

$$n = i$$

$$L_P = (\Delta z)(n) \tag{3.2.39}$$

This procedure has been utilized for the design and analysis of many industrial bioreactors [4] and has been observed to provide reliable estimates of design variables.

A third approach to industrial reactor modeling is to assume the reactor to be equivalent to a combination of different ideal reactors. A CSTR – PFR combination is the simplest example. It could also be a PFR – CSTR combination (of different volume ratios), a CSTR – PFR – CSTR combination and so on. The number of combinations possible is practically legion. Accordingly, this approach could demand large number of trials and could often turn out to be laborious.

A reactor combination of CSTR – 1 (of active volume V_{m1} and dead zone, V_{d1}) followed by a PFR (active volume = V_P) and CSTR – 2 (active volume = V_{m2} and dead zone = V_{d2}) is sketched in figure (3.3). It is assumed that there exists a bypass stream of flow rate, (RQ_0). The total volume of the bioreactor (V) is thus equal to $V_{m1} + V_{d1} + V_P + V_{m2} + V_{d2}$. Out of this, the active volume within which bioconversion occurs is only ($V_{m1} + V_P + V_{m2}$).

Since all the components are ideal reactors, the fractional substrate conversion (α) attained in such a reactor combination at a specific feed flow rate of Q_0 L/hr could be computed by solving the performance equation of each (with the known values of kinetic parameters). If this computed value of α agrees with that determined through pilot plant tests (in a pilot plant bioreactor), then this model could be accepted as final. Otherwise, the trial would have to be repeated with a different volume ratio ($V_{m1} + V_P + V_{m2}$) and a different value of R. If even after repeated trials, α (computed) does not agree with α (experimental), a different reactor combination has to be considered and computations repeated. Evidently, unless luck prevails, such a procedure could become too lengthy and often, laborious.

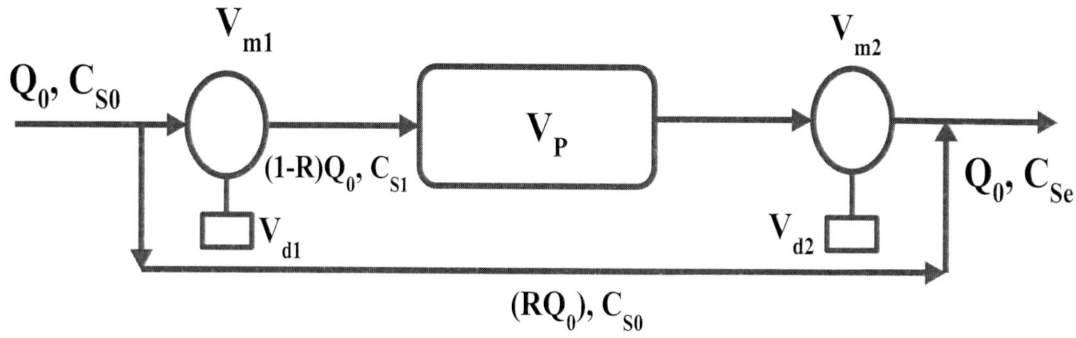

Figure 3.3

3.3 Design and Analysis of Moving Bed Biofilm Reactors

We have discussed on the basic characteristics of moving bed biofilm reactors in Chapter – 1 and Chapter – 2. They are basically tank reactors. The tank holds the substrate solution in which the particle – biofilm aggregates remain suspended. The slurry is agitated mechanically using a suitable impeller (propeller, turbine). Due to the agitation, the particle-biofilm aggregates are uniformly distributed within the liquid bulk and this also helps in keeping the aggregates (the discrete phase) in suspension.

In case of aerobic systems (three phase operation), air is sparged under pressure from below and this also promotes agitation of the slurry. The particle – biofilm aggregates tend to move (wobble) within the liquid (due to the agitation provided) and hence the name, *moving bed biofilm reactors*. In the strict sense, they are better called *slurry reactors*.

These bioreactors are restricted to low capacity installations. The tank size employed is relatively small so as to ensure effective contacting between the continuous phase (substrate solution) and the discrete phase (particle – biofilm aggregates). When operated as batch reactors (see figure 3.4), they may be approximated to ideal reactors with allowable error. In that case, the performance equation shall be

$$t = \int_{C_{Se}}^{C_{S0}} dC_S / \left[\eta(-r_{SP})(int)\right] \tag{3.3.1}$$

where η = effectiveness factor at $C_S = C_{SP}$

C_{SP} = substrate concentration in liquid at liquid – biofilm interface

$\left(-r_{SP}\right)\left(int\right)$ = intrinsic rate of bioconversion at $C_S = C_{SP}$

If resistance to substrate transfer from the liquid bulk to the surface of particle – biofilm aggregates is negligible, then

$$C_{SP} \cong C_S \tag{3.3.2}$$

After substituting the kinetic equation representing the intrinsic rate, the integral of equation (3.3.1) is to be evaluated numerically using Simpson's rule or trapezoidal rule.

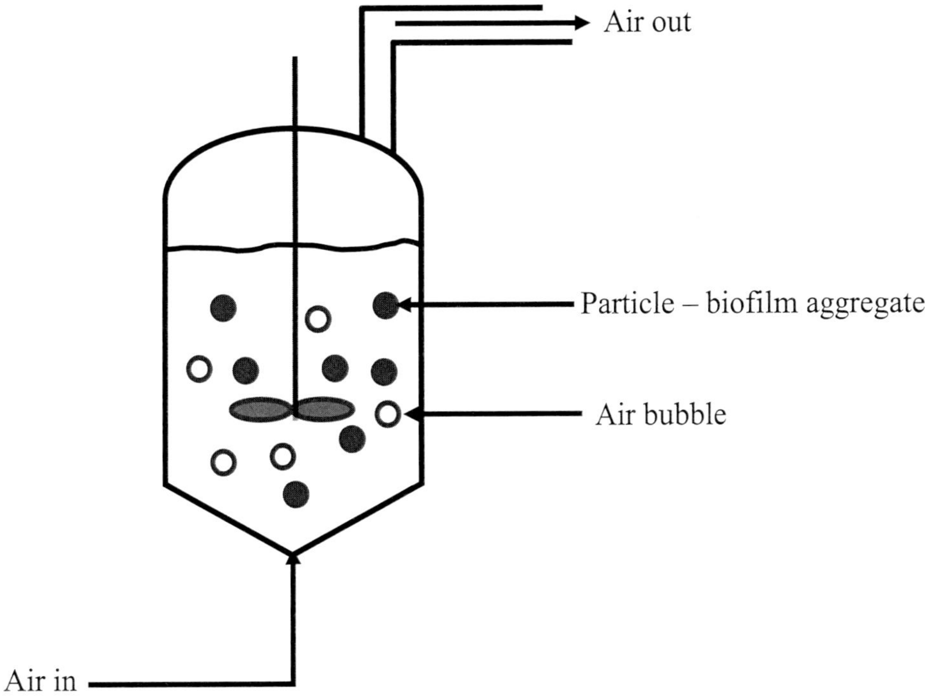

Figure 3.4: Three phase Moving Bed Biofilm Reactor (batch operation)

What has been illustrated in figure (3.4) is the three phase operation of a moving bed biofilm reactor. In this case, the process is aerobic and air is being sparged under pressure from the bottom at a specified flow rate. For systems of this kind, the fractional gas holdup $\left(\in_g\right)$ in the reactor tank becomes another operating parameter, apart from the fractional liquid holdup $\left(\in_L\right)$. Both of these parameters $\left(\in_L,\ \in_g\right)$ would have to be estimated from available experimental correlations (discussed subsequently).

In the case of two - phase systems (no air is supplied from outside, neither is there any evolution of gases during the process), if X is the mass fraction of catalyst particles (particle – biofilm aggregates) in the reaction mixture (commonly called, *the catalyst loading*), then its volume fraction in the slurry (which is nothing but the solid holdup, $\in_S$) is given by

$$\in_S = \left(X/\rho_{Sm}\right)/\left[\left(X/\rho_{Sm}\right)+\left(1-X\right)/\rho_L\right] \tag{3.3.3}$$

$$= 1/\left[1+\left(1-X\right)\rho_{Sm}/\left(X\rho_L\right)\right] \tag{3.3.4}$$

where ρ_{Sm} = density of particle – biofilm aggregates

ρ_L = density of substrate solution

And $\in_L = \left(1 - \in_S\right)$ (3.3.5)

As stated earlier, in the case of three phase systems, we have to invariably depend on the experimental correlations reported in the literature for the estimation of the fluid holdups in the reactor ($\in_L$ and $\in_g$). One such correlation proposed by Godlewska and Karcz [8] for the estimation of fractional gas holdup in three phase, agitated vessels is reproduced below :

$$\in_g = 0.00383 \left(P_{gLS} / V\right)^{0.521} \left(U_g\right)^{0.185} \left(X\right)^{0.0169} \quad (3.3.6)$$

where U_g = average superficial velocity of gas (air) through the reactor

$$= Q_g / \left(\pi D_t^2 / 4\right) \quad (3.3.7)$$

Q_g = volumetric flow rate of gas (air)

D_t = tank (reactor) diameter

The power input per unit volume $\left(P_{gLS} / V\right)$ may be computed from the correlation proposed by Patrick and Kennedy [9], discussed in Chapter – 2 (see equation – 2.5.17).

The fractional liquid holdup is then obtained by difference, as

$$\in_L = \left(1 - \in_S - \in_g\right) \quad (3.3.8)$$

If X is the catalyst lading (g/g) maintained, then

$$X = \left(\in_S \rho_{Sm}\right) / \left[\left(\in_S \rho_{Sm}\right) + \left(\in_L \rho_L\right) + \left(\in_g \rho_g\right)\right] \quad (3.3.9)$$

For any specified value of catalyst loading (X), the above two equations can be solved simultaneously for the two unknowns, $\in_S$ and $\in_L$.

Example 3.3

Xanthan gum is to be synthesized through aerobic microbial fermentation of lactose present in de-proteinised cheese whey using a culture of *Xanthomonas campestris* in a batch, moving bed biofilm reactor. The reactor tank is 0.8 m in diameter, fitted with four baffles and is agitated with a 500 rpm Rushton impeller (diameter = one-third of tank diameter). Cheese whey permeate (de-proteinized cheese whey) diluted to a lactose concentration of 9.0 g/L is filled upto 1.0 m height in the tank and 1.0 mm silica granules each surrounded by a 0.3 mm biofilm (cell mass concentration in the biofilm = 100 g/L) are kept in suspension in the substrate solution. A catalyst loading of 3% (by mass) is employed and compressed air is sparged from the bottom at the rate of 7200 L/hr (average density of air = 1.4 kg/m^3). The bioconversion reportedly follows Contois kinetics with μ_m = 0.29 hr^{-1}, K_C = 2.0 and overall yield coefficient for cell mass production (Y) = 0.5. Neglecting mass transfer resistance in the liquid bulk, determine the batch time required to affect 75% conversion of lactose to Xanthan gum. Take density of silica = 2200 kg/m^3 and that of microbial solution = 1030 kg/m^3. Effective diffusivity of substrate into the biofilm = 3.0 × 10^{10} m^2/s. Compare the results with t = 11.5 hr, observed during pilot plant tests.

Solution:

Assuming the bioreactor to be more or less equivalent to an ideal batch reactor, the performance equation of the reactor shall be that given in equation (3.3.1). However, since mass transfer resistance in liquid bulk is reportedly negligible, $C_{SP} \cong C_S$ and $(-r_{SP})(int)$ = $(-r_S)(int)$. Thus,

$$t = \int_{C_{Se}}^{C_{S0}} dC_S / \left[\eta(-r_S)(int)\right] \tag{i}$$

Contois kinetic equation (rewritten for attached growth of microbes) is that given in equation (2.4.32) of Chapter – 2 :

$$(-r_S)(int) = \mu_m(app)C_S / \left[K_C(app) + C_S\right] \tag{ii}$$

where $\mu_m(app) = (\mu_m / Y)x_f\, f(1 - \in_L - \in_g)/\in_L$ (iii)

$$K_C(app) = K_C\, x_f\, f(1 - \in_L - \in_g)/\in_L \tag{iv}$$

Substituting in equation (i), we get

$$\mu_m(app)\ t = \int_{2.25}^{9.0} F(C_S)dC_S \tag{v}$$

where $F(C_S) = (K_C(app) + C_S)/(\eta C_S)$

To note that C_{S0} = 9.0 g/L and C_{Se} = 9.0 (1 – 0.75) = 2.25 g/L.

Let us now proceed to compute the fractional fluid holdups ($\in_L$, $\in_g$) in the reaction mixture. For this, let us first estimate $\in_g$ from the experimental correlation reported by Godlewska and Karcz (equation – 3.3.6), coupled with the correlation reported by Patrick and Kennedy (equation – 2.5.17 of Chapter – 2). Thus,

$$Re_m = \text{mixing Reynolds number}$$
$$= \left(nD_a^2\rho_L / \mu_L\right)$$

Since n = (500/60) = 8.333 rps, $D_a = (D_t/3)$ = (0.8/3) = 0.2667 m, ρ_L = 1000 kg/m^3 and μ_L = 0.001 kg/(m.s),

$$Re_m = 592592.0$$

Since Re_m is greater than 10,000, the impeller is operating in fully developed turbulent zone. Therefore, from equation (2.5.19) of Chapter – 2,

P_L = power consumption for agitation of ungassed liquid

$= \left(K_T n^3 D_a^5 \rho_L\right)$

where K_T = 6.3. Therefore,

$$P_L = 4916.32 \text{ W}$$

Now, from equation (2.5.18),

$$P_{gL} = \left[P_L^2 n D_a^5 / Q_g^{0.53}\right]^{0.43}$$

where Q_g = 7200 L/hr = 0.002 m³/s. Therefore,

$$P_{gL} = 894.727 \text{ W}$$

Now, f = volume fraction of biofilm in particle – biofilm aggregate

$$= 1-\left(d_P / d_{Pm}\right)^3$$

Since d_P = 1.0 mm and d_{Pm} = 1.0 + 2(0.3) = 1.6 mm,

$$f = 0.75586$$

Also, $\rho_{Sm} = f\rho_m + (1-f)\rho_S$

Since ρ_m = 1030 kg/m³ and ρ_S = 2200 kg/m³,

$$\rho_{Sm} = 1315.64 \text{ kg/m}^3$$

Therefore, from equation (2.5.17),

$$P = P_{gLS} = \left(1.26 P_{gL}\right) - X\left(\rho_{Sm} / \rho_L\right)^{6.9}$$

where X = 0.03, $\left(\rho_{Sm} / \rho_L\right)$ = (1315.64)/(1000) = 1.31564. Hence,

$$P_{gLS} = 1127.157 \text{ W}$$

The volume of reaction mixture (V) is,

$$V = \left(\pi D_t^2 / 4\right)H = [\pi (0.8)^2/4]\,(1.0) = 0.50265 \text{ m}^3$$

Accordingly, (P_{gLS} / V) = 2242.429 W/m³

Now, from equation (3.3.6),

$$\in_g = 0.00383\left(P_{gLS} / V\right)^{0.521}\left(U_g\right)^{0.185}\left(X\right)^{0.0169}$$

where $U_g = Q_g / \left(\pi D_t^2 / 4\right) = 0.004$ m/s. Therefore,

$$\in_g = 0.07237$$

Now, by solving equations (3.3.8) and (3.3.9) simultaneously, we get

$$\in_S = 0.0214$$

$$\in_L = 0.90623$$

To note that, $\rho_g = 1.4$ kg/m³ (given). Now,

$$\mu_m(app) = (0.29 / 0.5)(100)(0.75586)(0.0214)/(0.90623)$$
$$= 1.035242 \text{ g/(L.hr)}$$

$$K_C(app) = (2.0)(100)(0.75586)(0.0214)/(0.90623)$$
$$= 3.5698 \text{ g/L}$$

Equation (v), therefore, reduces to

$$(1.035242)t = \int_{2.25}^{9.0} F(C_S)\,dC_S \qquad \text{(vii)}$$

where $F(C_S) = (3.5698 + C_S)/(\eta C_S)$ (viii)

Let us now evaluate the integral of equation (vii) numerically using trapezoidal rule. For this, the values of $F(C_S)$ at different values of C_S are computed and are listed in tables (3.3.1) and (3.3.2).The values of effectiveness factor (η) are computed from the correlation proposed by Gottifreddi and Gonzo (equation – 2.4.27 of Chapter – 2) as,

$$(1/\eta)^2 = (1/\eta_d)^2 + \exp\left[\varnothing_b - (1/\eta_d)^2\right] \quad \text{(ix)}$$

where
$$\varnothing_b = (6/5)(\varnothing')^2/(1+\beta)^2$$

$$\varnothing = L^*\sqrt{\mu_m(app)/\left[D_e K_C(app)\right]}$$

$$L^* = \left(d_{Pm}^3 - d_P^3\right)/\left(6d_{Pm}^2\right)$$

Substituting the values of d_P and d_{Pm}, we get

$$L^* = 0.20156 \text{ mm}$$

Since $D_e = 3.0 \times 10^{-10}$ m²/s,

$$\varnothing' = 0.1044$$

And $\beta = C_S/K_C(app) = (C_S/3.5698)$ (x)

$$\varnothing_b = (0.01309)/(1+\beta)^2 \quad \text{(xi)}$$

From equation (2.4.22),

$$\eta_d = (13.54)\left[(1+\beta)/\beta\right]\sqrt{\beta - \ln(1+\beta)} \quad \text{(xii)}$$

The values of η are thus computed from equations (ix) to (xii). Incidentally, for all values of C_S greater than 5.0 g/L, the value of effectiveness factor (η) is seen to be very close to unity. Hence, at these high values of C_S, the parameter $F(C_S)$ could be computed straightaway without resorting to the computation of η. These results are tabulated in table (3.3.2).

Table – 3.3.1

C_S, g/L	β	η_d	η	$F(C_S)$
2.25	0.63028	13.1756	0.9975	2.59294
2.50	0.70	13.5354	0.9977	2.43342
2.75	0.77035	13.8868	0.9979	2.3029
3.0	0.84038	14.2331	0.99807	2.19416
3.25	0.9104	14.57326	0.9982	2.102163
3.5	0.98045	14.90826	0.99833	2.02331
3.75	1.05048	15.2377	0.99845	1.95498
4.0	1.1205	15.5620	0.99855	1.8952
4.25	1.19054	15.8817	0.99864	1.84246

C_S, g/L	β	η_d	η	$F(C_S)$
4.5	1.26057	16.1967	0.99872	1.79558
4.75	1.3306	16.50726	0.99879	1.75365
5.0	1.40	16.8152	0.99886	1.715907

Table – 3.3.2

C_S, g/L	$F(C_S)$
5.25	1.67996
5.50	1.64905
5.75	1.620835
6.0	1.594967
6.25	1.571168
6.50	1.54920
6.75	1.52886
7.0	1.50997
7.25	1.492386
7.5	1.47597
7.75	1.46062
8.0	1.446225
8.25	1.4327
8.5	1.419976
8.75	1.407977
9.0	1.39664

Now, as per trapezoidal rule,

$$I = \int_{2.25}^{9.0} F(C_S)dC_S = (h/2)\left[f_0 + 2(f_1 + f_2 + \ldots f_{n-1}) + f_n\right]$$

where h = step size = 0.25 g/L, f_0 = 2.59294, f_1 = 2.43342, f_2 = 2.3029 and so on. Finally, $f_n = f_{28}$ = 1.39664. Therefore,

$$(1.035242)t = I = 11.7121 \text{ g/L}$$

or
$$t = (11.7121/1.035242) = \mathbf{11.313 \ hr.}$$

This is the batch time required to affect 75 % conversion of lactose to Xanthan gum. It compares closely with the experimental value (from pilot plant tests) of 11.5 hr. In the present case therefore, the assumption of ideal performance of bioreactor does not induce serious error.

Continuous flow moving bed biofilm reactors are also common in industrial practices. They can be operated continuously at relatively larger capacities than batch reactors. Occasional shutdown and re-startup (including sterilization) shall not be required.

Example (2.9) of Chapter – 2 illustrates the performance analysis of a two phase, continuous flow MBBR, with special reference to lactic acid synthesis from de-proteinised cheese whey. However, in this example, the bioreactor has been assumed to be equivalent to an ideal CSTR. This could often be too much of an approximation. Design and analysis of flow reactors of this kind are better performed based on the tanks-in-series model [2]. Such an analysis is illustrated in Example (3.4) given below.

Example 3.4

Rework Example (2.9) of Chapter – 2, based on tanks-in-series model. Take the volume of bioreactor, V $= 3.0$ m^3 and neglect the mass transfer resistance in the liquid bulk. Compare the results obtained with those reported from pilot plant tests, given below :

Feed flow rate, m3/hr	Fractional substrate conversion
0.30	0.82
0.35	0.70

Solution:

Let us proceed to analyze the bioreactor performance by assuming it to be equivalent to three ideal CSTRs (of equal volume) in series ($n = 3$). Also, since mass transfer resistance in liquid bulk is negligible, $C_{SP} \cong C_S$.

Case – 1 :

Here, the feed flow rate (Q_0) = 0.3 m^3/hr. Therefore,

$$\tau(tot) = (V \in_L)/Q_0 = (3.0)(1-0.02)/(0.3)$$

$$= 9.8 \text{ hr}$$

Accordingly, the space time of each ideal CSTR in the cascade is,

$$\tau = \tau(tot)/3 = 3.266 \text{ hr}$$

Now, the performance equation for CSTR – 1 is,

$$3.266 = (C_{S0} - C_{S1})(K_S + C_{S1}) / [\eta \mu_m (app) C_{S1}]$$

Substituting $C_{S0} = 9.0$ g/L, $K_S = 0.22$ g/L and $\mu_m(app) = 0.8 \text{ g}/(\text{L.hr})$ and rearranging, we get

$$(C_{S1})^2 + [2.61333\,\eta - 8.78] C_{S1} - 1.98 = 0 \qquad \text{(i)}$$

The above equation can be solved only by trial, since the value of at η (at $C_S = C_{S1}$) is not known at the outset. Thus, let

$$\eta = 1.0$$

Then, equation (i) would reduce to

$$(C_{S1})^2 - 6.16667\ C_{S1} - 1.98 = 0$$

On solving, C_{S1} = 6.47257 g/L

Let us now compute the value of η from Gottifreddi and Gonzo's correlation (equation – 2.4.27 of Chapter – 2). Thus,

$$\beta = (C_{S1}/K_S) = (6.47257/0.22) = 29.42$$

As computed in Example (2.9),

$$L^* = 0.22245$$

$$\varnothing' = 0.4084$$

$$\eta_d = (3.4628)\left[(1+\beta)/\beta\right]\sqrt{\beta - \ln(1+\beta)}$$

And, $$(1/\eta)^2 = (1/\eta_d)^2 + \exp\left[0.2/(1+\beta)^2 - (1/\eta_d)^2\right]$$

In the present case, since $\beta = 29.42$,

$$\eta_d = 18.25876$$

And $$\eta = 0.99989$$

Since the value of η is very close to 1.0, the computed value of C_{S1} is final and no further trials are required. To note that C_{S1} is the substrate concentration in the exit stream from CSTR – 1.

Now, similarly for CSTR – 2, the performance equation shall be

$$(C_{S2})^2 + [2.61333\eta + 0.22 - C_{S1}]C_{S2} - 0.22C_{S1} = 0$$

Substituting $C_{S1} = 6.47257$ g/L, we get

$$(C_{S2})^2 + [2.61333\eta - 6.25257]C_{S2} - 1.423965 = 0$$

For solving by trial, let η = 1.0. Then, the above equation gets reduced to

$$(C_{S2})^2 - 3.63924C_{S2} - 1.423965 = 0$$

On solving, $$C_{S2} = 3.9956 \text{ g/L}$$

The value of is now computed following the same procedure as before. Thus,

$$\beta = (3.9956/0.22) = 18.1618$$

$$\eta_d = 14.24798$$

And $$\eta = 0.9997$$

Here also, the computed value of η is very close to unity and hence, we can avoid further trial.

Finally, for CSTR – 3 (the last reactor in the cascade), the performance equation shall be

$$(C_{Se})^2 + [2.61333\eta + 0.22 - C_{S2}]C_{Se} - 0.22C_{S2} = 0$$

On substituting $C_{S2} = 3.9956$ g/L, the above equation reduces to

$$\left(C_{Se}\right)^2+\left[2.61333\,\eta-3.7756\right]C_{Se}-0.879=0$$

Putting η = 1.0 (as the first trial), we get

$$\left(C_{Se}\right)^2-1.16227C_{Se}-0.879=0$$

On solving, $C_{Se}=1.6842$ g/L

Computation of η is now performed as before

$$\beta=(1.6842/0.22)=7.65545$$

$$\eta_d=9.179515$$

And $\eta=0.99865$

Once again, η is very close to 1.0 and no further trials are required. The fractional substrate conversion attained (α) is, therefore,

α = (9.0 – 1.6842)/(9.0) = **81.2867 %**

It can be seen that the above-computed value of α is very close to the reported experimental value (from pilot plant tests) of 82 %. This confirms the accuracy of the mathematical model developed (tanks-in-series model, with n = 3).

Case – 2 :

This corresponds to a feed flow rate (Q_0) of 0.35 m^3/s. Therefore,

$$\tau(tot)=8.4 \text{ hr}$$

And $\tau=\tau(tot)/3=2.8$ hr

Now, proceeding similarly as in Case-1, the performance equation for CSTR – 1 in the series cascade is

$$\left(C_{S1}\right)^2+\left[2.24\,\eta+0.22-C_{S0}\right]C_{S1}-0.22C_{S0}=0$$

Substituting $C_{S0}=9.0$ g/L, we get

$$\left(C_{S1}\right)^2+\left[2.24\,\eta-8.78\right]\,C_{S1}-1.98=0$$

This equation once again is to be solved by trial. Thus, let η = 1.0. Then, the above equation shall reduce to

$$\left(C_{S1}\right)^2-6.54\,C_{S1}-1.98=0$$

On solving, $C_{S1}=6.38$ g/L

As in Case-1, the value of effectiveness factor (η) is now computed from Gottifreddi and Gonzo's correlation. Thus,

$$\beta=(6.83/0.22)=31.04545$$

$$\eta_d=18.77$$

And $\eta=0.9999$

Since is very close to 1.0 (as assumed), we can discontinue further trials.
Now, for CSTR – 2 of the cascade,

$$(C_{S2})^2 + [2.24\eta + 0.22 - C_{S1}]\ C_{S2} - 0.22\ C_{S1} = 0$$

Substituting $C_{S1} = 6.83$ g/L, we get

$$(C_{S2})^2 + [2.24\eta - 6.61]\ C_{S2} - 1.5026 = 0$$

As the first trial, taking η = 1.0, we obtain,

$$(C_{S2})^2 - 4.37\ C_{S2} - 1.5026 = 0$$

On solving, $C_{S2} = 4.69$ g/L

Computation of η (at $C_S = C_{S2}$) is now done as before :

$$\beta = (4.69\ /\ 0.22) = 21.31818$$

$$\eta_d = 15.4712$$

And $\eta = 0.99979$

The value of η is very close to 1.0 and no further trials are demanded. Finally, for CSTR – 3,

$$(C_{Se})^2 + [2.24\eta + 0.22 - C_{S2}]\ C_{Se} - 0.22\ C_{S2} = 0$$

Taking $\eta = 1.0$ (as the first trial) and substituting $C_{S2} = 4.69$ g/L, we get

$$(C_{Se})^2 - 2.23\ C_{Se} - 1.0318 = 0$$

On solving, $C_{Se} = 2.6233$ g/L
Computing η as in earlier cases,

$$\beta = (2.6233\ /\ 0.22) = 11.9241$$

$$\eta_d = 11.48566$$

And $\eta = 0.99939 \approx 1.0$

No further trials are thus required. The fractional substrate conversion attained (α) is, therefore,

$$\alpha = (9.0 - 2.6233)\ /\ (9.0) = \mathbf{70.852\ \%}$$

This value of α (computed) is also very close to the experimental value reported, such as 70.0 %, thereby confirming the reliability of the developed mathematical model. The results are summarized below :

Feed flow rate, m³/hr	α(computed), %	α(experimental),%
0.30	81.2867	82.0
0.35	70.85	70.0

Incidentally, if we assume the bioreactor to be equivalent to a single ideal CSTR of volume 3.0 m³, then the computed value of fractional substrate conversion (α) shall be 78.29

% in Case – 1 and 69.18 % in Case – 2, thereby demonstrating that tanks-in-series model predicts the bioreactor performance more accurately. Also, substrate conversion provided by the series- cascade is higher than that provided by a single CSTR of same volume.

3.4 Design and Analysis of Packed Bed Bioreactors

Among the heterogeneous column reactors, packed bed reactors are probably the oldest. They could be of two types:

a. Packed bed biofilm reactors
b. Packed bed immobilized enzyme/cell bioreactors

We have briefly discussed on these in Chapter 1. In packed bed biofilm reactors, the particle- biofilm aggregates form a packed bed or fixed bed of height L_P inside the vertical column. The bed is supported on a porous plate at the bottom, which also serves as the liquid distributor (see figure 3.5). The feed solution is admitted from the bottom and it moves up through the bed. As the feed solution flows upward, the substrate diffuses into the biofilm, where it is converted into product(s) and, the product(s) diffuses out into the liquid bulk. The product solution is discharged from the top.

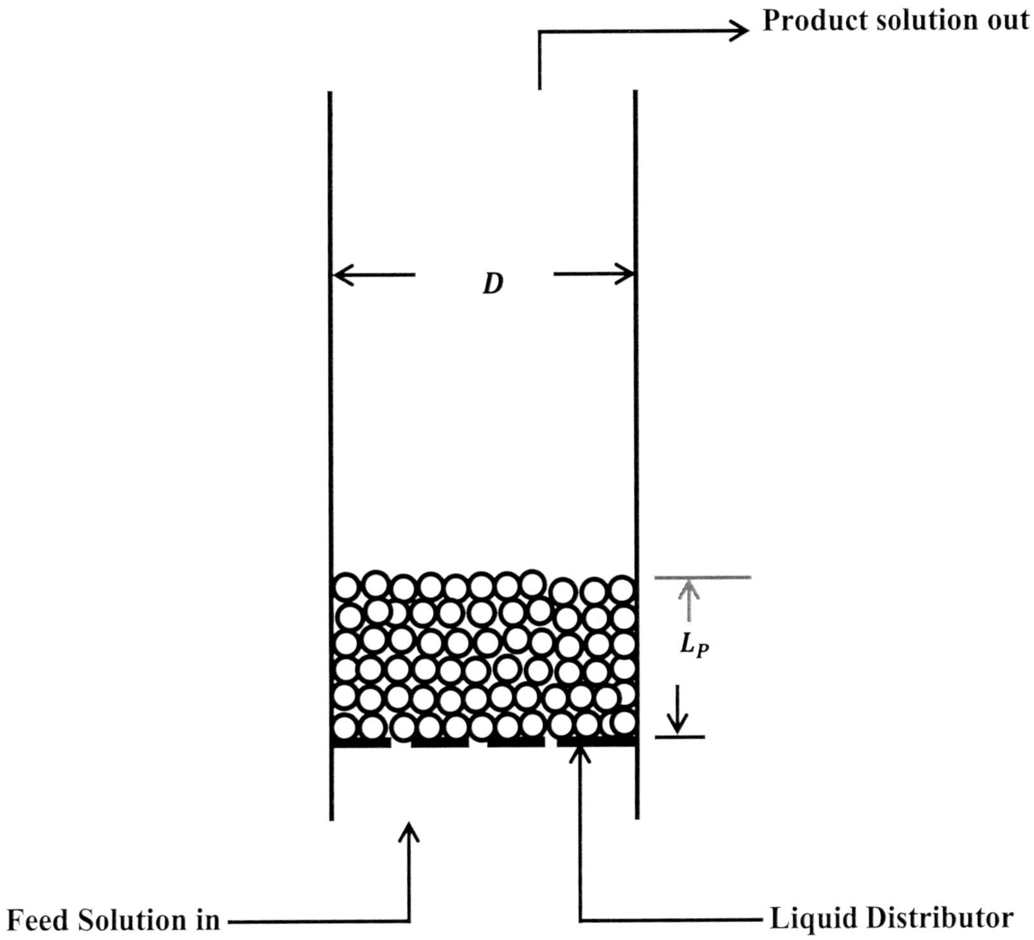

Figure 3.5: Schematic of Packed Bed Biofilm reactor (upflow mode of operation). $U_L < U_{mf}$ each particle rests on the other. $\varepsilon_P = 0.3$ to 0.4.

When operated in the downflow mode, the bioreactor is often called the *trickle bed bioreactor* (the term *trickle bed filter* or *trickling filter* is also used particularly in wastewater treatment circles, though this is an incorrect, often misleading, terminology). Here, the feed solution is admitted from the top and it flows down (trickles down) the bed under gravity (at a low velocity). When used for aerobic wastewater treatment, the bioreactor is constructed in large sizes and the column is kept open at the top. As a result, atmospheric oxygen (air) diffuses into the substrate solution and that meets the oxygen requirement of the process. Pumping of compressed air using huge air compressors is not required and this, coupled with the downflow mode of operation (feed solution is not required to be pumped) brings down the operating cost of the bioreactor.

The construction and operation of packed bed immobilized enzyme/cell bioreactors are exactly analogous, except that the packed bed is now composed of support particles accommodating the immobilized enzyme or the microbial cells. Also, the upflow mode of operation is more commonly employed here.

It is to be kept in mind that unlike fluidized bed, semifluidized bed or sludge bed (UASB) bioreactors, packed bed bioreactors are restricted to low capacities (low feed flow rates). At high feed flow rates, the bed tends to get fluidized.

Three phase operation with co-current flow (upward flow) of feed solution and air (often, sterile air) is also practiced in many packed bed biofilm reactors handling aerobic processes. Production of Xanthan gum from de-proteinised cheese whey (discussed in example 3.3) is an example. Trickle bed reactors with co-current downflow of both liquid and gas are also common in many petrochemical and pharmaceutical industries.

Alternately, gases produced during the process (such as carbon dioxide gas) move up through the packed bed, thereby formulating a three phase, countercurrent operation. For example, in the trickle bed bioreactors discussed above, though no air is sparged into the reactor, the carbon dioxide gas formed during the aerobic digestion of dissolved organic matter in the wastewater (feed solution) moves up through the packed bed (as the feed solution trickles down), thereby constituting a three phase, countercurrent operation of the bioreactor.

In the case of three phase operation, it becomes invariable to resort to experimental correlations available in literature for the estimation of fractional liquid holdup $(\in_{PL})$ and fractional gas holdup $(\in_{Pg})$ in the bioreactor.

For the estimation of liquid holdup $(\in_{PL})$ in trickle beds that operate with downflow of both liquid and gas, correlations have been proposed Sato et.al [10] and Midoux et.al [11]. They have employed a dimensionless pressure drop parameter (β), which is defined as

$$\beta = \left\{ \frac{\left(\dfrac{-\Delta P}{L_P} \right)_{liq}}{\left(\dfrac{-\Delta P}{L_P} \right)_{gas}} \right\}^{\frac{1}{2}} \tag{3.4.1}$$

The pressure drop (or, the pressure gradient) across the packed bed for single phase flow (flow of either liquid or gas) could be estimated with good accuracy from the well-established Ergun's equation [12,13], which is reproduced below:

$$f_b = (150 / Re_b) + 1.75 \tag{3.4.2}$$

where f_b is the bed friction factor such that

$$f_b = \left[(-\Delta P)/L_P\right]\left[d_P/(\rho_L U_L^2)\right]\left[\in_P^3/(1-\in_P)\right] \tag{3.4.3}$$

$$Re_b = \text{bed Reynolds number} = (d_p U_L \rho_L / \mu_L) \tag{3.4.4}$$

For gas flow through the packed bed U_L, μ_L, ρ_L are to be replaced by U_g, μ_g, ρ_g respectively.

The correlation proposed by Sato et al.[10] is,

$$\in_{PL} = (0.4\beta)(a_b)^{1/3} \tag{3.4.5}$$

where a_b = specific surface area of the bed, m^2/m^3 of the bed

= (total surface area of particles in the bed) / (total volume of the bed)

$$= (6/d_P)(1-\in_P) \tag{3.4.6}$$

The above correlation is reportedly valid for $0.1 \le \beta \le 20$.

Midoux et al. [11] have proposed a modified correlation, which, they report, is applicable over a larger range of β-parameter, such as $0 \le \beta \le 80$:

$$\in_{PL} = \varnothing / (1+\varnothing) \tag{3.4.7}$$

$$\text{where } \varnothing = 0.66(\beta)^{0.81} \tag{3.4.8}$$

In the case of packed beds involving co-current upward flow of both gas and liquid (both gas and liquid are admitted from the bottom of the bed), Kumar et. al. [14] have reported the following correlation for the estimation of fractional liquid holdup ($\in_{PL}$) in the bed:

$$\in_{PL} = 1.489(Re_{bL})^{0.142}(Re_{bg})^{-0.222}\left[\in_P/(1-\in_P)\right]^{0.547}\left[d_P/D\right]^{0.054}\left[g\mu_L^4/(\rho_L\sigma_L^3)\right]^{0.054} \tag{3.4.9}$$

$$\text{where } Re_{bL} = \text{liquid phase Reynolds number} = (d_P U_L \rho_L / \mu_L) \tag{3.4.10}$$

$$Re_{bg} = \text{gas phase Reynolds number} = (d_P U_g \rho_g / \mu_g) \tag{3.4.11}$$

σ_L = interfacial tension, N/m

The systems considered are air-water and air - glycerol solutions, the packed beds being composed of randomly packed ceramic spheres, 3.87 to 8.5 mm in diameter (bed porosity or voidage = $\in_P$ = 0.35 to 0.40). The above correlation is valid for liquid flow rates in the range of 2.17 to 43.0 kg/(m^2.s) and gas flow rates ranging from 0.265 to 2.4 kg/(m^2.s).

The simplest mathematical model that could be used to describe (analyze) the performance of a packed bed bioreactor is that which assumes true plug flow through the reactor bed (see Example 3.5). No doubt, a more rigorous analysis shall be that based on the PFDR approach (discussed in Section 3.2). That is, the bioreactor is assumed equivalent to a plug flow dispersion reactor (PFDR) with a pre-determined value of the axial dispersion coefficient

(D_{LP}). Typically, the value of D_{LP} varies from 0.007 to 0.008 m²/s in most two phase, packed bed reactors (see also equations (3.2.10) to (3.2.12) of Section 3.2).

Example 3.5

A trickle bed bioreactor composed of 50 mm stones (each surrounded by a thin biofilm of 0.3 mm thickness) is to be used for the aerobic treatment of wastewater from a fertilizer manufacturing unit. The effluent from the fertilizer unit with a BOD of 50 mg/L is fed from the top of the column at 100 $m^3/(m^2.d)$ and its BOD is to be reduced to 10 mg/L. Compute the required packed height of the bioreactor, if the bioconversion follows Monod's kinetics with $\mu_m = 60.0 d^{-1}$, $K_S = 6.5$ mg/L and overall yield coefficient for cell mass production (assumed constant) = 0.50. The average voidage (porosity) of the bed is 0.60 and cell mass concentration in biofilm = 1000 mg/L. Take effective diffusivity of substrate into biofilm = 3.0×10^{-10} m^2/s and liquid-to-particle mass transfer coefficient = 90.0 m/d.

Solution:

Though, strictly speaking, the bioreactor exhibits a three phase operation (the carbon dioxide gas formed during the process moves up countercurrent to the trickling liquid solution), we shall assume here that the total amount of carbon dioxide formed is not substantially large (a part of the gas does get dissolved in the wastewater as well) and its movement through the bed does not significantly disturb the overall operation of the bioreactor. In other words, we may, with allowable error, assume that the bioreactor exhibits more or less a two-phase operation.

Further, let us also assume that the operation of the bioreactor is close to plug flow and accordingly, its performance equation shall be that given in equation (3.1.12). Thus,

$$\left[\tau_P \mu_m (app)\right] = \int_{10}^{50} \left[\left(k_S + C_{SP}\right) / \left(\eta C_{SP}\right)\right] dC_S \qquad \text{(i)}$$

where

$$\tau_P = \left(V \in_{PL}\right) / Q_0$$

$$\in_{PL} \cong \in_P = 0.60$$

$$Q_0 = (100A)\, m^3 / d$$

$$K_S = 6.5\, mg / L$$

A = cross-sectional area of reactor column

From eq. (2.4.9), $$\mu_m (app) = \left(\mu_m / Y\right) x_f\, f\left(1 - \in_P\right) / \in_{PL}$$

where $$\mu_m = 60.0, \quad x_f = 1000\, mg / L$$

$$f = 1 - \left(d_P / d_{Pm}\right)^3 = 1 - (50 / 50.6)^3 = 0.035$$

Therefore, $$\mu_m (app) = 2800\, mg / (L.d)$$

Equation (i), now reduces to (to note that $V = A\, L_P$ m³),

$$(16.8)L_P = \int_{10}^{50} F(C_S)\,dC_S \qquad \text{(ii)}$$

where $$F(C_S) = (6.5 + C_{SP})/(\eta C_{SP}) \qquad \text{(iii)}$$

The integral of equation (ii) is to be evaluated using Simpson's rule or trapezoidal rule or any other well-established numerical technique. For this, the values of $F(C_S)$ are to be computed at different values of C_S ranging from $C_S = C_{Se} = 10\ mg/L$ to $C_S = C_{Se} = 50\ mg/L$.

In this connection, let us estimate the effectiveness factor (η) from Gottifreddi and Gonzo's correlation (equation 2.4.27 of chapter 2) as,

$$(1/\eta)^2 = (1/\eta_d)^2 + \exp\left[\varnothing_b - (1/\eta_d)^2\right] \qquad \text{(iv)}$$

where $\varnothing_b = (6/5)(\varnothing')^2/(1+\beta)^2$

$\varnothing' = L^*\sqrt{\mu_m(app)/[K_S D_e]}$

$L^* = (d_{Pm}^3 - d_P^3)/(6d_{Pm}^2)$

Substituting $d_P = 50$ mm and $d_{Pm} = 50.6$ mm, we get

$$L^* = 0.296 \text{ mm}$$

Since $D_e = 3.0\times10^{-10}\ \text{m}^2/\text{s}$,

$\varnothing' = 1.2067$

And $\varnothing_b = (1.7473)/(1+\beta)^2$

Now, $\beta = C_{SP}/K_S = (C_{SP}/6.5)$ (v)

From equation (2.4.22),

$$\eta_d = (1.17197)\left[(1+\beta)/\beta\right]\sqrt{\beta - \ln(1+\beta)} \qquad \text{(vi)}$$

Finally, $(1/\eta)^2 = (1/\eta_d)^2 + \exp\left[(1.7473)/(1+\beta)^2 - (1/\eta_d)^2\right]$ (vii)

The values of η are thus computed from equations (v) to (vii). However, for this, we need to know the value of C_{SP} (substrate concentration at liquid-biofilm interface or at the surface of each particle-biofilm aggregate) in advance. This can be accomplished from equation (3.2.15) with the help of the known value of liquid-to-particle mass transfer coefficient (k_{LP}). The specific interfacial area for mass transfer (a) is obtained from equation (2.5.5) of chapter-2 as,

$$a = s_P(1-\epsilon_P)/\epsilon_{PL}$$

$$= (6/d_{Pm})(1-\epsilon_P)/\epsilon_{PL}$$

Since $d_{Pm} = 50.6$ mm and $\epsilon_{PL} = \epsilon_P = 0.60$,

$$a = 79.05138\ \text{m}^2/\text{m}^3$$

Therefore from equation (3.2.15),

$$(k_{LP}a)C_S - (k_{LP}a)C_{SP} = \eta\mu_m(app)C_{SP}/(6.5 + C_{SP})$$

Substituting $k_{LP} = 90.0$ m/d and the known values of a and $\mu_m(app)$ and rearranging, we get

$$C_{SP}^2 + \left[6.5 + (0.393556)\eta - C_S\right]C_{SP} - (6.5C_S) = 0 \qquad \text{(viii)}$$

The computation of C_{SP} and thereby η, thus demands a trial and error procedure. At first, a value of C_{SP} is assumed, then η is computed from equations (v) to (vii). Equation (viii) is now solved C_{SP} and it is checked whether the computed value of C_{SP} agrees with that assumed earlier. If not, a new value of C_{SP} is assumed and the trials are repeated. We shall illustrate this with an example :

Step 1: $C_S = C_{Se} = 10\,mg/L$

Step 2: Taking $\eta = 1.0$, equation (viii) becomes

$$C_{SP}^2 - (3.10644)C_{SP} - 65 = 0$$

On solving, $C_{SP} = 9.7637\,mg/L$

This is our assumed value of C_{SP}.

Step 3: Now, η is computed as ,

$$\beta = (9.7637)/(6.5) = 1.5021$$

$$\eta_d = 1.49308 \qquad \text{[from equation (vi)]}$$

And, $\eta = 0.87954$ [from equation (vii)]

Step 4: Equation (viii) becomes

$$C_{SP}^2 - (3.15385)C_{SP} - 65 = 0$$

On solving, $C_{SP} = 9.79195\,mg/L$

Since this value of C_{SP} does not differ significantly from that assumed earlier, we can discontinue further trial. Thus, from equation (iii),

$$F(C_S) = (6.5 + 9.79195)/\left[(0.87954)(9.79195)\right] = 1.89168$$

In this way, the values of $F(C_S)$ at all values of substrate concentration (C_S) ranging from 10 mg/L to 50 mg/L are computed and results are listed in table (3.5.1).

Table 3.5.1

Substrate concentration (C_S), mg/L	Substrate concentration at aggregate surface (C_{SP}), mg/L	Effectiveness factor (η)	$F(C_S)$
10.0	9.7919	0.8795	1.89168
12.0	11.7718	0.8998	1.7249
14.0	13.7553	0.9156	1.60832
16.0	15.740	0.928	1.52258

Substrate concentration (C_S), mg/L	Substrate concentration at aggregate surface (C_{SP}), mg/L	Effectiveness factor (η)	$F(C_S)$
18.0	17.730	0.9379	1.4571
20.0	19.720	0.946	1.4055
22.0	21.7114	0.9526	1.36398
24.0	23.704	0.958	1.3298
26.0	25.6976	0.9627	1.3014
28.0	27.692	0.9666	1.2773
30.0	29.6868	0.97	1.25665
32.0	31.6823	0.9728	1.2388
34.0	33.6782	0.9753	1.22318
36.0	35.6746	0.9775	1.20943
38.0	37.6713	0.9794	1.19724
40.0	39.6683	0.981	1.1864
42.0	41.6655	0.9825	1.1766
44.0	43.663	0.9838	1.16776
46.0	45.66	0.985	1.15975
48.0	47.6585	0.986	1.15252
50.0	49.6565	0.9869	1.14583

Now, from Simpson's rule,

$$I = \int_{10}^{50} F(C_S)\,dC_S = \left(\frac{h}{3}\right)\left[f_0 + 4(f_1 + f_3 + \ldots f_{n-1}) + 2(f_2 + f_4 + \ldots f_{n-2}) + f_n\right]$$

where h = step size = 2.0 mg/L. $f_0 = 1.89168$, $f_1 = 1.7249$, $f_2 = 1.60832$ and so on. Finally, $f_n = 1.14583$.

Therefore, $I = 52.92394$ mg/L

And from equation (ii),

$$L_P = (52.92394 / 16.8) = \mathbf{3.15\ m}$$

This is the required packed height of the bioreactor.

As has been discussed earlier, the PFDR approach is more accurate and more reliable for bioreactor design. We leave to the readers to perform the design of above trickle bed biofilm reactor by assuming it to be equivalent to a PFDR, as an interesting exercise. Numerical examples based on PFDR approach have also been presented in the subsequent sections in connection with the design and analysis of fluidized bed and semifluidized bed bioreactors.

3.5 Design and Analysis of Fluidized bed Bioreactors

As stated in the earlier section, packed bed bioreactors are restricted to low capacity installations. For operation at high capacities (when the feed solution is to be processed at high volumetric flow rates), fluidized bed bioreactors are to be preferred.

As in the case of packed bed bioreactors, fluidized bed bioreactors are also column reactors that utilize particle – biofilm aggregates or solid particles within which the microbial cells or the enzyme have been immobilized. These particles remain suspended in an ascending stream of the feed solution that is admitted from the bottom of the column through the liquid distributor (see figure 3.6). In the case of aerobic systems (gas- liquid-solid or three phase systems), air is also admitted from the bottom and it moves up the column in the form of tiny, discrete bubbles. The feed solution forms the continuous phase. Alternately, gases may be produced during the process (for example, biogas formed during the anaerobic digestion of organic matter) and these gases move up in the form of discrete bubbles, thereby constituting a three phase fluidized bed (see figure 3.7).

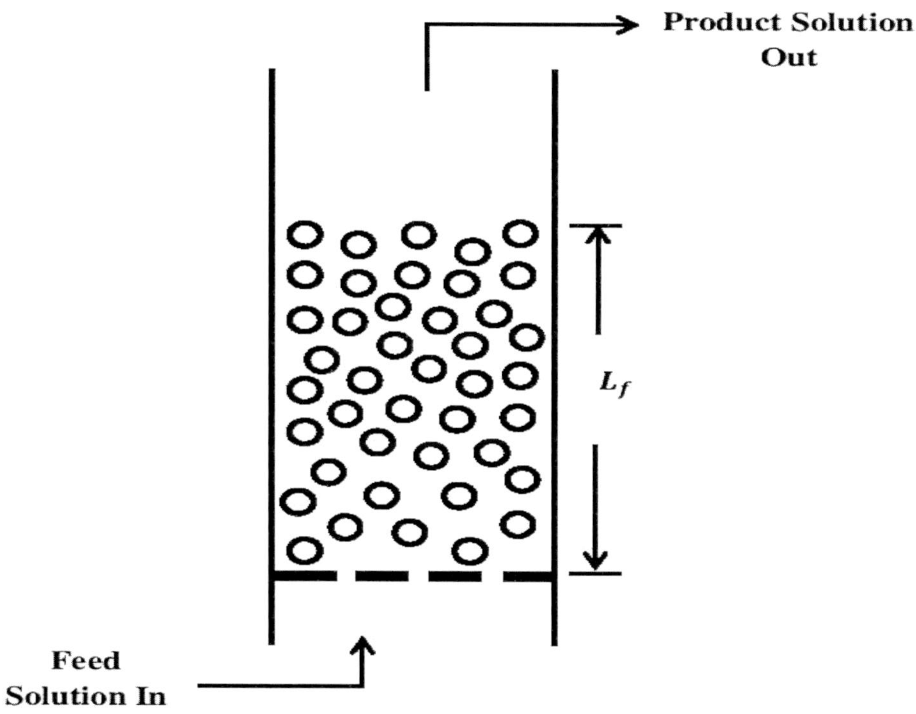

Figure 3.6: Two phase Fluidised Bed Biofilm Reactor. Fully fluidized (expanded) bed, $U_{mf} < U_L < V_t$, $L_f > L_{mf}$, $\varepsilon_f > \varepsilon_{mf}$. Particles suspended in ascending fluid stream.

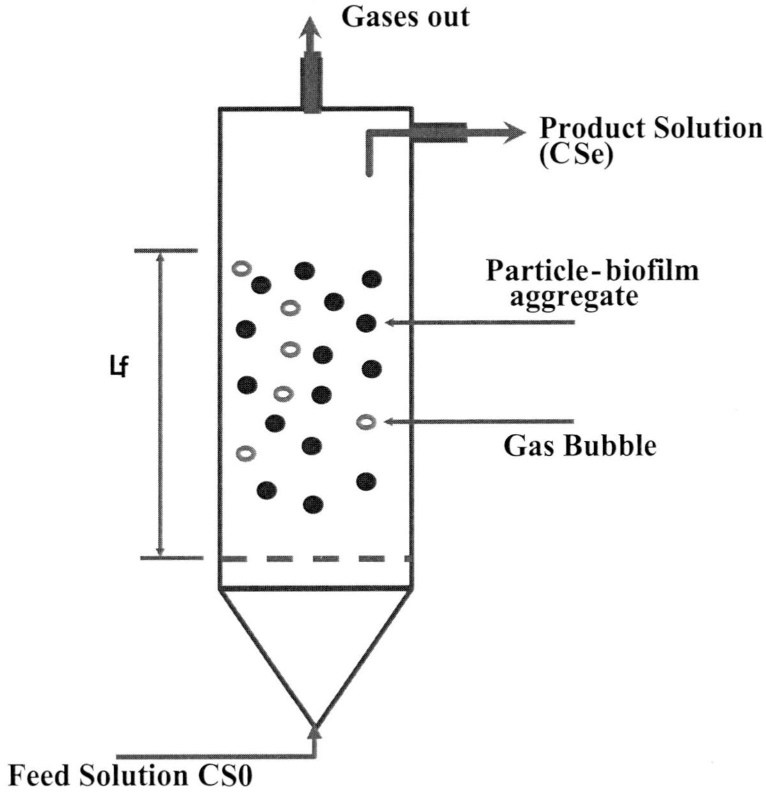

Figure 3.7: Three phase fluidized bed biofilm reactor (gases formed during the process).

The superficial velocity (U_L) or the volumetric flow rate (Q_0) of the feed solution employed in this bioreactor is much higher than that employed in packed bed bioreactors. In a packed bed, all the particles are in contact with each other and there is no relative motion between the particles. As the upward velocity of the feed solution (U_L) is increased, a stage is reached (at $U_L = U_{mf}$) when the loosest arrangement of particles gets formed within the column (the height of the bed increasing from L_P to L_{mf}). Every particle is just detached from the other and is not in contact with each other. This is called the *minimum fluidized* or *incipient fluidized bed.* The fluid velocity (U_{mf}) at which this occurs is called the *minimum fluidization velocity* or *incipient fluidization velocity*. When the upward velocity of the feed solution is further increased, the bed starts expanding and its height increases to L_f (see figure 3.6). This is called the *fully fluidized bed* and its height L_f and voidage ($\in_f$) depends on the upward velocity (or flow rate) of the fluid (feed solution) employed. If the velocity of the ascending fluid reaches or exceeds the terminal free settling velocity of the particles (V_t), then all the particles would get carried off the column and the column starts functioning more as a conveyor, rather than a bioreactor. For the effective operation of a fluidized bed bioreactor therefore, the upward velocity of the feed solution must be higher than the minimum fluidization velocity (U_{mf}), but lower than the terminal free settling velocity (V_t) of the particles. A popular choice is

$$(1.2)U_{m_f} \leq U_L \leq (0.8)V_t \tag{3.5.1}$$

Once fully fluidized, the pressure drop across the bed remains more or less constant and shall not increase with further increase in fluid velocity or fluid flow rate. This is an exclusive characteristic of fluidized beds and the bioreactor could be operated at high feed flow rates (at high capacities) without sacrificing much on the operating cost.

3.5.1 Estimation of Hydrodynamic parameters

Computation of terminal free settling velocity (V_t) of particles, we have already discussed in Chapter-2 (see section-2.5). The minimum fluidization velocity (U_{m_f}) could either be determined experimentally (by recording the pressure drop across the expanded bed at different fluid velocities (U_L) and then plotting pressure drop versus U_L) or computed from the approximate correlation proposed by Wen and Yu [15], given below:

$$Re_{mf} = [(33.67)^2 + 0.0408\ Ar_m]^{1/2} - (33.67) \tag{3.5.2}$$

where Re_{mf} = Reynolds number at minimum fluidization

$$= (d_P U_{mf} \rho_L) / \mu_L \tag{3.5.3}$$

Ar_m = modified Archimedes number (also called Galileo number)

$$= d_P^3\ (\rho_S - \rho_L) g \rho_L / \mu_L^2 \tag{3.5.4}$$

In the case of particle-biofilm aggregates, d_P and ρ_S are to be replaced by d_{Pm} and ρ_{Sm} respectively.

The above correlation is applicable to two - phase fluidized beds only (either liquid - fluidized bed or gas - fluidized bed). It is an approximate correlation (since has been developed based on many simplifying assumptions), but is still used in most design calculations, mainly due to the fact that for design purposes, we need only an approximate estimate of U_{mf}, since the actual operating velocity employed U_L is much larger than this minimum fluidization velocity.

It maybe stated here that in the special case of *circulating fluidized bed bioreactors* however, the reactor is operated with feed velocities higher than the carry – over velocity. No doubt, this forms a special case and in all conventional fluidized bed bioreactors, the feed velocity is selected based on the criterion given in equation (3.5.1).

In the case of three phase fluidized beds, a large number of experimental correlations have been reported by various authors for the estimation of the minimum fluidization velocity of the liquid, U_{Lmf}. They have been developed for co-current fluidized beds in which the liquid forms the continuous phase and the gas (either air that is supplied from outside or gases that are formed during the bioconversion process) moves up in the form of tiny, discrete bubbles. Examples are correlations proposed by Begowich and Watson [16], Fortin [17], Costa et.al [18], Song et.al [19], Nacef et.al [20], Zhang et.al [21], Ruiz et.al [22] and Jena et.al [23]. Being experimentally developed, all of these correlations are highly empirical in nature and each is applicable to a specific range of particle size, particle density, gas velocity and liquid properties (density, viscosity). A few of them are further limited by the column diameter employed as well. The *dimensional* correlation reported by Fortin [17] is reproduced below :

$$U_{Lmf} = 0.427\left(U_g\right)^{-0.198}\left(d_P\right)^{1.539}\left(\rho_S - \rho_L\right)^{0.775} \tag{3.5.5}$$

where U_g = average superficial velocity of gas through the bed, m/s.

As stated earlier, d_P and ρ_S are to be replaced by d_{Pm} and ρ_{Sm} in the case of particle-biofilm aggregates.

The voidage or void fraction ($\in_f$) of the fluidized bed shall be invariably higher than that of packed bed or static bed. This, once again, is a hydrodynamic parameter, the magnitude of which depends on the liquid velocity/gas velocity employed, apart from the particle size and particle density. In the case of two phase, liquid - fluidized beds in which *particulate fluidization* occurs, a reliable estimate of $\in_f$ (which shall be equal to the fractional liquid holdup, $\in_{fL}$) is possible from the correlation proposed by Richardson and Zaki [24] way back in 1954. Their correlation is presented below:

$$\in_f = \in_{fL} = (U_L / U_i)^{1/n} \tag{3.5.6}$$

where the parameter U_i is defined as

$$U_i = V_t \exp\left(-2.303 d_P / D\right) \tag{3.5.7}$$

$$n = \left[a_0 + b_0\left(d_p / D\right)\right]\left(Re_T\right)^m \tag{3.5.8}$$

Re_T = modified Reynolds number

$$= \left(D V_t \rho_L / \mu_L\right) \tag{3.5.9}$$

D = column diameter

V_t = terminal free settling velocity of each particle or particle-biofilm aggregate (see equation 2.5.27 of Chapter-2)

The values of correlation constants (a_0, b_0 and m) depend on the magnitude of the modified Reynolds number $\left(Re_T\right)$ and are listed in Table – 3.5.1.

Table – 3 .5.1 Values of Correlation constants of Equation (3.5.8)

Modified Reynolds number, Re_T	a_0	b_0	m
< 0.2	4.65	20.0	0.0
0.2 – 1.0	4.40	18.0	– 0.03
1.0 – 200	4.40	18.0	– 0.01
200 – 500	4.40	0.0	– 0.10
Above 500	2.40	0.0	0.0

It must be kept in mind that Richardson and Zaki's correlation, discussed above, has been developed for liquid – fluidized beds in which particulate or homogeneous fluidization occurs. In the case of gas – fluidized beds in which aggregative fluidization occurs (with violent movement of gas bubbles), the above correlation is hardly dependable.

The voidage (the total fluid holdup, $\in_f$) of a three phase fluidized bed is the sum of the fractional liquid hold ($\in_{fL}$) and the fractional gas holdup ($\in_{fg}$), the latter being of much lower magnitude in most cases. Once again the literature is practically flooded with experimental correlations proposed by different investigators for the estimation of these two parameters. An extensive survey on these correlations has been reported by Muroyama and Fan [25]. The correlation proposed by the Dakshinamurthy et.al [26] that is based on large amount experimental data collected on particles of various sizes and densities, is given below :

$$\in_f = k\ (U_L / V_t)^m \left(U_g \mu_L / \sigma_L\right)^{0.08} \tag{3.5.10}$$

where σ_L = interfacial tension, N/m.

The values of correlation constants (k, m) depend on the particle Reynolds number (Re_P) and are listed below:

If $Re_P < 500$, $k = 2.12$, $m = 0.41$

If $Re_P > 500$, $k = 2.65$, $m = 0.60$ (3.5.11)

Similarly, Saberian – Broudjenni et.al [27] have proposed correlations for the estimation of total fluid holdup ($\in_f$) and fractional gas holdup ($\in_{fg}$) in three phase fluidized beds, based on extensive experimentation :

$$\in_f = 0.4149\left(U_L / U_{mf}\right)^{0.27}\left[1 + 0.07\left(Re_{Pg}\right)^{0.34}\right] \tag{3.5.12}$$

$$\in_{fg} = \in_f \left(U_g - U_N\right) / \left(U_g + U_L\right) \tag{3.5.13}$$

$$\in_{fL} = \left[\in_f - \in_{fg}\right] \tag{3.5.14}$$

where $U_N = 0.017\left[\rho_L U_g^2\right]^{0.45}$ (3.5.15)

Re_{Pg} = modified Reynolds number = $\left(d_p U_g \rho_L / \mu_L\right)$ (3.5.16)

U_{mf} = minimum fluidization velocity of liquid in liquid fluidized bed (to be computed from Wen and Yu's correlation, equation 3.5.2).

More recently, Jena et.al. [28] have reported a modified correlation for the estimation of fractional gas holdup ($\in_{fg}$) in three-phase fluidized beds, that contains as many as four dimensionless groups :

$$\in_{fg} = (Mo)^{0.05} (Fr_L)^{-0.093} \left(Fr_g\right)^{0.401} \left(\rho_s / \rho_L\right)^{-0.0906} \left(d_P / D\right)^{0.081} (Eo)^{0.157} \left(L_S / D\right)^{0.026} \tag{3.5.17}$$

where Mo = Morton number = $\left[g\left(\mu_L{}^4\right) / \left(\rho_L \sigma_L{}^3\right)\right]$ (3.5.18)

Fr_L = liquid phase Froude number = $\left(U_L^2 / Dg\right)$ (3.5.19)

Fr_g = gas phase Froude number = $\left(U_g^2 / Dg\right)$ (3.5.20)

Eo = Eötvös number = $\left(\rho_L - \rho_g\right) D^2 g / \sigma_L$ (3.5.21)

L_S = height of static bed (prior to fluidisation)

The above correlation is valid for $\left(\rho_S / \rho_L\right) = 2.12$ to 2.48, $(d_P / D) = 0.0155$ to 0.063, Fr_L and $Fr_g = 0.00046$ to 0.025 and $Eo = 1370$ to 1528.

Ramesh and Murugesan [29] have also conducted elaborate experimentation and have come forward with an alternate correlation for the estimation of $\in_{fg}$, as given below:

$$\in_{fg} = 0.17\left(Fr_g\right)^{0.33}\left(Re_L\right)^{-0.065}\left(Ar_m\right)^{0.125}\left(Mo\right)^{0.05} \quad \text{for } Re_L < 100 \quad (3.5.22)$$

$$= 0.11\left(Fr_g\right)^{0.35}\left(Re_L\right)^{0.2}\left(Ar_m\right)^{0.11}\left(Mo\right)^{0.075} \quad \text{for } Re_L > 100 \quad (3.5.23)$$

where Re_L = modified liquid phase Reynolds number

$$= \left(DU_L\rho_L / \mu_L\right) \quad (3.5.24)$$

Ar_m = modified Archimedes number (defined in equation 3.5.4)

For the estimation of the fractional liquid holdup ($\in_{fL}$) in three phase fluidised beds, the correlation proposed by Kim et.al [30] is reproduced below:

$$\in_{fL} = 1.504\left(Fr_L\right)^{0.234}\left(Fr_g\right)^{-0.086}\left(Re_L\right)^{-0.082}\left(Ar_m\right)^{0.125}\left(We_L\right)^{0.092} \quad (3.5.25)$$

where We_L = Weber number = $\left(DU_L^2\rho_L / \sigma_L\right)$ (3.5.26)

The liquid phase Froud number (Fr_L) and the gas phase Froud number (Fr_g) are those defined earlier in equations (3.5.19) and (3.5.20) respectively.

Kim and coworkers have conducted experiments on air – water, air – acetone and air – sugar solutions systems with particles ranging in size from 1.0 to 6.0 mm (particle density = 2300 to 2950 kg/m^3), the superficial gas velocity (U_g) being maintained at 0.007 to 0.161 m/s. The liquid velocity (U_L) ranged from 0.027 to 0.102 m/s.

An alternate correlation for the estimation of $\in_{fL}$ in co-current three phase fluidized beds has been reported by Kato and coworkers [31] as given below :

$$\in_{fL} = k \ (U_L / V_t)^m \quad (3.5.27)$$

where $$k = 1 - \left[9.7N^{0.092}\right] / \sqrt{350 + \left(Re_P\right)^{1.1}} \quad (3.5.28)$$

$$N = \left(U_g^4\rho_L\right) / \left(g\sigma_L\right) \quad (3.5.29)$$

$$m = \left(b+1\right) / \left(a + 2.7b\right) \quad (3.5.30)$$

$$a = 5.1 + 86.2N^{0.285} \quad (3.5.31)$$

$$b = \left(Re_P\right)^{0.9}\left[0.1 + 0.443N^{0.165}\right] \quad (3.5.32)$$

All the above listed correlations for the estimation of fractional fluid holdup are for co-current three phase fluidized beds, in which the liquid forms the continuous phase and gas moves up in the form of tiny, discrete bubbles.

The magnitude of axial dispersion coefficient in fluidized beds (D_{Lf}) is relatively higher than that in packed beds, mainly due to the higher degree of backmixing affected. The value of D_{Lf} typically varies from 0.03 to 0.032 m^2/s. Experimental correlations have also been proposed in the literature for the estimation of this parameter. For example, Kim and Kim [32] have reported the following correlation for the estimation of axial dispersion coefficient in co-current, three phase, fluidized beds (liquid forming the continuous phase and the gas moving up in the form of discrete bubbles):

$$(d_PU_L / D_{Lf}) = (20.19)\ (d_P / D)^{1.66}\ [U_L / (U_L + U_g)]^{1.03} \quad (3.5.33)$$

3.5.2 Performance Analysis (Mathematical modeling)

Once the hydrodynamic parameters such as the minimum fluidization velocity $\left(U_{mf} \text{ or } U_{Lmf}\right)$ of the liquid (continuous phase) and the fractional fluid holdup ($\in_{fL}$, $\in_{fg}$) are computed from selected correlations, we can proceed to the design of the bioreactor. Once again, the simplest method of performance analysis of a fluidized bed bioreactor (biofilm reactor or immobilized cell/enzyme bioreactor) is that based on the assumption that the bioreactor is more or less equivalent to an ideal CSTR (completely backmixed reactor). In such a case, the performance equation shall be

$$\tau_m = \left[V_m \in_{fL} / Q_0\right] = \left(C_{S0} - C_{Se}\right) / \left[\eta_e\left(-r_{Se}\right)(int)\right] \tag{3.5.34}$$

where η_e = value of effectiveness factor at $C_S = C_{Se}$

$\left(-r_{Se}\right)(int)$= intrinsic rate of bioconversion at $C_S = C_{Se}$

It has been nevertheless assumed that mass transfer resistance in liquid bulk (resistance to substrate transport from liquid bulk to particle surface / biofilm surface) is negligible and $C_{SPe} \approx C_{Se}$.

The PFDR approach shall be, no doubt, more accurate and more desirable. The procedure shall be more or less the same as that discussed in Section (3.2) except for a few modifications. Unlike the packed bed reactor, the height of the fluidized bed reactor (L_f) is a hydrodynamic parameter, the magnitude of which depends on the liquid velocity (U_L) employed. The height of the expanded bed increases with increase in liquid velocity or liquid flow rate. Since the height and thereby the volume of the reactor (reaction zone) increases with increase in feed flow rate (Q_0), the fractional substrate conversion (α) also increases with increase in Q_0. This is an exquisite advantage of fluidized bed bioreactors and consequently, the bioreactor can be operated at high capacities (at high feed flow rates), simultaneously affecting high degree of substrate conversion. In reactors of constant volume such as tank reactors and packed bed reactors, increase in feed flow rate (Q_0) causes decrease in residence of time of fluid elements within the reaction zone and thereby lowers the fractional substrate conversion (α) attained.

Since the height of the fluidized bed (L_f) is fixed by the feed flow rate (Q_0) employed, what is required to be computed is the fractional substrate conversion (α) attained at this flow rate. The numerical scheme discussed in equations (3.2.18) to (3.2.39) of Section (3.2), therefore, gets slightly modified as outlined below:

Step 1: Compute the hydrodynamic parameters such as minimum fluidization velocity (U_{mf} or U_{Lmf}) and fluid holdup ($\in_{fL}$, $\in_{fg}$) from selected experimental correlations. To note that the operating liquid velocity (U_L) is to be selected based on the criterion given an equation (3.5.1). Now, compute feed flow rate (Q_0) as

$$Q_0 = \left(\pi D^2 / 4\right) U_L \tag{3.5.35}$$

Step 2: Compute the height of the fluidized bed (L_f) at the above feed flow rate from a solid balance as

$$L_S\left(1 - \in_P\right) = L_f\left(1 - \in_f\right) \tag{3.5.36}$$

where L_S and $\in_P$ are the height and voidage of the static bed (prior to fluidization).

Step 3: Assume a value of the fractional substrate conversion (α). For example, let $\alpha = 0.80$.

Step 4: Compute the substrate concentration in the exit stream (C_{Se}) as

$$C_{Se} = C_{S0}(1-\alpha) \tag{3.5.37}$$

Step 5: Select the step size (Δz). For example, let $\Delta z = 1.0$ mm $= 0.001$ m.

Step 6: Put $i = 1$

$$C_S(1) = C_{Se}$$

$$W(1) = 0.0$$

Step 7: Put $j = 1$

$$C_S(j) = C_S(i)$$

$$W(j) = W(i)$$

Step 8: Compute $C_{SP}(j)$ and $\eta(j)$ based on the procedure outlined in Step – 5 of the scheme outlined in Section (3.2) of Chapter – 2.

Step 9: Compute $F(j)$ from equation (3.2.20) of Section (3.2), after replacing D_L by D_{Lf}.

Step 10: Compute $m(j)$ as

$$m(j) = (\Delta z)F(j) \tag{3.5.38}$$

If $j = 4$, GO TO Step 12.

Step 11: This step is same as Step - 9 of the scheme described in Section (3.2).

Step 12: Compute $C_S(i+1)$ and $W(i+1)$ as indicated in Step -10 of the scheme described in Section (3.2).

Step 13: Compute $L_f(cal)$ as,

$$L_f(cal) = (i)(\Delta z) \tag{3.5.39}$$

Step 14: If $L_f(cal) \geq L_f$, GO TO Step – 15. Otherwise, put $i = i+1$ and repeat the computations starting from Step – 7.

Step 15: Compute $C_{S0}(cal)$ as, $C_{S0}(cal) = C_S(i+1)$ (3.5.40)

Step 16: If $C_{S0}(cal) < C_{S0}$, select a lower value of fractional substrate conversion (α) and repeat the computations starting from Step - 4. Otherwise, proceed to Step -17.

Step 17:Print α, L_f, Q_0, C_{S0}.

We have started the computations from the top of the fluidized bed, where $z = 0$, $C_S = C_{Se}$. Computations are continued downward until $z = L_f$. If the value of $C_S(z)$ at this point (at $z = L_f$) is very close to C_{S0}, then the assumed value of α is correct. Otherwise, a lower value of α is assumed and the computations repeated. The fractional substrate conversion (α) attained at the specified feed flow rate (Q_0) and feed concentration (C_{S0}) is thus estimated by trial. This also specifies the height of the fluidized bed / expanded bed (L_f) at that feed flow rate.

3.5.3 Advantages and Industrial Applications

The specific advantages of fluidized bed bioreactors are summarized below:

(a) Fluidized bed bioreactors provide high degree of fractional substrate conversion at high capacities (high feed flowrates)

(b) The fractional substrate conversion (α increases with increase in feed flow rate (Q_0) since the height (and thereby the volume) also increases with increase in Q_0 (an exclusive characteristic of these reactors)

(c) Once fully fluidized (once in steady operation), the pressure drop across the bed does not increase further with increase in feed flow rate; the reactor can be henceforth operated at high capacities with meagre increase in the operating cost.

(d) Every particle/ particle-biofilm aggregate is completely surrounded by the substrate solution and hence there is efficient contacting between the feed solution and the catalyst particles (no problems of channeling or bypass flow as that occur in packed bed reactors).

The operating cost of fluidized bed bioreactors is invariably higher than that of packed bed reactors (since they employed larger fluid flow rates) and there could be problem of entrainment of particles in the outgoing product solution. To eliminate entrainment loss of particles, a disengagement of the top of the fluidized disengagement space of larger diameters often provided at the top of the fluidized bed. As the product solution enters this space, its velocity decreases (due to increase in cross sectional area) and the entrained particles settle down back to the fluidized bed. No doubt, entrainment loss is serious only with fine particles and particles of low density.

Successful studies on the application of fluidized bed bioreactors for the synthesis of biodiesel from vegetable oils have been reported by De et.al [33, 34, 35]. They have conducted studies on trans- esterification of neem oil with methyl acetate using immobilized lipase enzyme as the catalyst, to produce the biodiesel, neem oil methyl ester (NOME) and have reported that when neem oil and methyl acetate are used in stoichiometric molar proportion of 1:3, 80% conversion of neem oil to biodiesel could be attained at a feed flow rate of 23000 L/hr, when the expanded bed height of the bioreactor (L_f) is 3.28 m. The performance of the bioreactor is thus truly encouraging.

Similarly, performance characteristics of these bioreactors when used for wastewater treatment (both aerobic and anaerobic) have also been quite attractive. Biswas and Narayanan [36] have reported that when employed for the removal of toxins like phenolics, cresols and arsenic from industrial effluents, these bioreactors provide as much as 80% BOD destruction at a feed flow rate of 21,600 L/hr, when the expanded bed height of the aerobic bioreactor is 2.43 m (column diameter = 0.5m). The same authors have reported that when used for aerobic dephenolization of coke oven wastewater, around 70% dephenolization could be affected at a feed flow rate of 20,000 L/hr in a fluidized bed bioreactor of 0.5m in diameter and 3.9 m in bed height.

Large scale synthesis of lactic acid from de-proteinized cheese whey and diluted molasses in fluidized bed biofilm reactors has also been successful. Narayanan [37] has reported that as much as 78% conversion of sucrose (present in molasses) to lactic acid is possible in a fluidized bed biofilm reactor (employing a biofilm composed of a culture of *Enterococcus faecalis*) of expanded bed height 3.0 m (column diameter = 0.5m) at a feed flow rate of 10,800 L/hr.

Fluidized bed reactors have been extremely popular in chemical process industries. For example, they are being employed in petroleum refineries for the catalytic cracking and catalytic reforming of petroleum oils for more than half a century. They have also been recommended for the bio- desulfurisation of petroleum oils [38].

LPO (liquid phase oxygen) utilization technology (discussed in Chapter 1) has also been successfully adapted to the operation of three phase fluidized bed biofilm reactors, handling aerobic processes. With a 2.0-3.0 mM concentration of hydrogen peroxide in the feed solution, significant BOD destruction has been found possible within a reasonably low reactor volume [39]. The operation of the bioreactor gets simplified to two phase (since no air is being sparged in) and the operating cost of huge air compressors gets dispensed with. Kinetic equation on this has been discussed in Chapter 2.

Fluidized bed bioreactors of modified design have also been successfully designed and launched in a number of process industries. An excellent example is fluidized bed columns of diverging- converging geometry (also called, variable area design)[40]. The geometry of the reactor column is that shown schematically in figure (3.8). The column is composed of a number of segments, each segment being of minimum diameter (D_1), maximum diameter (D_2) and segment length (L_S). Each segment is thus equivalent to two frustums of cone joined base to base. Any number of such segments could be joined together, to form a tall column. If n is the number of segments employed, then the total height (L) of the column shall be, $L=\left(nL_S\right)$. The angle of convergence/ divergence (θ) maintained is only 5^0 ($\tan\theta=1/12$). From simple geometry, it can be easily deduced that

$$\tan(\theta)=\left(D_2-D_1\right)/L_S \qquad (3.5.41)$$

The dimensions D_1, D_2 and L_S are so chosen that ($\tan\theta$) becomes close to (1/12). This type of geometry has been successfully utilized for the improved design of

(a) heat exchangers [41,42],
(b) condensers [43, 44],
(c) falling film absorbers [45,46],
(d) flat plate solar collectors [47],
(e) solar parabolic trough concentrating collectors [48],
(f) tubular ultrafiltration units [49].

In all the cases, the system has been observed to exhibit significantly augmented performance efficiency (the heat transfer coefficient / mass transfer coefficient in the system is 300-350% higher than that in the conventional system of straight, cylindrical geometry, but of same surface area per unit length) with relatively marginal increase (12-15%) in the pressure drop penalty (operating cost). The phenomenon has been substantiated both mathematically as well as experimentally.

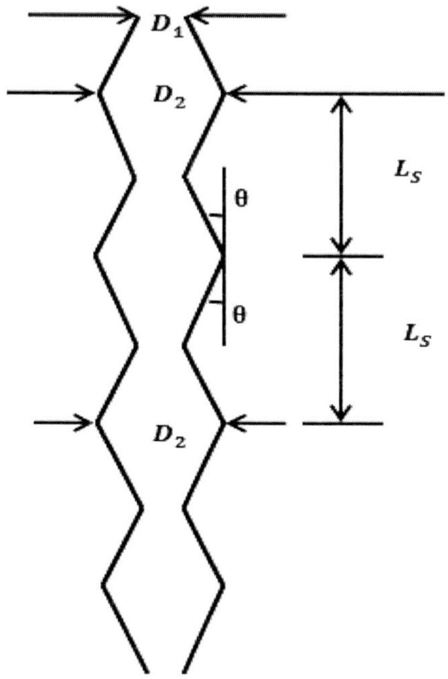

Figure 3.8: Schematic of diverging – converging column

With respect to synthesis of lactic acid from molasses and cheese whey, Basak et. al [50] have reported that the fractional conversion of sucrose / lactose attained in such reactors is 20 – 22% higher than that obtained in a conventional fluidized bed biofilm reactor of cylindrical geometry, but of same volume per unit length (at the same feed flow rate, same biomass concentration in biofilm and with support particles of same size). In addition, the operating cost of bioreactors of such modified geometry has been observed to be only marginally higher (12 – 15% higher) than that of conventional fluidized beds. De et. al [33,35] have also demonstrated, both mathematically as well as experimentally, that fluidized bed bioreactors of such diverging – converging geometry provide 20 – 25% higher yield of biodiesel, when used for biodiesel synthesis from neem oil and methyl acetate (with lipase enzyme immobilized in silica granules as the catalyst).

The fabrication cost of bioreactors of this category shall be, no doubt, higher.

Example 3.6:

Biodiesel (jatropha oil butyl ester) is to be manufactured in a fluidized bed bioreactor that uses lipase enzyme immobilised in 1.5 mm silica granules as a catalyst (density of silica = 2100 kg/m^3). A 1:3 (by mole) mixture of oil and butyl acetate is admitted from the bottom at 32000 L/hr and the fractional conversion of oil desired is 79%.

(a) Determine the required dimensions of the bioreactor (diameter, bed height) by assuming it to be the equivalent to an ideal CSTR,

(b) Compute the bioreactor dimensions, by assuming it to be equivalent to a PFDR, with axial dispersion coefficient = 0.0315 m^2/s.

Assume that the kinetics of transesterification follows equation (2.3.90) of Chapter-2 with r_{max} = 1.975 mol/(L.s), K_{m1} = 1.125 mol/L, K_{m2} = 15.83 mol/L and K_{Si} = 1.1 mol/L. Neglect mass transfer resistance to substrate transfer in liquid bulk and take effective diffusivity of substrate into particle = 2.95×10^{10} m^2/s. Property values of oil and acetate are given below:

	Jatropha oil	Butyl acetate
Density, kg/m^3	909	882
Kinematic viscosity, centistokes	28.43	0.7766
Molecular weight, g/mol	900	116

Solution:

Let us first select the operating velocity of the liquid (substrate solution). For this, we have to first quantify the physical and transport properties (density, viscosity, concentration) of the feed solution. Since the feed solution contains oil and butyl acetate in the molar ratio of 1:3, molar concentration of oil in the feed solution shall be,

$$C_{S1}(0) = 1/\left[(M_A/\rho_A)+(3M_B/\rho_B)\right]$$

where M_A = molecular weight of jatropha oil = 900.0 g/mol

M_B = molecular weight of butyl acetate = 116.0 g/mol

ρ_A = density of oil = 909 kg/m^3

ρ_B = density of butyl acetate = 882 kg/m^3

Therefore, $C_{S1}(0) =$ 0.72225 mol/L.

Now, the density of the feed solution is,

$$\rho_L = (M_A+3M_B)/\left[(M_A/\rho_A)+(3M_B/\rho_B)\right]$$

$$= C_{S1}(0)(M_A+3M_B) = 901.368 \text{ kg/m}^3$$

The viscosity of the feed solution can be estimated from the empirical formula,

$$\mu_L = \rho_L\left[\exp(X)-0.8\right](10^{-6}) \quad \text{(i)}$$

where $X = exp\left[0.0688(VBN)-0.755\right]$ (ii)

(VBN) = viscosity blending number of liquid mixture (feed solution)

$$= x_A(VBN_A)+(1-x_A)(VBN_B) \quad \text{(iii)}$$

VBN_A, VBN_B = viscosity blending number of oil and that of butyl acetate, respectively

x_A = mass fraction of oil in liquid mixture

$= M_A/(M_A+3M_B) = 0.72115$

$$VBN_A = (14.534)\, ln\,[ln(\gamma_A) + 0.8)] + 10.975$$

where γ_A = kinematic viscosity of oil (in centistokes) = 28.43 centistokes

Therefore, VBN_A = 28.6549

Similarly, VBN_B = – 0.46127 (since γ_B = 0.7766 cs)

Accordingly, from equation (iii), (VBN) = 20.5359

Then, from equation (ii), $X = 1.9307$

And, from equation (i), $\mu_L = 0.005493$ kg/(m.s)

Let us now proceed to compute the minimum fluidization velocity (U_{mf}) from Wen and Yu's correlation (equation 3.5.2). Thus, from equation (3.5.4),

$$Ar_m = \left(d_p\right)^3 \left(\rho_S - \rho_L\right) g \rho_L / \left(\mu_L^2\right)$$

Since d_p = (1.5 × 10^{3}) m, ρ_S = 2100 kg/m^3, ρ_L, μ_L = computed above,

$$Ar_m = 1184.32$$

Therefore, from equation (3.5.2), $Re_{mf} = 0.71$

and $U_{mf} = 0.003$ m/s

To select the operating liquid velocity (U_L), we have to estimate the terminal settling velocity of each particle (V_t) as well. Thus, from equation (2.5.27) of Chapter – 2,

$$V_t^2 = (4/3)\left(\rho_S - \rho_L\right) g\, d_P / \left(\rho_L f_D\right) = 0.026 / f_D \qquad \text{(iv)}$$

Also, $Re_P = (246.1409) V_t$ (v)

The above two equations are to be solved simultaneously, by trial. Thus, let

$$V_t = 0.1 \text{ m/s}$$

Re_P (from equation-v) = 24.614

ln (Re_P) = 3.2033

Now, from equation (2.5.30) and table (2.1) of Chapter-2,

$\ln\left(f_D\right)$ = 3.32 – 0.929 (3.18257) + 0.0429 (3.18257)^2 + 0.0014 (3.18257)^3

or $f_D = 2.294$

Then, V_t (from equation-iv) = 0.10646 m/s

The value of V_t does not differ substantially from that assumed at the outset. However, we could refine our computation by proceeding to one more trial. Thus, let

$$V_t = 0.107 \text{ m/s}$$

Then, $Re_P = 26.337$

$$\ln\left(Re_P\right) = 3.2709$$

$$\ln\left(f_D\right) = 0.789267$$

$$f_D = 2.2018$$

And $V_t = 0.10866$ m/s

Since the assumed and computed values of V_t are fairly close to each other, we may discontinue further trials and select an average value of V_t such as, V_t = 0.1078 m/s as final.

It may be noted that here also an accurate value of terminal settling velocity is not essential, since we shall be selecting an operating liquid velocity (U_L) much below it (as per the criterion given in equation-3.5.1). Thus, let us select the superficial liquid velocity as

$$U_L = 0.045 \text{ m/s}$$

which very well satisfies equation (3.5.1). Accordingly, the required cross-sectional area (A) of the reactor column shall be

$$A = (Q_0 / U_L)$$

$$= (32/3600) / (0.045) = 0.1975 \text{ m}^2$$

And $D = \left[(4.0)(0.1975)/\pi\right]^{1/2} = 0.50$ m

This is the required diameter of the bioreactor column.

To compute the required volume of the bioreactor, we have to first estimate the fractional liquid holdup ($\in_{fL}$) in the fluidised bed. Being a liquid fluidised bed ($\in_f = \in_{fL}$), the value of $\in_{fL}$ could be estimated with reliable accuracy from Richardson and Zaki's correlation (equation-3.5.6). Thus, from equation (3.5.9),

$$Re_T = (DV_t \rho_L / \mu_L) = 8844.6632$$

Now, with reference to table (3.5.1), $n = a_0 = 2.4$

And, $$U_i = V_t \exp[-2.303 d_P / D]$$

$$= (0.1078) \exp[-2.303 (1.5 \times 10^{3}) / 0.5]$$

$$= 0.10706 \text{ m/s}$$

Therefore, from equation (3.5.6), $\in_f = \in_{fL} = 0.6989$

(a) Based on the assumption that the bioreactor is equivalent to an ideal CSTR (completely backmixed) and the mass transfer resistance in liquid bulk is negligible, the performance equation of the bioreactor shall be that given an equation (3.5.33):

$$\tau_m = (V_m \in_{fL} / Q_0) = (C_{S1}(0) - C_{S1}(e)) / [\eta_e (-r_{se})(int)] \qquad \text{(vi)}$$

Since 79% conversion of oil is desired,

$$C_{S1}(e) = C_{S1}(0)[1 - 0.79]$$

$$= (0.72225)(0.21) = 0.15167 \text{ mol/L}$$

Since the intrinsic rate follows equation (2.3.90) of Chapter 2,

$$(-r_{se})(int) = [r_{max} C_{S1} C_{S2}] / [K_{m1} C_{S1} + K_{m2} C_{S2} + K_{m2} (C_{S2}^2 / K_{Si}) + C_{S1} C_{S2}] \qquad \text{(vii)}$$

where $C_{S1} = C_{S1}(e) = 0.15167$ mol/L, $C_{S2} = C_{S2}(e) = 3\ C_{S1}(e) = 0.455$ mol/L.

To note that the oil and the acetate are fed to the bioreactor in the stoichiometric proportion of 1:3 (by mole) and hence, $C_{S2}(e) = 3\ C_{S1}(e)$. Substituting the numerical values of kinetic constants, we get

$$(-r_{se})(int) = 47.0815 \text{ mol/(L.hr)}$$

Therefore, from equation (vi),

$$V_m = (0.55648 / \eta_e) \quad \text{(viii)}$$

Now, to compute the value of the effectiveness factor (η_e) at $C_{S1} = C_{S1}(e)$, let us rewrite the expression for intrinsic rate [namely, equation-(vii)] after substituting $C_{S1} = C_{S1}(e)$ and $C_{S2}(e) = 3\ C_{S1}(e)$, as

$$(-r_{se})(int) = (r_{max} / K_1) C_{S1}(e) / [(K_2 / K_1) + C_{s1}(e)] \quad \text{(ix)}$$

where $K_1 = 1 + (3\ K_{m2} / K_{Si})$,

$K_2 = (K_{m1}/3) + K_{m2}$

On substituting the numerical values of the kinetic constants, we get,

$$(-r_{se})(int) = (0.04471)\ C_{S1}(e) / [\ 0.36685 + C_{S1}(e)] \quad \text{(x)}$$

This is of the same form as the classical Michaelis- Menten equation. We can, therefore, use equation (2.4.28) for the computation of η_e. Thus,

$\beta = (C_{S1}(e) / 0.36685) = (0.15167 / 0.36685) = 0.41343$

$L^* = d_P / 6 = 0.25$ mm

$\phi' = 5.08144$

$\eta_d = 0.247$

η (from equation -2.4.28) = 0.996

Therefore, from equation (viii), $V_m = 0.5587\ m^3$

$L_f = (V_m / A) = (0.5587 / 0.1975) =$ **2.83 m**

(b) The PFDR approach, as stated earlier, is the more accurate and realistic approach to bioreactor design. In this case, the performance equation is to be solved numerically using a numerical algorithm such as the modified fourth order Runge-Kutta method. The steps involved are those summarised in equations (3.5.36) to (3.5.38). We shall present here a sample computation with the step size of 0.1m. However, it must be kept in mind that the lower the step size chosen, the higher shall be the accuracy of the iterative procedure (numerical scheme) handled. A step size of not less than 1mm or 0.001 m is essential to obtain reliable design results. Here, we have shown computation with Δz = 0.1m as an illustration (since the change in value of C_S within an increment of 0.001m shall be too small). The procedure is outlined below:
The performance equation after rearranging in terms of $W(z)$ (where $(z) = (dC_{S1} / dz)$), is

$$(dW / dz) = F(z) = (U_L / D_{Lf}) W(z) + [r_{max}(app) / D_{Lf}] \eta(z) C_{S1}(z) / [K_m(app) + C_{S1}(z)]$$

where $r_{max}(app) = (r_{max} / K_1) = 0.04471$ mol/(L.s)

$K_m(app) = (K_2 / K_1) = 0.36685$ mol/L

Substituting the above values of kinetic constants and also, U_L = 0.045m/s and D_{Lf} = 0.0315 m²/s, the above equation reduces to

$$F(z) = (1.42857)\ W(z) + (1.419365)\ \eta(z)C_{S1}(z)/\ [\,0.36685 + C_{S1}(z)] \tag{xi}$$

The value of $\eta(z)$ is computed from equation (2.4.28) or (2.4.29). For this,

$$\beta = \left(C_{S1}(z)/0.36685\right)$$

$\phi' = 5.08144$ (computed earlier)

$\eta_d = (0.27831)\left[(1+\beta)/\beta\right]\sqrt{\beta - \ln(1+\beta)}$

$\eta = 1 - (0.19679)\ [\,\eta_d \coth(\eta_d) - 1\,]$, for $\eta_d \le 1.0$

$\quad = \eta_d - (0.19679)\ [\coth(\eta_d) - 1\,]$, for $\eta_d \ge 1.0$

The numerical scheme now proceeds as follows:

Step – 1: Δz = step size = 0.1 m

Step – 2: $i = 1$

$C_{S1}(1) = C_{S1}(e) = 0.15167$ g/L

$W(i) = W(1) = 0.0$

Step – 3: $j = 1$

$C_{S1}(j) = C_{S1}(i) = 0.1567$ g/L

$W(j) = W(i) = 0.0$

Step – 4: Computation of effectiveness factor, $\eta(j)$:

$\beta(j) = (0.1567/0.36685) = 0.41343$

$\eta_d(j) = 0.247$ (computed earlier)

$\eta(j) = 0.996$ (computed earlier)

Step – 5: From equation (xi), $F(j) = F(1) = 0.4135$ mol/(L.m^2)

$m(1) = m(j) = (0.1)\ (0.4135) = 0.04135$ mol/(L.m)

Step – 6: $j = 2$

$C_{S1}(j) = C_{S1}(2) = C_{S1}(i) + (\Delta z/2)W(i) = C_{S1}(i) = 0.15167$ mol/L

$W(j) = W(2) = W(i) + (1/2)m(1) = (0.04135\ /\ 2) = 0.020675$ mol/(L.m)

Since $C_{S1}(j)$ remains same as 0.15167 mol/L, $\eta(j) = \eta(2) = 0.996$ (computed earlier).

Step – 7: From equation (xi),

$F(j) = F(2) = 0.443047$ mol/(L.m^2)

$m(2) = m(j) = (0.1)\ (0.443047) = 0.0443047$ mol/(L.m)

Step – 8: $j = 3$

$$C_S(j) = C_S(3) = C_S(i) + (\Delta z/2)W(i) + (\Delta z/4)m(1)$$

= 4.0 + (0.1/4) (0.04135) = 0.1527 mol/L

$$W(j) = W(3) = W(i) + (1/2)m(2) = (0.0443047/2) = 0.02215 \text{ mol/(L.m)}$$

Step – 9: Computation of $\eta(j)$:

$$\beta(j) = (0.1527/0.36685) = 0.4162566$$

$$\eta_d(j) = 0.2473556$$

$$\eta(j) = 0.996$$

Step – 10: $F(j) = F(3) = 0.447147$ mol/(L.m^2)

$m(3) = m(j) = 0.0447147$ mol/(L.m)

Step – 11: $j = 4$

$$C_{S1}(j) = C_{S1}(4) = C_{S1}(i) + (\Delta z)W(i) + (\Delta z/2)m(2)$$

= (0.15167) + (0.1/2) (0.0443047) = 0.153885 mol/L

$$W(j) = W(4) = W(i) + m(3) = 0.0447147 \text{ mol/(L.m)}$$

Step – 12: Computation of $\eta(j)$:

$\beta(j)$ = (0.153885/0.36685) = 0.41947

$$\eta_d(j) = 0.2477$$

$$\eta(j) = 0.996$$

Step – 13: $F(j) = F(4) = 0.481644$ mol/(L.m^2)

$m(4) = 0.0481644$ mol/(L.m)

Step – 14:

$$C_{S1}(i+1) = C_{S1}(2) = C_{S1}(i) + (\Delta z)\ W(i) + (\Delta z/6)\left[\ m(1) + m(2) + m(3)\right]$$

= (0.15167) + (0.1/6) (0.130369) = 0.15384 mol/L

$$W(i+1) = W(2) = W(i) + (1/6)\left[\ m(1) + 2\ m(2) + 2\ m(3) + m(4)\right]$$

= 0.044592 mol/(L.m)

To note that $C_{S1}(2)$ is the value of C_{S1} at 0.1m below the top of the fluidized bed, since we have started our computations from the top of the bed. This value of $C_{S1}(2)$ is much less than $C_{S1}(0)$, namely 0.72225 mol/L. Therefore, we have to put $i = (i+1)$ and repeat the computations starting from Step - 3.

Computations were performed using a step size of $\Delta z = 0.001$m, starting from the top of the bed, proceeded downward until the computed value of $C_{S1}(i+1)$ slightly exceeded 0.72225 mol/L. This occurred at $i = 2370$. Therefore,

$$L_f = (2370)\ (0.001) = \mathbf{2.37\ m}$$

If the initial height of the static bed (prior to fluidisation) is L_S and its voidage 0.35, then

$$L_S\ (1 - 0.35) = L_f\ (1 - \in_f) = 2.37\ (1 - 0.69689)$$

or $L_S = 1.1$ m

Thus, if a static bed of silica granules accommodating immobilized lipase enzyme of height 1.1m and voidage 0.35 is fluidized at a feed flow rate of 32000 L/hr, it shall form a fluidized bed of height 2.37m and shall affect 79% conversion of jatropha oil to biodiesel.

Example 3.7

Aerobic treatment of a sample of wastewater is to be performed to reduce its BOD from 20 mg/L to 4 mg/L in a fluidized bed biofilm reactor composed of 2.5 mm silica particles each surrounded by a thin microbial film of thickness 0.3 mm. Cell mass concentration in microbial film = 2000 mg/L. Density of microbial solution = 1030 kg/m^3 and that of silica = 2100 kg/m^3. The feed solution is admitted from the bottom at the rate of 21600 L/hr and air is sparged at a superficial velocity of 0.025 m/s (which may be assumed constant throughout the process). Compute the required column diameter and the expanded bed height, if bioconversion reportedly follows

Monod's kinetic equation with $\mu_m = 30\ d^{-1}$, $K_S = 5.4$ mg/L and overall yield coefficient for cell mass production (Y) = 0.57. Perform the design computations based on the following two approaches, such as

(a) by assuming the bioreactor to be equivalent to an ideal CSTR (completely back mixed reactor),

(b) by considering the bioreactor's performance as equivalent to that of a PFDR (plug flow dispersion reactor).

Take interfacial tension = 0.073 N/m and effective diffusivity of substrate into biofilm = 3.5 × 10^{10} m^2/s.

Solution:

The bioreactor under consideration is a three phase, fluidized bed biofilm reactor in which the liquid (wastewater) forms the continuous phase and gas moves up in the form of tiny, discrete bubbles as discussed in example (3.6) of section (3.4). The total gas flow through the bed, in fact, involves that of the air sparged and that of the carbon dioxide produced. Part of the oxygen in the air gets dissolved and gets consumed and an equivalent amount of CO_2 is continuously produced. We shall, therefore, assume here, with allowable error, that the superficial velocity of gas (more precisely, gas mixture) through the bed remains more or less constant, such that $U_g = 0.025$ m/s ≈ constant.

To perform the design of the bioreactor, let us first estimate the hydrodynamic parameters such as the operating liquid velocity and the phase holdups (fractional gas holdup, fractional liquid holdup) in the fluidized bed. For this, we have to resort to selected experimental correlations.

For example, let us first compute the minimum fluidization velocity of the liquid $\left(U_{Lmf}\right)$ from the dimensional correlation proposed by Fortin (equation 3.5.5), after replacing d_P by d_{Pm}, and ρ_S by ρ_{Sm}. Thus,

$$d_{Pm} = (d_P + 2\delta) = 2.5 + 2(0.3) = 3.1\text{mm}$$

Therefore, $f = 1 - (d_P / d_{Pm})^3 = 0.4755$

$$\begin{aligned}\rho_{Sm} &= \rho_S(1-f) + (\rho_m f)\\ &= (2100)\,(1-0.4755) + (1030)\,(0.4755)\\ &= 1591.215\ \text{kg/m}^3\end{aligned}$$

Since $U_g = 0.025$ m/s and $\rho_L = 1000$ kg/m³ (assumed), from equation (3.5.5),

$$U_{Lmf} = 0.017175\ \text{m/s}$$

To select the operating liquid velocity (U_L), we have to determine terminal free settling velocity (V_t) of each particle-biofilm aggregates in the liquid as well. This demands, as shown earlier in Example 3.7, a trial and error procedure. Thus, from equation (2.5.27) of Chapter – 2,

$$V_t^2 = (4/3)(\rho_{Sm} - \rho_L)g\,d_{Pm} / (\rho_L f_D) = [0.023948 / f_D] \tag{i}$$

Also, $Re_P = \left(d_{Pm}V_t\rho_L / \mu_L\right) = (3100)\ V_t$ (ii)

To start the trials, let us assume, $V_t = 0.2$ m/s

Then, from equation (ii), $Re_P = (3100)\,(0.2) = 620$

$\ln\left(Re_P\right) = 6.4297$

Now, with reference to table (2.1) and equation (2.5.30) of Chapter-2,

$$\ln\left(f_D\right) = -0.6522$$

or $f_D = 0.52089$

Now, from equation (i), $V_t = 0.214$ m/s

Since the above computed value of V_t does not differ significantly from the assumed value, we may discard further trials and safely take an average value of V_t such as $V_t = 0.207$ m/s as the final value. Now, based on the criterion given in equation (3.5.1), let us select the operating liquid velocity as,

$$U_L = 0.03\ \text{m/s}$$

The required cross-sectional area (A) of the column shall be, therefore,

$$A = \left(Q_0 / U_L\right) = (21.6/3600) / (0.03) = 0.2\ \text{m}^2$$

The column diameter, $D = \sqrt{4(0.2)/\pi} = 0.5$ m

Let us now proceed to compute the phase holdups. For the estimation of the total fluid holdup in the bed, let us select the correlation proposed by Dakshinamurthy and co-workers, namely equation (3.5.10). Thus, since Re_P (computed above) is greater than 500, $k = 2.65$ and $m = 0.6$. Therefore, from equation (3.5.10),

$$\in_f = k\left(U_L / V_t\right)^m \left(U_g \mu_L / \sigma_L\right)^{0.08}$$

$$= 2.65\ (0.03/0.207)^{0.6}\ [(0.025)\ (0.001)\ /\ (0.073)]^{0.08}$$

$$= 0.44$$

Now, let us use the correlation proposed by Ramesh and Murugesan (equation 3.5.22 or 3.5.23) for the estimation of the fractional gas holdup in the bed. Thus,

$$Re_L = (0.5)\ (0.03)\ (1000)\ /\ (0.001) = 15{,}000$$

Since $Re_L > 100$, from equation (3.5.23),

$$\in_{fg} = 0.11\left(Fr_g\right)^{0.35}\left(Re_L\right)^{0.2}\left(Ar_m\right)^{0.11}\left(Mo\right)^{0.075}$$

where Mo = Morton number = $[\, g\,(\mu_L{}^4)\,/\,(\rho_L \sigma_L{}^3)] = 2.519 \times 10^{11}$

Fr_g = gas phase Froude number = $\left(U_g^2 / Dg\right) = 0.00012755$

Ar_m = modified Archimedes number = $d_P^3\ (\rho_S - \rho_L)\, g\ \rho_L\, /\, \mu_L{}^2 = 17961.1$

Therefore, $\in_{fg} = 0.036$

The fractional liquid holdup in the bed is, therefore,

$$\in_{fL} = \left(\in_f - \in_{fg}\right) = \left(0.44 - 0.036\right) = 0.404$$

All the hydrodynamic parameters have been thus evaluated.

(a) If we assume the bioreactor to be perfectly backmixed (ideal CSTR approximation), then the performance equation shall be that given in equation (3.5.33). Thus,

$$\tau_m = \left[V_m \in_{fL} / Q_0\right] = \left(C_{S0} - C_{Se}\right) / \left[\,\eta_e\left(-r_{Se}\right)(int)\right] \tag{iii}$$

where $\left(-r_{Se}\right)(int) = \mu_m(app)\, C_{Se} / (K_S + C_{Se})$

$C_{Se} = 4.0$ mg/L

$$\mu_m(app) = \left(\mu_m / Y\right)\left(x_f\, f\right)\left(1 - \in_f\right) / \in_{fL}$$

$$= (30/24)\ (2000)\ (0.4755)\ (1 - 0.44)\ /\ [(0.57)\ (0.404)]$$

$$= 2890.83 \text{ mg/(L.hr)}$$

Therefore, $\left(-r_{Se}\right)(int) = 1230.14$ mg/(L.hr)

Let us compute the effectiveness factor (η_e) from equation (2.4.28) or (2.4.29). Thus,

$$L^* = \left[d_{Pm}{}^3 - d_P{}^3\right] / (6\, d_{Pm}{}^2) = 1.24568 \text{ mm}$$

$$\varnothing' = L^*\ [\mu_m(app) / \left(D_e K_S\right)]^{1/2} = 5.064$$

To note that $D_e = 3.5 \times 10^{-10}$ m²/s. Now,

$\beta = (4.0\ /\ 5.4) = 0.7407$

$\eta_d = (0.279268)\left[(1+\beta)/\beta\right]\sqrt{\beta - \ln(1+\beta)} = 0.283365$

And $\eta = \eta_e = 0.9947$ (from equation – 2.4.28)

Therefore, from equation (iii), $V_m = 0.6991$ m^3

Since $A = \pi(0.5)^2/4 = 0.19635$ m^2,

$L_f = (V_m / A) = \mathbf{3.56\ m}$

(b) More accurately, if we proceed to design the reactor as being equivalent to a PFDR, then the procedure shall be same as that discussed in the earlier example (Example 3.7).

To apply the PFDR approach, let us first evaluate the axial dispersion coefficient (D_{Lf}) from the correlation proposed by Kim and Kim (equation 3.5.33). Thus, since U_L = 0.03 m/s, U_g = 0.025 m/s, d_{pm} = 0.0031m and D = 0.5m, from equation (3.5.33),

$$D_{Lf} = 0.03973 \text{ m}^2/\text{s}$$

The performance equation of PFDR in terms of $W(z)$ (where $(z) = (dC_S / dz)$), is

$$(dW/dz) = F(z) = (U_L / D_{Lf})W(z) + \left[\mu_m(app)/D_{Lf}\right]\eta(z)C_S(z)/\left[K_S + C_S(z)\right]$$

Substituting the values of U_L , D_{Lf} and those of kinetic constants, the above equation reduces to

$$F(z) = (0.755)\ W(z) + (20.2116)\ \eta(z)C_S(z)/[5.4 + C_S(z)] \qquad \text{(iv)}$$

The value of $\eta(z)$ is computed from equation (2.4.28). In this connection,

$$\beta = (C_S(z)/5.4)$$

$$\phi' = 5.064 \text{ (computed earlier)}$$

$$\eta_d = (0.279268)\left[(1+\beta)/\beta\right]\sqrt{\beta - \ln(1+\beta)}$$

The different steps involved in the numerical solution are now listed below:

Step – 1: Δz = step size = 0.1 m

Step – 2: $i = 1$

$C_S(i) = C_S(1) = C_{Se}$ = 4.0 mg/L

$W(i) = W(1) = 0.0$

Step – 3: $j = 1$

$C_S(j) = C_S(i) = 4.0$ mg/L

$W(j) = W(i) = 0.0$

Step – 4: Computation of effectiveness factor, $\eta(j)$:

$\beta(j) = (4.0/5.4) = 0.7407$

$\eta_d(j) = 0.283365$ (computed earlier)

$\eta(j) = 0.9947$ (computed earlier)

Step – 5: From equation (iv), $F(j) = F(1) = 8.555$ mg/(L.m^2)

$m(1) = m(j) = (0.1)\ (8.555) = 0.8555$ mg/(L.m)

Step – 6: $j\ =\ 2$

$C_S(j) = C_S(2) = C_S(i) + (\Delta z/2)W(i) = C_S(i) = 4.0$ mg/L

$W(j) = W(2) = W(i) + (1/2)m(1) = 0.42775$ mg/(L.m)

Since $C_S(j)$ remains same as 4.0 mg/L, $\eta(j) = \eta(2) = 0.9947$ (computed earlier).

Step – 7: From equation (iv),

$F(j) = F(2) = 8.87795$ mg/(L.m^2)

$m(2) = m(j) = (0.1)(8.87795) = 0.887795$ mg/(L.m)

Step – 8: $j = 3$

$C_S(j) = C_S(3) = C_S(i) + (\Delta z/2)W(i) + (\Delta z/4)m(1).$

$= 4.0 + (0.1/4)\ (0.8555) = 4.0213875$ mg/L

$W(j) = W(3) = W(i) + (1/2)m(2) = (0.887795/2) = 0.4438975$ mg/(L.m)

Step – 9: Computation of $\eta(j)$:

$\beta(j) = (4.0213875/5.4) = 0.7447$

$\eta_d(j) = 0.283775$

$\eta(j) = 0.9947$

Step – 10: $F(j) = F(3) = 8.916535$ mg/(L.m^2)

$m(3) = m(j) = 0.89165$ mg/(L.m)

Step – 11: $j = 4$

$C_S(j) = C_S(4) = C_S(i) + (\Delta z)W(i)\ +\ (\Delta z/2)m(2)$

$= (4.0) + (0.1/2)\ (0.887795) = 4.044389$ mg/L

$W(j) = W(4) = W(i) +\ m(3)\ = 0.89165$ mg/(L.m)

Step – 12: Computation of $\eta(j)$:

$\beta(j) = 0.74896$

$$\eta_d(j) = 0.284216$$

$$\eta(j) = 0.9947$$

Step – 13: $F(j) = F(4) = 9.28257$ mg/(L.m²)

$$m(4) = 0.928257 \text{ mg/(L.m)}$$

Step – 14: The substrate concentration at 0.1 m below the top of the bed is now given by,

$$C_S(i+1) = C_S(2) = C_S(i) + (\Delta z)W(i) + (\Delta z/6)[m(1) + m(2) + m(3)]$$

$$= 4.0439 \text{ mg/L}$$

$$W(i+1) = W(2) = W(i) + (1/6)[\, m(1) + 2\, m(2) + 2\, m(3) + m(4)]$$

$$= 0.89044 \text{ mg/(L.m)}$$

Computations are to be continued downward until the computed value of $C_S(i+1)$ becomes equal to or slightly exceeds C_{S0} (namely, 20 mg/L).

Computations have been performed using a step size of Δz = 0.001 m and it has been observed that convergence is attained at i = 2456 (at i = 2456, the value of $C_S(i+1)$ slightly exceeded 20 mg/L). Therefore,

$$L_f = (0.001)\,(2456) = \mathbf{2.456\ m}$$

This is the expanded bed height within which the required degree of BOD destruction has been affected.

3.6 Design and Analysis of Semi-fluidized bed Bioreactors

Semifluidized bed bioreactors combine the advantages of packed bed bioreactors and fluidized bed bioreactors. They thus form an attractive compromise between fixed bed and fluidized bed operations. However, the terminology, *semifluidized bed* could often be confusing and even misleading. A semifluidized bed does not mean *half fluidized bed.* In fact, the fluid velocity employed in semifluidized bed reactors is substantially larger than that employed in conventional fluidized bed reactors.

Semifluidization is achieved by fixing an additional porous plate (called top restraint) at a height of L_{sf} metres from the bottom liquid distributor (see figure 3.9). As the feed solution is admitted from the bottom at a substantially high velocity, the bed expands, but the expansion of the bed is restricted by the top restraint. Those particles (or particle-biofilm aggregates) which reach the top restraint, accumulate below it, thereby forming a packed bed of height L_P there. The pore size of the top restraint is less than the average size of particles and hence they are retained by the top restraint. The product solution, however, passes through the pores of the plate and is discharged from the top.

Once steady state has been attained at a particular feed flow rate, the column shall be composed of a packed bed of height L_P at the top and a fluidized bed of height L_f below it (see figure 3.9). This is called the *semifluidized bed.* The total height (L_{Sf}) of the semifluidized bed, which is the distance between the top restraint and the bottom liquid distributor, shall be, therefore, equal to the sum of these two. Thus,

$$L_{Sf} = \left(L_P + L_f\right) \tag{3.6.1}$$

The system is thus equivalent to a fluidized bed followed by a packed bed (something like a CSTR-PFR combination). However, it must be kept in mind that the packed section of the column is one that gets formed during the process and hence, its height L_P and voidage $\in_P$ are hydrodynamic parameters, whose magnitudes depend on the fluid flow rate employed (Q_0) and the physical properties of the fluid and the particles (ρ_L, ρ_S).

The ratio of the total height of the semi fluidized bed (L_{Sf}) to that of the initial static bed (L_S) is called the *bed expansion ratio* (R). Thus,

$$R = (L_{Sf} / L_S) \tag{3.6.2}$$

where L_S = height of the static bed (prior to fluidization)

One of the popular choices is, R = 2.0. That is, L_{Sf} is chosen as twice the initial static bed height.

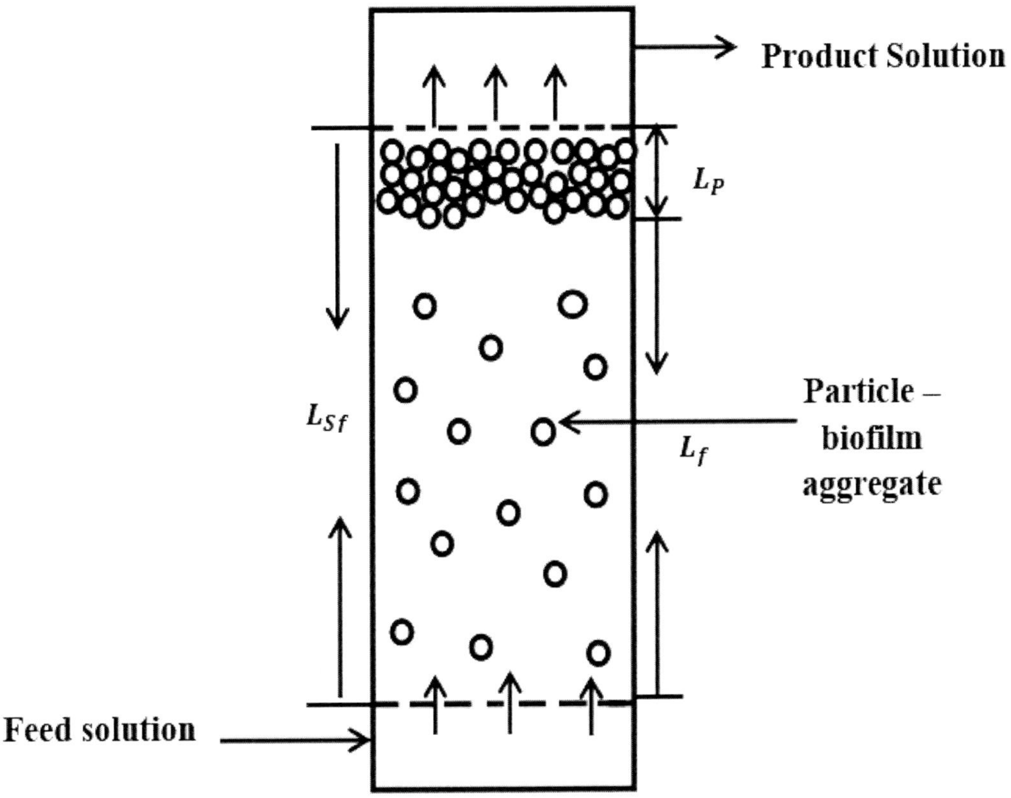

Figure 3.9: Schematic of two phase semifluidized bed biofilm reactor

In the case of three phase operation (G-L-S semifluidized beds), the gas (say, air) is also admitted from the bottom and it moves up the fluidized section as well as the packed section in the form of tiny, discrete bubbles.

3.6.1 Estimation of Hydrodynamic parameters

The hydrodynamic parameters associated with the design and operation of semi fluidized bed bioreactors are the minimum semifluidization velocity of the liquid, $U_{SL}(min)$, the height of the fluidized section (L_f) and that of the packed section (L_P) and the fractional fluid holdup in both sections $\left(\in_{fL}, \in_{fg}, \in_{PL}, \in_{Pg}\right)$. Once again, the literature is flooded with a large number of experimental correlations reported by different authors for the estimation of these parameters [51]. Obviously, correlations are to be selected keeping their limitations very much in mind.

The minimum semifluidization velocity of the liquid (the continuous phase) is the liquid velocity at which the first particle enters the top packed section of the column or reactor. More precisely, $U_{SL}(min)$, is the liquid velocity at which the top of the expanded bed just touches the top restraint. Experimental correlations have been proposed by many investigators for the estimation of the minimum semifluidization velocity. One such correlation proposed by Kurian and Rao [52] is reproduced below:

$$(U_{SL}(min) - U_{mf}) / (V_t - U_{mf}) = (0.61)[(R-1)/R]^{-1.2} \tag{3.6.3}$$

where U_{mf} = minimum fluidization velocity in conventional fluidized bed (from Wen and Yu's correlation, equation – 3.5.2).

V_t = terminal free settling velocity of particle – biofilm aggregates (see equation – 2.5.27 of Chapter 2).

Alternate correlations have been proposed by Roy and Sharma [53[and Roy and Sharat Chandra [54].

In the case of three phase (G –S –L) semifluidized beds, the minimum semifluidization velocity of liquid, $U_{SL}(min)$, would depend on the average superficial velocity of the gas (U_g) as well. The correlation proposed by Jena [55] is given below :

$$U_{SL}(min) = 0.037(U_g)^{0.267}(R)^{0.0497}(d_{Pm})^{0.228}(\mu_L)^{0.157} \tag{3.6.4}$$

All of the above correlations are summarized in table (3.6.1).

The operating superficial liquid velocity (U_{SL}) is to be then chosen in such a way that it does exceed $U_{SL}(min)$, but falls much below V_t. In other words,

$$U_{SL}(\min) < U_{SL} < V_t \tag{3.6.5}$$

A popular choice is

$$(1.15)U_{SL}(\min) < U_{SL} < (0.75)V_t \tag{3.6.6}$$

If $U_{SL} = V_t$, then all the particles (all particle – biofilm aggregates) shall enter the top packed section and the voidage of the bottom fluidized section shall become equal to 1.0.

TABLE – 3.6.1 Correlations for estimation of minimum semifluidisation velocity. Liquid phase: water, Solid phase: glass beads, polymer granules, pulverized ore, Gas phase (for S.No.4 only): air

S. No.	Correlation	Reference
1.	$\left[U_{SL}(\min) - U_{mf}\right] / \left[V_t - U_{mf}\right] = 0.61[R/(R-1)^{-1.2}$ $U_{mf} = \left(Re_{mf}\mu_L\right)/\left(d_P\rho_L\right)$	52

S. No.	Correlation	Reference
	$Re_{mf} = \sqrt{(33.67)^2 + 0.0408 Ar_m} - (33.67)$ $Ar_m = d_P^3 (\rho_S - \rho_L) g \rho_L / (\mu_L)^2$	
2.	$\left[U_{SL}(min)/U_{mf}\right] = 1.875 (D/d_P)^{0.266} (\rho_S/\rho_L)^{-0.228} (R)^{0.585}$	53
3.	$\left[U_{SL}(min)/U_{SL}(\max)\right] = 0.473 (D/d_P)^{-0.2} (\rho_S/\rho_L)^{0.17} (R)^{0.38}$ $U_{SL}(\max) = (1.85 X 10^4) d_P^{0.65} (\rho_S - \rho_L)^{0.55} (\rho_L)^{-0.45} (\mu_L)^{-0.1}$	54
4.	$U_{SL}(\min) = 0.037 (U_g)^{-0.267} (R)^{0.0497} (d_P)^{0.228} (\mu_L)^{-0.157}$ (for G – L – S semifluidized bed)	55

In the case of liquid – solid semifluidized beds, the fractional liquid holdup in the fluidized section, $\in_{fL}$ (which is equal to the total voidage of this section, $\in_f$) can be estimated, with reasonable accuracy, from the correlation proposed by Richardson and Zaki [24], given in equation (3.5.6) of Section (3.5).

In the case of the packed section, it has been observed that its voidage ($\in_P$) does not vary materially with liquid flow rate and other system parameters and consequently, we may assume, with allowable error,

$$\in_{PL} = \in_P = \text{constant} = 0.35 \text{ to } 0.40. \qquad (3.6.7)$$

The fractional fluid holdups in the fluidized section of three phase semifluidized bed bioreactors $(\in_f, \in_{fL}, \in_{fg})$ could be evaluated from any of the experimental correlations that have been proposed for three phase fluidized beds, such as those presented in equations (3.5.10) to (3.5.27) of Section – 3.5. The only difference is that U_L is to be replaced by U_{SL}. For the convenience of the readers, we have summarized some of the selected correlations in this connection in table (3.6.2).

TABLE – 3.6.2A Selected correlations for the estimation of fractional fluid holdup in the fluidized section of gas – liquid – solid semifluidized beds

S. No	Correlation	Range of applicability	Reference
1.	$\in_f = 2.12 (U_{SL}/V_t)^{0.41} (We_m)^{0.08}$ for $Re_P < 500$ $= 2.65 (U_{SL}/V_t)^{0.6} (We_m)^{0.08}$ for $Re_P > 500$ $We_m = [\mu_L U_g / \sigma_L]$	Gas: air, nitrogen. Liquid: water, kerosene. Solid: glass beads, iron shots, lead shots & sand.	26

S. No	Correlation	Range of applicability	Reference
2.	$\in_f = 0.371(U_{SL})^{0.271}(U_g)^{0.041}(\rho_S-\rho_L)^{-0.316}$ $(d_P)^{-0.268}(\mu_L)^{0.055}(D)^{-0.033}$ $\in_{fg} = 0.048(U_g)^{0.72}(d_P)^{0.168}(D)^{-0.125}$	d_P = 4.6 – 6.2 mm ρ_S = 1170 – 2240 kg/m^3 $U_g \leq$ 0.173 m/s $U_{SL} \leq$ 0.12 m/s D = 72 – 152 mm	16
3.	$\in_f = 1-0.631\exp\left[-0.418U_{SL}-0.004U_g\right]$ $\in_{fL} = 0.006(U_{SL})^{0.246}(U_g)^{-0.059}$	d_P = 0.25 mm ρ_S = 2500 kg/m^3 U_g = 0.0128 – 0.02 m/s U_{SL} = 0.0039 – 0.156 m/s	56
4.	$\in_f = 0.4149\left[U_{SL}/U_{mf}\right]^{0.27}\left[1+0.07\left(Re_{Pg}\right)^{0.34}\right]$ $\in_{fg} = \in_f\left(U_g - U_N\right)/\left(U_g + U_{SL}\right)$ $U_N = 0.017\left[\rho_L U_g^{\ 2}\right]^{0.45}$ $Re_{Pg} = \left(d_P U_g \rho_L / \mu_L\right)$	Gas: N_2, CO_2, Helium. Liquid: water, gas oil, kerosene, cyclohexane. Solid: glass beads, alumina granules.	27

TABLE – 3.6.2B

Sl No.	Correlation	Range of applicability	Reference
5.	$\in_{fg} = 0.17(Fr_g)^{0.33}(Re_L)^{-0.065}(Ar_m)^{0.125}(Mo)^{0.05}$, for $Re_L < 100$ $= 0.11(Fr_g)^{0.35}(Re_L)^{0.2}(Ar_m)^{0.11}(Mo)^{0.075}$, for $Re_L > 100$ $Fr_g = \left(U_g^{\ 2}/Dg\right)$, $Re_L = \left(DU_{SL}\rho_L/\mu_L\right)$ $Ar_m = \left(d_P^{\ 3}(\rho_S-\rho_L)g\,\rho_L/\mu_L^{\ 2}\right)$, $Mo = \left(g\,\mu_L^{\ 4}\right)/\left(\rho_L\,\sigma_L^{\ 3}\right)$	d_P = 4.11 – 6.8 mm ρ_S = 2030 – 2480 kg/m^3 ρ_L = 1000 – 1150 kg/m^3 $U_g \leq$ 0.13 m/s $U_{SL} \leq$ 0.13 m/s σ_L = 0.049 – 0.072 N/m μ_L = 0.0008 – 0.015 kg/(m.s)	29

Sl No.	Correlation	Range of applicability	Reference
6.	$\in_{fL} = 1.504(Fr_L)^{0.234}(Fr_g)^{-0.086}(Re_L)^{-0.082}(We_L)^{0.092}$ $Fr_L = \left(U_{SL}^2 / Dg\right)$, $We_L = \left[D U_{SL}^2 \rho_L / \sigma_L\right]$	$d_P = 1.0 - 6.0$ mm $\rho_S = 2300 - 2950$ kg/m^3 $\rho_L = 962 - 1230$ kg/m^3 $\mu_L = 0.001 - 0.07$ kg/(m.s) $U_g = 0.007 - 0.161$ m/s $U_{SL} = 0.027 - 0.12$ m/s	30
7.	$\in_{fg} = (Mo)^{0.05}(Eo)^{0.157}(Fr_L)^{-0.093}(Fr_g)^{0.401}$ $(\rho_S/\rho_L)^{-0.0906}(d_P/D)^{0.081}(L_S/D)^{0.026}$ $Eo = \left(\rho_L - \rho_g\right)D^2 g / \sigma_L$ $Mo = \left(g\,\mu_L^4\right) / \left(\rho_L\,\sigma_L^3\right)$	$1.1078 \times 10^{-11} \leq Mo \leq 3.34 \times 10^{-10}$ $Eo =$ 1370.66 – 1527 $Fr_L =$ 0.00459 – 0.0252 $Fr_g =$ 0.000458 – 0.0252 $\rho_S / \rho_L =$ 2.124 – 2.481 $d_P / D =$ 0.0155 – 0.0629 $L_S / D = 1.71 - 3.0$	28

TABLE – 3.6.2C

Sl No.	Correlation	Range of applicability	Reference
8.	$\varepsilon_{fg} = 0.98\left(Mo\right)^{0.0265}\left(Fr_L\right)^{0.0438}\left(Fr_g\right)^{0.382}\left(\rho_S / \rho_L\right)^{-0.529}$ $\left(d_P / D\right)^{0.0339}\left(\varphi_S\right)^{-0.0217}$ φ_S = particle sphericity	$d_P = 1.0 - 13.6$ mm $\rho_S = 1245 - 2960$ kg/m^3 $\rho_L = 995 - 1620$ kg/m^3 $\mu_L = 0.00085 -$ 0.032 kg/(m.s) $U_g = 0.00375 -$ 0.1375 m/s $U_{SL} = 0.008 - 0.032$ m/s $\varphi_S = 0.33 - 1.0$	57

As in the case of liquid – solid systems, in the case of three phase semifluidized beds also, the total fluid holdup in the packed section ($\in_P$) may be treated as more or less constant and equal to 0.35 to 0.40. For estimating the fractional gas holdup in the packed section of three phase semifluidized beds, Chern and coworkers [58] have proposed the following *dimensional* experimental correlation:

$$\in_{Pg} = (33.7838 \in_P)\, U_{SL}^{0.35} U_g^{-0.977} \tag{3.6.8}$$

And the fractional liquid holdup is then obtained by difference:

$$\in_{PL} = \in_P - \in_{Pg} \tag{3.6.9}$$

As stated earlier, the height of the packed section (L_P) and that of the fluidised section $\left(L_f\right)$ are also hydrodynamic parameters and their relative magnitudes depend on the liquid velocity used (U_{SL}), the size and density of the discrete phase (d_P, ρ_S or d_{Pm}, ρ_{Sm}) and the properties of the feed solution (ρ_L, μ_L). In the case of three phase semifluidized beds, these could also be functions of the voidages of the two sections ($\in_P, \in_f$). Selected experimental correlations reported in the literature are summarized in table (3.6.3). For example, the correlation reported by Singh et. al [59] is given below:

$$C_L = 0.974 + 0.324 \ln\left(C_V\right) \tag{3.6.10}$$

where C_V = flow parameter = $(U_{SL} - U_{mf}) / (V_t - U_{mf})$ (3.6.11)

C_L = capacity parameter = $(L_{Sf} - L_S) / (L_{Sf} - L_P)$ (3.6.12)

U_{mf} = minimum fluidization velocity in conventional fluidized bed (from Wen and Yu's correlation, equation – 3.5.2).

In the case of three phase systems, Chern et al. [58] have reported that the height of the packed section formed is governed by the total voidage of the fluidized section and also that of the packed section. The correlation proposed by them is given below:

$$(L_S - L_P)/(L_{Sf} - L_P) = (1 - \in_f)/(1 - \in_P) \tag{3.6.13}$$

Once L_P is known (from the above equations), the height of the fluidised section (L_f) is obtained by difference, as

$$L_f = \left(L_{Sf} - L_P\right) \tag{3.6.14}$$

TABLE – 3.6.3 Correlations for the estimation of the height of packed section of semifluidized beds. Liquid phase: water. Solid phase: glass beads, polymer granules, pulverized ore. Gas phase (for S.No.5 only): air

S.No	Correlation	Reference
1.	$C_V = 0.61(C_L)^{1.2}$	52
2.	$\left(L_P / L_S\right) = 2.21\left[U_{SL} / U_{SL}(\max)\right]^{2.08} \left(D / d_p\right)^{0.23} \left(\rho_S / \rho_L\right)^{0.17} \left(D/L_S\right)^{0.15} (R)^{-0.88}$	53
3.	$C_L = 0.974 + 0.324 \ln (C_V)$	59
4.	$C_V = (C_L)^{3.15}$ for $1.0 < (1/C_L) < 1.3$	60
5.	$(L_S - L_P)/(L_{Sf} - L_P) = (1 - \in_f)/(1 - \in_P)$ (for G-L-S semifluidized bed)	58

An increase in liquid velocity (U_{SL}) tends to increase the height of the packed section (L_P) and relatively lower that of the fluidized section. This is because at high liquid velocities, more particles or particle – biofilm aggregates are carried to the top packed section and consequently, its height increases. Popularly, in many industrial semifluidized bed bioreactors is, $(L_P/L_f) = 0.5$.

3.6.2 Performance Analysis (Mathematical modeling)

The simplest mathematical model for the performance analysis of semifluidized bed bioreactors is the one that assumes the bioreactor to be equivalent to a CSTR – PFR combination. In other words, the fluidized section of the bioreactor is assumed equivalent to an ideal CSTR (completely backmixed) and the packed section is being assumed equivalent to a PFR. This obviously involves a degree of approximation. Wherever such an approximation is acceptable, the performance equation of the fluidized section shall be

$$\tau_m = \left(C_{S0} - C_{Sb}\right) / \left[\eta_b \left(-r_{Sb}\right)(int)\right] \tag{3.6.15}$$

where τ_m = space time = $\left[V_m \in_{fL} / Q_0\right]$ (3.6.16)

C_{Sb} = concentration of substrate in the solution leaving the fluidized section and entering the packed section (substrate concentration at the interface between the packed section and the fluidized section)

η_b = value of effectiveness factor at $C_S = C_{Sb}$

$\left(-r_{Sb}\right)(int)$ = intrinsic rate of bioconversion at $C_S = C_{Sb}$

Q_0 = volume flow rate of feed solution entering the reactor

After substituting the kinetic equation representing the intrinsic rate of bioconversion, the above performance equation would have to be solved for C_{Sb} by trial, since the effectiveness factor (η_b) is not known at the outset and is a nonlinear function in C_{Sb} .

Similarly, based on the PFR approximation, the performance equation for the packed section shall be

$$\tau_P = \int_{C_{Se}}^{C_{Sb}} \left[\frac{1}{\eta\left(-r_S\right)(int)}\right] dC_S \tag{3.6.17}$$

where $\tau_P = \left[V_P \in_{PL} / Q_0\right]$ (3.6.18)

The integral of above equation is to be evaluated numerically using Simpson's rule or trapezoidal rule. The value of C_{Se} can be estimated, once again, only by trial. It may be noted here that the substrate concentration changes from C_{S0} to C_{Sb} in the fluidized section and then from C_{Sb} to C_{Se} in the packed section. It has been observed [61 - 64] that in bioreactors of this kind, the major share of bioconversion occurs in the packed section (though it is of smaller volume than the fluidized section).

Once the substrate concentration in the product solution leaving the bioreactor (C_{Se}) has been thus estimated (by trial), the fractional substrate conversion attained in the bioreactor (α) is obtained as

$$\alpha = \left(C_{S0} - C_{Se}\right) / C_{S0} \tag{3.6.19}$$

A numerical example illustrating the above procedure has been discussed at the end of this section (Example – 3.8).

A more accurate approach to the design and analysis of semifluidized bed bioreactors is, no doubt, the PFDR approach (discussed already in the previous sections under packed bed and fluidized bed bioreactors). In the present case however, we shall assume that the bioreactor is equivalent to two plug flow dispersion reactors in series, PFDR-1 representing the packed section and PFDR-2 representing the fluidized section. Accordingly, the performance equation for PFDR-1(packed section) shall be

$$-U_{SL}(dC_S/dz)+D_{LP}\,(d^2C_S/dz^2)=\eta(-r_S)(int) \tag{3.6.20}$$

where D_{LP} = axial dispersion coefficient in the packed section.

As stated earlier, the magnitude of D_{LP} usually ranges from 0.007 to 0.008 m^2/s (see also correlations given in section – 3.2).

The boundary conditions are,

BC-1: At $z=0$ (at the top outlet of the bioreactor),

$$C_S=C_{Se} \tag{3.6.21}$$

BC-2: At $z=L_P$ (at the bottom of the packed section or top of the fluidized section),

$$C_S=C_{Sb} \tag{3.6.22}$$

Similarly, the performance equation for PFDR – 2 (the fluidized section) is

$$-U_{SL}(dC_S/dz)+D_{Lf}\,(d^2C_S/dz^2)=\eta(-r_S)(int) \tag{3.6.23}$$

The boundary conditions shall be,

BC-1: At $z=0$ (at the top of the fluidised section or at the interface between the packed section and the fluidized section),

$$C_S=C_{Sb} \tag{3.6.24}$$

BC-2: At $z=L_f$ (at the bottom of the fluidised section or at the inlet to the bioreactor),

$$C_S=C_{S0} \tag{3.6.25}$$

The above two performance equations are to solved successively using the same numerical procedure (that involves a modified form of fourth order Runge – Kutta method) discussed in Section (3.2) and that has been used for the design and analysis of packed bed bioreactors and fluidized bed bioreactors (discussed in the earlier sections). The major difference is that here we have to solve two differential equations successively (one after the other). It is also important to keep in mind that the reduced expression for the intrinsic kinetics of bioconversion shall be different for the packed section from that for the fluidized section (since the voidages of the two are different).

For the convenience of the readers, we shall briefly summarize the procedure here with an example. Let the bioconversion intrinsically follow Monod's kinetics such that

$$(-r_S)(int)=\mu_m(app)C_S/(K_S+\ C_S) \tag{3.6.26}$$

where $\mu_m(app) = (\mu_m / Y)(x_f f)(1-\in_P) / \in_{PL}$, for the packed section (3.6.27)

$= (\mu_m / Y)(x_f f)(1-\in_f) / \in_{fL}$, for the fluidized section (3.6.28)

Let us also consider a liquid – solid system and let us further assume that the mass transfer resistance to substrate transport in the liquid bulk is negligible so that $C_{SP} \approx C_S$.

Now, putting $W(z) = (dC_S / dz)$ in equation (3.6.20) and on rearrangement (after inserting the expression for intrinsic rate as well), we get

$$F(z) = (dW / dz) = (U_{SL} / D_{LP}) W(z) + \left[\mu_m(app)\ \eta\ C_S / D_{LP}\right] / (K_S + C_S) \tag{3.6.29}$$

where $\mu_m(app)$ is that defined in equation (3.6.27).

Similarly, the performance equation of the fluidized section, namely equation (3.6.23), yields on similar re-arrangement,

$$F(z) = (U_{SL} / D_{Lf}) W(z) + \left[\mu_m(app)\ \eta\ C_S / D_{Lf}\right] / (K_S + C_S) \tag{3.6.30}$$

where $\mu_m(app)$ is that defined in equation (3.6.28).

The stepwise procedure shall be then as follows (to note that in this case also, computations are started from the top of the reactor column where $C_S = C_{Se}$ and proceeded downwards) :

Step 1: Specify the bed expansion ratio R. For example, let

$$R = 2.0 \tag{3.6.31}$$

This fixes the total height of the semifluidized bed (L_{Sf}), since

$$L_{Sf} = R\ (L_S) \tag{3.6.32}$$

where L_S = height of the static bed (prior to fluidization).

Step 2: Compute all the hydrodynamic parameters such as the operating liquid velocity (U_{SL}), the height of the packed section (L_P) and that of the fluidized section (L_f), the fractional liquid holdup in each section ($\in_{fL}, \in_{PL}$), from selected experimental correlations discussed earlier. The volumetric flow rate of the feed solution (Q_0) may also be kept computed as

$$Q_0 = (\pi D^2 / 4)\ U_{SL} \tag{3.6.33}$$

Step 3: Select the step size (Δz). For example, let $\Delta z = 1.0$ mm $= 0.001$ m.

Step 4: Compute the value of $\mu_m(app)$ for the packed section from equation (3.6.27).

Step 5: Assume a value of the fractional substrate conversion (α). For example, let $\alpha = 0.80$.

Step 6: Compute the substrate concentration in the product stream as

$$C_{Se} = C_{S0}(1-\alpha) \tag{3.6.34}$$

Step 7: Now, proceed to solve the performance equation for the packed section (PFDR – 1):

Put $i = 1$

$$C_S(1) = C_{Se}$$

$$W(1) = 0.0$$

Step 8: Put $j = 1$

$$C_S(j) = C_S(i)$$

$$W(j) = W(i)$$

Step 9: Compute the effectiveness factor, $\eta(j)$ from any of the correlations discussed in Chapter – 2 and $F(j)$ from equation (3.6.29).

Step 10: Compute $m(j)$ as

$$m(j) = (\Delta z) F(j) \tag{3.6.35}$$

If $\boldsymbol{j} = 4$, GO TO Step 12.

Step 11: Put $j = j+1$

If $j = 2$, $C_S(j) = C_S(i) + (\Delta z/2) W(i)$

If $j = 3$, $C_S(j) = C_S(i) + (\Delta z/2)\ W(i) + (\Delta z/4)\ m(j-2)$

If $j = 4$, $C_S(j) = C_S(i) + (\Delta z) W(i) + (\Delta z/2) m(j-2)$

If $j < 4$, $W(j) = W(i) + (1/2) m(j-1)$

If $j = 4$, $W(j) = W(i) + m(j-1)$

GO TO Step – 9 (repeat the computations starting from Step 9).

Step 12: Compute $C_S(i+1)$ and $W(i+1)$ as

$$C_S(i+1) = C_S(i) + (\Delta z)\ W(i) + (\Delta z/6)\left[m(1) + m(2)\right) \tag{3.6.36}$$

$$W(i+1) = W(i) + (1/6)\left[m(1) + 2m(2) + 2m(3) + m(4)\right] \tag{3.6.37}$$

Step 13: Compute $L_P(cal)$ as,

$$L_P(cal) = (i)(\Delta z) \tag{3.6.38}$$

Step 14: If $L_P(cal) \geq L_P$, GO TO Step – 15. Otherwise, put $i = i+1$ and repeat the computations starting from Step – 8.

Step 15: Compute C_{Sb} as

$$C_{Sb} = C_S(i+1) \tag{3.6.39}$$

Print C_{Sb}.

Step 16: Compute the value of $\mu_m(app)$ for the fluidised section from equation (3.6.28).

Step 17: Now, proceed to solve the performance equation for the fluidized section (PFDR – 2) similarly. Thus,

Put $i = 1$

$$C_S(1)=C_{Sb}$$

$$W(1)=0.0$$

Step 18: Put $j=1$

$$C_S(j)=C_S(i)$$

$$W(j)=W(i)$$

Step 19: Compute $\eta(j)$ from any of the correlations discussed in Chapter – 2 and $F(j)$ from equation (3.6.30).

Step 20: Compute $m(j)$ from equation (3.6.35) of Step – 10.

If $j=4$, GO TO Step – 22.

Step 21: Put $j = j+1$.

If $j=2$, $C_S(j)= C_S(i)+(\Delta z/2)W(i)$

If $j=3$, $C_S(j)= C_S(i)+(\Delta z/2)\, W(i)+(\Delta z/4)\, m(j-2)$

If $j=4$, $C_S(j)= C_S(i)+(\Delta z)\, W(i)+(\Delta z/2)\, m(j-2)$

If $j<4$, $W(j)=W(i)+(1/2)m(j-1)$

If $j=4$, $W(j)=W(i)+m(j-1)$

GO TO Step – 19 (repeat the computations starting from Step-19).

Step 22: Compute $C_S(i+1)$ and $W(i+1)$ as

$$C_S(i+1)=C_S(i)+(\Delta z)\, W(i)+(\Delta z/6)\left[m(1)+m(2)+m(3)\right] \tag{3.6.40}$$

$$W(i+1)= W(i)+(1/6)\left[m(1)+2\,m(2)+2\,m(3)+m(4)\right] \tag{3.6.41}$$

Step 23: Compute $L_f(cal)$ as,

$$L_f(cal)=(i)(\Delta z) \tag{3.6.42}$$

Step 24: If $L_f(cal)\ge L_f$, GO TO Step – 25.

Otherwise, put $i=i+1$ and repeat the computations starting from Step – 18.

Step 25: Compute $C_{S0}(cal)$ as

$$C_{S0}(cal)=C_S(i+1) \tag{3.6.43}$$

Step 26: If $C_{S0}(cal)\ge C_{S0}$, proceed to Step – 27. Otherwise, re-choose α and repeat the computations starting from Step – 6.

Step 27: Print α. Print also the values of Q_0 , U_{SL}, $L_f, L_P, L_{Sf}, C_{Sb}, C_{S0}, C_{Se}$.

The performance analysis of three phase semifluidized bed bioreactors can also be performed following the same procedure as described above. The only difference shall be in the estimation of the hydrodynamic parameters such as the minimum semifluidization velocity of the liquid, U_{SL} (min), the (L_P / L_f) ratio and the fractional fluid holdups in the packed and fluidized sections ($\in_P, \in_{PL}, \in_{Pg}, \in_f, \in_{fL}, \in_{fg}$). These are functions of the average gas velocity (U_g) and the properties of the gas / air stream (ρ_g, μ_g) as well. Appropriate experimental correlations are to be selected from those listed in tables (3.6.1) to (3.6.3). Once all the hydrodynamic parameters have been thus estimated from selected correlations, the rest of the design procedure (numerical solution of performance equations) shall be exactly the same as that described above.

3.6.3 Advantages and Industrial Applications

Semifluidized bed bioreactors have become fairly popular in many process industries.

Studies on lactic acid synthesis from cheese whey permeate (deproteinized cheese whey) and molasses in semifluidized bed bioreactors have been reported by Das and Narayanan [61]. The findings have been substantiated both through rigorous mathematical modeling and simulation and pilot plant tests. The bioreactor has been observed to provide substantially large conversion of sucrose / lactose (present in molasses / cheese whey) and thereby provide a high yield of lactic acid within a low reactor volume and at large capacities (large feed flow rates). For example, 80% conversion of sucrose (present in molasses) to lactic acid is reportedly accomplished in a reactor column of diameter 0.5 m and total height (L_{Sf}) 1.0 m, at a feed flow rate of about 78,000 L/hr. The observation is similar in the case of cheese whey permeate fermentation as well. This speaks of the special advantage of employing semifluidisation technology.

Interestingly, the major share of substrate conversion is contributed by the packed section, though it occupies only around one-third of total reactor volume. This could be partly due to the fact that the flow of substrates solution through the packed section is close to plug flow (the magnitude of axial dispersion coefficient (D_{LP}) being quite low). It has been reported by all investigators [61-64] that the substrate concentration decreases from C_{So} to C_{Sb} in the fluidized section, but this decrease is relatively low, in spite of the fact that the height of the fluidized section (L_f) is much larger. On the other hand, the substrate concentration decreases sharply from C_{Sb} to C_{Se} in the packed section.

Another interesting observation is that the fractional substrate conversion (α) increases with increase in feed flow rate (Q_0). This is a specific characteristic exhibited by semifluidised bed bioreactors only, since in bioreactors of all other categories, the fractional conversion decreases with increase in substrate flow rate (due to the fact that the residence time of fluid elements within the reaction zone decreases).

This phenomenon, however, is not difficult to explain. As the feed flow rate (Q_0) increases, the height of the packed section (L_P) also increases, as more and more particles / particle-biofilm aggregates enter the packed section at the top. The height of the fluidized section (L_f) decreases relatively. In other words, there occurs a re-arrangement of reaction zones within the bioreactor. Since the major share of bioconversion occurs in the packed section, as its height increases (with increase in feed flow rate), the ultimate fractional conversion of substrate attained also increases. This phenomenon has been substantiated by other investigators as well, such as by Biswas and Narayanan [62], Narayanan [63] and De et. al [64].

The operating cost of the semifluidized bed biofilm reactors shall be nevertheless higher than that of the conventional fluidized bed biofilm reactors due to the presence of the top restraint and the additional packed section at the top, both of which offer added resistance to the flow of substrate solution.

This type of design is also not suitable if the biocatalyst particles have a tendency to undergo deactivation and are to be reactivated and recycled continuously. In the case of fluidized bed bioreactors, it can be operated in the circulating mode (as circulating fluidized bed reactors) and in that case, continuous reactivation or regeneration of catalyst particles and recycle could be made possible [65].

On the other hand, relatively lower size support particles can be handled in semifluidized bed biofilm reactors (as compared to fluidized bed bioreactors) without any serious concern over entrainment loss. The top restraint helps in inhibiting escape of particles or particle-biofilm aggregates in the outgoing product solution.

Synthesis of Xanthan gum through aerobic fermentation of cheese whey permeate in three phase semifluidized bed biofilm reactors has been equally successful [63]. More than 85% conversion of lactose to Xanthan gum could be reportedly attained at substrate flow rates exceeding 1600 L/min, within a reactor column of height 1.0 m (L_{Sf} = 1.0 m) and diameter, 0.5 m. As in the case of lactic acid synthesis, here also the fractional substrate conversion, α (fractional conversion of lactose to Xanthan gum) has been observed to be increasing with increase in feed flow rate (due to reasons explained earlier).

Application of semi-fluidized bed biofilm reactors to biological treatment of industrial effluents has been surveyed by Jena et.al [66]. They have highlighted the specific advantages of employing semifluidization technology in such cases. Biswas and Narayanan [62] have demonstrated that in the case of aerobic treatment of industrial effluents for the removal of toxins such as p-cresol, when a fluidized bed biofilm reactor demands a significantly large reactor volume (L_f exceeding 10 meters), a semifluidized biofilm reactor could provide the same degree of toxin removal within a much lower reactor volume (L_{Sf} = 2–3 meters).

Successful case studies on anaerobic digestion of municipal sewage mixed with wastewater from pharmaceutical industry [51,62] have demonstrated that more than 85% BOD destruction could be achieved in a semifluidized bed biofilm reactor of height 1.5 m (L_{sf} = 1.5 m) and diameter 0.5m, at a feed flow rate as high as 80,000 L/hr (80.0 m^3/hr).

Studies on aerobic treatment of wastewater in semifluidized bed biofilm reactors with LPO (liquid phase oxygen) technology utilization have been reported by Narayanan [67]. The results reported are truly encouraging. As discussed earlier, due to LPO utilization, the operating cost of huge air compressors gets eliminated completely and the concentration of hydrogen peroxide (nascent oxygen supplier) required is not large.

De et.al [64] have studied biodiesel synthesis from neem oil and methyl acetate in semifluidized bed bioreactors (composed of immobilized lipase catalyst). They report that by applying semifluidization technology, high degree trans-esterification of neem oil (more than 80%) can be achieved at fairly high capacities (feed flow rates exceeding 30,000 L/hr) and within a much lower reactor volume (L_{Sf} = 1.2 m and D = 0.5m). The oil and acetate were used in the stoichiometric molar proportion of 1:3.

Large scale use of semifluidized bed reactors / bioreactors in process industries can be, therefore, considered as truly promising.

The specific advantages of semifluidized bed bioreactors may be summarized as given below:

(1) These bioreactors provide large degree of bioconversion at significantly high capacities (feed flow rates) within a reasonably low reactor volume.
(2) The fractional substrate conversion or the product yield attained increases with increase in feed flow rate, in spite of the fact that the residence time of fluid elements within the reaction zone does get diminished. This is an exclusive, attractive characteristic exhibited by this class of bioreactors only.
(3) The reactor can be thus operated at high capacities, simultaneously affecting high degree of bioconversion, without increasing the reactor volume (keeping the volume requirement of the reactor still reasonably low).
(4) Due to the presence of the top restraint, there shall be hardly any entrainment loss of particles / particle-biofilm aggregates in the outgoing product solution.
(5) The operating cost of these bioreactors shall be invariably high, since the top restraint and the top packed section offer additional resistance to fluid flow. Also, this type of design shall not be suitable for handling biocatalyst particles that get deactivated fast and are required to be continuously re-activated and recycled.

Example 3.8:

Microbial fermentation of de-proteinized cheese whey using a modified strain of Lactobacillus family (the Monod kinetic constants for which are $\mu_m = 8.7\ hr^{-1}$, $K_S = 0.22$ g/L) is to be conducted in a 1.2 m high semifluidized bed biofilm reactor, composed of 1.5 mm support particles each surrounded by a 0.3 mm thick biofilm (average density of microbial solution = 1030 kg/m^3, density of support particles = 2100 kg/m^3). The feed solution with a lactose concentration of 9.0 g/L is admitted from the bottom at the rate of 50,200 L/hr. Compute the fractional substrate conversion attained, if the overall yield coefficient for cell mass production = 0.65 (could be assumed constant) and the bed expansion ratio chosen = 2.0. Assume that the voidage of the packed section formed is more or less constant and is equal to 0.40 and take effective diffusivity of substrate (lactose) into the biofilm = 3.0 × 10^{-10} m^2/s. Compute the fractional substrate conversion, based on the following two approaches :

(a) if the reactor could be assumed equivalent to a CSTR – PFR combination,
(b) if it is assumed to be equivalent to two PFDRs in series. Take axial dispersion coefficient in the packed section to be 0.0075 m^2/s and that in the fluidized section 0.0315 m^2/s.

Solution:

Let us first evaluate the hydrodynamic parameters affecting the operation of the bioreactor. Thus, let us, at the outset, evaluate the minimum semifluidization velocity of the liquid (continuous medium), $U_{SL}(min)$, from the correlation proposed by Kurian and Rao (equation - 3.6.3). For this, we have to first evaluate U_{mf} from Wen and Yu's correlation (equation 3.5.2) and V_t from the generalized law of settling (equation - 2.5.27 of Chapter 2). Thus,

$$Ar_m = \left[d_{Pm}{}^3 \left(\rho_{Sm} - \rho_L \right) g \rho_L \right] / \mu_L{}^2$$

where d_{Pm} = 1.5 + 2(0.3) = 2.1 mm

$$\rho_{Sm} = (1-f)\rho_S + (f\,\rho_m)$$

$$f = 1-\left(d_P / d_{Pm}\right)^3 = 1 - (1.5 / 2.1)^3 = 0.63557$$

Therefore, $\rho_{Sm} = 2100\ (1 - 0.63557) + (1030)\ (0.63557) = 1419.94\ \text{kg/m}^3$

Now, taking $\rho_L = 1000\ \text{kg/m}^3$ and $\mu_L = 0.001\ \text{kg/(m.s)}$,

$$Ar_m = 38112.83$$

From equation (3.5.2), therefore,

$$Re_{mf} = 18.1824$$

And $U_{mf} = \left(Re_{mf}\,\mu_L\right) / \left(d_{Pm}\,\rho_L\right) = 0.008658$ m/s

Now, from the generalized law of settling (equation – 2.5.27 of Chapter - 2),

$$V_t = \left[(4/3)\left(\rho_{Sm} - \rho_L\right) g\, d_{Pm} / \left(\rho_L\, f_D\right) \right]^{1/2}$$

or $V_t^2 = (0.011523) / f_D$ (i)

Also, $Re_P = \left(d_{Pm} V_t\, \rho_L / \mu_L\right) = (2100)\ V_t$ (ii)

The above two equations are to be solved by trial. Thus, let

$$V_t = 0.12\ \text{m/s}$$

Then from equation (ii), $Re_P = (2100)\ (0.12) = 252.0$

Now, with reference to table (2.1) and equation (2.5.30) of Chapter – 2,

$$f_D = 0.75526$$

Then from equation (i), $V_t = 0.1235$ m/s

Since the above – computed value of V_t does not differ significantly from the assumed value, we may choose an average value of $V_t = 0.12175$ m/s as final. Therefore, from equation (3.6.3),

$$U_{SL}(\min) = 0.03868\ \text{m/s}$$

To note that the bed expansion ratio = R = 2.0 (specified in the problem). Let us, therefore, choose an operating liquid velocity of 0.07 m/s. In other words,

$$U_{SL} = 0.07\ \text{m/s}$$

The above choice does satisfy the criterion specified in equation (3.6.4). Since liquid flow rate is 50.2 m^3/hr,

$$A = (Q_O / U_{SL}) = (50.2/3600)/(0.07) = 0.2\ \text{m}^2$$

And $$D = \sqrt{4A/\pi} = 0.5\ \text{m}.$$

This is the required column diameter. The height of the packed section formed (L_P) and that of the fluidized section (L_f) are also to be computed from selected experimental correlations. Thus, in this case, let us compute them based on the correlation proposed by Singh et.al (equation – 3.6.10). Thus,

$$C_L = 0.974 + 0.324 \ln(C_V) \tag{iii}$$

where $C_V = (U_{SL} - U_{mf})/(V_t - U_{mf}) = 0.5424$

Therefore, $C_L = 0.974 + 0.324 \ln(0.5424) = 0.77579$

Since $C_L = (L_{Sf} - L_S)/(L_{Sf} - L_P)$, and $L_S = (L_{Sf}/R) = (1.2/2.0) = 0.6$ m,

$L_P = 0.4266$ m

And $L_f = (1.2 - 0.4266) = 0.7734$ m.

It is given that, $\in_P = \in_{PL} = 0.40$.

The fractional liquid holdup in the fluidized section ($\in_{fL}$), which, in this case, is equal to its total voidage ($\in_f$), may be estimated from Richardson and Zaki's correlation(equation – 3.5.6). Thus,

$$\in_{fL} = \in_f = [U_{SL}/U_i]^{1/n} \tag{iv}$$

where, $U_i = V_t \exp[-2.303\, d_{Pm}/D] = 0.12058$ m/s

Now, $Re_T = (D V_t \rho_L / \mu_L) = 60875.0$

Since Re_T is greater than 500, from table (3.1), $n = a_0 = 2.4$. Therefore, from equation (iv),

$$\in_{fL} = \in_f = 0.797$$

(a) If we assume that the bioreactor is equivalent to a CSTR – PFR combination, then the performance equation for the fluidized section (ideal CSTR) shall be

$$\tau_m = [V_m \in_{fL}/Q_0] = (C_{S0} - C_{Sb})(K_S + C_{Sb})/[\eta_b\, \mu_m(app)\, C_{Sb}] \tag{v}$$

where $V_m = (\pi D^2/4) L_f = 0.15185$ m^3

$Q_0 = 50.2$ m^3/hr

$K_S = 0.22$ g/L

$\mu_m(app) = (\mu_m/Y)(x_f f)(1 - \in_f)/\in_{fL} = 1083.37$ g/(L.hr)

To note that $\mu_m = 8.7$ hr^{-1}, Y = 0.65 and $x_f = 500$ g/L. Equation (v), therefore, becomes

$$C_{Sb}^2 + (2.6118\,\eta_b - 8.78) C_{Sb} - 1.98 = 0 \tag{vi}$$

The above equation can be solved only by trial, since the value of η_b (at $C_S = C_{Sb}$) is not known at the outset. Thus, let

$$\eta_b = 1.0$$

Now, on solving equation (vi), $C_{Sb} = 6.4748$ g/L

Let us now compute the value η_b at $C_S = C_{Sb} = 6.4748$ g/L. For this, we shall use equations (2.4.28) and (2.4.29). Thus,

$$L^* = [d_{Pm}^3 - d_P^3]/(6 d_{Pm}^2) = 0.22245 \text{ mm}$$

$$\varnothing' = L^* \; [\mu_m(app) / (D_e K_S)]^{1/2} = 15.0209$$

To note that $D_e = 3.0 \times 10^{-10}$ m²/s.

$$\beta = (6.4748/0.22) = 29.43$$

$$\eta_d = (0.09415)\left[(1+\beta)/\beta\right]\sqrt{\beta - \ln(1+\beta)} = 0.4965$$

And $\qquad \eta = \eta_b = 0.9946$ (from equation – 2.4.28)

Since the value of η_b is very close to 1.0, no further trials are required and the above – computed value of C_{Sb} is final.

Now, the performance equation for the packed section (assumed equivalent to a PFR) is,

$$\tau_P = \left[V_P \in_{PL} / Q_0\right] = \int_{C_{Se}}^{C_{Sb}} \left[\frac{(K_S + C_S)}{\eta\, \mu_m(app) C_S}\right] dC_S \qquad \text{(vii)}$$

The above integral is to be evaluated numerically using Simpson's rule or trapezoidal rule. However, it may be observed that since η is close to 1.0 at $C_S = C_{Sb}$, it shall necessarily have a higher magnitude and shall be much more close to 1.0 at $C_S = C_{Se}$. If η is taken equal to 1.0, then the integral of equation (vii) can be integrated analytically to yield,

$$\tau_P\, \mu_m(app) = (C_{Sb} - C_{Se}) + K_S \ln(C_{Sb}/C_{Se}) \qquad \text{(viii)}$$

Since $V_P = (\pi D^2/4) L_P = 0.08376$ m³,

$$\tau_P = \left[V_P \in_{PL} / Q_0\right] = 0.0006674 \text{ hr} = 0.04 \text{ min}$$

Also, $\mu_m(app) = (\mu_m/Y)(x_f f)(1 - \in_P)/\in_{PL} = 6380.145$ g/(l.hr)

Now, on solving equation(viii) by trial, we get

$$C_{Se} = 2.43 \text{ g/L}$$

Let us now compute the value of η at $C_S = C_{Se}$ as follows:

$$\varnothing' = 36.45226$$

$$\beta = (2.43/0.22) = 11.04545$$

$$\eta_d = 0.12376$$

And $\eta = 0.99986$ (from equation – 2.4.28)

Thus, the assumption that $\eta \approx 1.0$ shall not cause any serious error and the above – computed value of C_{Se} ($C_{Se} = 2.43$ g/L) can be treated as final. The fractional substrate conversion attained is, therefore,

$$\alpha = (C_{S0} - C_{Se})/C_{S0}$$

$$= (9.0 - 2.43) / (9.0) = 73.0\ \%$$

To note that the conversion attained in the fluidized section is (9.0 – 6.4748)/(9.0) = 28 %, while that attained in the packed section is, (6.4748 – 2.43) / (6.4748) = 62.47 %. Thus, the major share of substrate conversion occurs in the packed section, though its height and void volume are relatively lower than those of the fluidized section.

(b) For a more accurate analysis of the bioreactor performance, the PFDRs – in – series approach is to be employed, which assumes the bioreactor to be equivalent to two PFDRs in series, PFDR – 1 representing the packed section and PFDR - 2 representing the fluidized section.

The performance equation for each section and the associated boundary conditions are those given in equations (3.6.20) to (3.6.25) and the procedure of numerical solution is also that presented earlier below these equations. We shall present here one sample iteration, as an example. As in examples (3.7) and (3.8), we shall illustrate computations with a larger step size of 0.1 m. In real-life design calculations, a step size as low as 0.001m (1.0 mm) must have to be employed. The hydrodynamic parameters have already been computed in part (a) of this problem and therefore, we are not repeating them here. The different steps involved in the numerical solution are thus as follows :

Step – 1: Δz = step size = 0.1 m

Step – 2: Let $C_{Se} = 2.43$ g/L. To note that this is the value of C_{Se} computed in part (a). We have purposefully selected this value as the first approximation to reduce the number of iterations required. In cases otherwise, computations are to be performed starting with a random, but reasonably chosen, value of C_{Se}.

Step – 3: The performance equation of the packed section (PFDR – 1) is now solved:

$$i = 1$$

$$C_S(1) = C_{Se} = 2.43 \text{ g/L}$$

$$W(i) = W(1) = 0.0$$

Step – 4: $j = 1$

$$C_S(j) = C_S(i) = 2.43 \text{ g/L}$$

$$W(j) = W(i) = 0.0$$

Step – 5: Computation of effectiveness factor, $\eta(j)$:

$$\beta(j) = C_S(j)/K_S = (2.43/0.22) = 11.04545$$

$$\eta_d(j) = 0.12376$$

$$\eta(j) = 0.99986$$

The above value of $\eta(j)$ is that computed in part (a). We have, therefore, avoided the elaborations here.

Step – 6: From equation (3.6.29), $F(j) = 215.8485$ g/(L.m^2)

$$m(1)= m(j)= (0.1)\ (216.2508) = 21.58485 \text{ g/(L.m)}$$

Step – 7: $j = 2$

$$C_S(j) = C_S(i) + (\Delta z/2)W(i) = C_S(i) = 2.43 \text{ g/L}$$

$$W(j) = W(i) + (1/2)m(1) = (21.58485 / 2) = 10.7924 \text{ g/(L.m)}$$

Since $C_S(j)$ remains same as 2.43 g/L, $\eta(j)=$ 0.99986 (computed earlier).

Step – 8: From equation (3.6.29),

$$F(j) = (215.8485) + (0.07 / 0.0075)\ (10.7924) = 316.57756 \text{ g/(L.m}^2)$$

$$m(2) = m(j) = (0.1)\ (316.57756) = 31.6577 \text{ g/(L.m)}.$$

Step – 9: $j = 3$

$$C_S(j) = C_S(i) + (\Delta z/2)W(i) + (\Delta z/4)m(1)$$

$$= 2.43 + (0.1/4)\ (21.58485) = 2.9696 \text{ g/L}$$

$$W(j) = W(i) + (1/2)m(2) = (31.6577/2) = 15.8288 \text{ g/(L.m)}$$

Step – 10: Computation of $\eta(j)$:

$$\beta(j) = C_S(j)/K_S = (2.9696 / 0.22) = 13.4982$$

$$\eta_d(j) = 0.137096$$

$$\eta(j) = 0.99983$$

Step – 11: $F(j) = (219.9655) + (147.73546) = 367.7 \text{ g/(L.m}^2)$

$$m(3) = m(j) = (0.1)\ (367.7) = 36.77 \text{ g/(L.m)}$$

Step – 12: $j = 4$

$$C_S(j) = C_S(i) + (\Delta z)W(i) + (\Delta z/2)m(2)$$

$$= 2.43 + (0.1/2)\ (31.6577) = 4.01288 \text{ g/L}$$

$$W(j) = W(i) + m(3) = 36.77 \text{ g/(L.m)}$$

Step – 13: Computation of $\eta(j)$:

$$\beta(j) = (4.01288/0.22) = 18.24036$$

$$\eta_d(j) = 0.16$$

$$\eta(j) = 0.999766$$

Step – 14: $F(j) = 567.1543$ g/(L.m^2)

$$m(4) = 56.71543 \text{ g/(L.m)}$$

Step – 15: $C_S(i+1) = C_S(i) + (\Delta z) W(i) + (\Delta z/6)\left[m(1) + m(2) + m(3)\right]$

$$= 2.43 + (0.1/6)\ (90.01255) = 3.9302 \text{ g/L}$$

$$W(i+1) = W(i) + (1/6)\left[\ m(1) + 2\ m(2) + 2\ m(3) + \ m(4)\right]$$

$$= 35.85928 \text{ g/(L.m)}$$

Step – 16: $L_P(cal) = (i)(\Delta z) = 0.1$ m

Since the above value of $L_P(cal)$ is less than L_P (namely, 0.4266 m), we put $i = (i+1) = 2$ and repeat the computations starting from Step – 4.

When computations were performed on a high speed computer with $\Delta z = 0.001$m and $C_{Se} = 2.75$ g/L, the value of $C_S(i+1)$ obtained at $L_P(cal) = 0.427$m, is 6.3953 g/L. Therefore,

$$C_{Sb} = 6.3953 \text{ g/L}$$

The procedure was repeated for the fluidized section, starting from $C_S(i) = C_S(1) = C_{Sb}$ = 6.3953 g/L and proceeded downwards. At $L_f(cal) = 0.774$ m, the calculated value of $C_S(i+1)$ only slightly exceeded 9.0 g/L. This, thus, confirmed the assumed value of C_{Se} = 2.75 g/L. The fractional substrate conversion (α) attained is, therefore,

$$\alpha = (9.0 - 2.75)\ /\ (9.0) = \mathbf{69.44\ \%}$$

This value is lower than that computed under part (a), based on ideal reactor performance (which is understandable as well). Here also,

α (attained in packed section) = (6.3953 – 2.75) / (6.3953) = 57 %

α (attained in fluidized section) = (9.0 – 6.3953) / (9.0) = 28.94 % The major share of substrate conversion is thus contributed by the packed section.

3.7 Design and Analysis of Inverse Fluidized Bed Bioreactors

Inverse fluidized bed bioreactors are also relatively new entries into the family of commercial bioreactors. The terminology, *inverse fluidized* often causes confusion, since it represents fluidization of discrete phase (solid particles or particle-biofilm aggregates) in a down -flowing stream of fluid (usually, liquid solution). To many, this may appear to be impossible and could even be labelled as an Utopian concept. However, today, inverse fluidization is being practiced successfully in a number of commercial systems. Inverse fluidization can be achieved in two ways. In other words, inverse fluidized beds (or, inverse fluidized bed bioreactors) can be broadly classified into two categories :

(a) In the first category (Category I), solid particles that constitute the discrete phase are so chosen that they are lighter than the fluidizing medium (liquid solution). For example, plastic beads made of polypropylene (PP) or low density polyethene (LDPE) are lighter than water and aqueous solutions and hence, when they are used as support particles in biofilm reactors, the resultant particle-biofilm aggregates shall remain suspended in the descending stream of liquid (feed solution). In other words, since ρ_S (or ρ_{Sm}) < ρ_L , the particle-biofilm aggregates (the discrete phase) shall remain suspended (fluidized) in the feed solution and thereby form a fluidized bed of height L_f inside the column, even though the liquid solution is flowing down under gravity (executing downflow). All inverse fluidised bad biofilm reactors fall under this category (see figure 3.10). In such biofilm reactors, particles of larger size (d_p = 20 to 25 mm) could be employed.

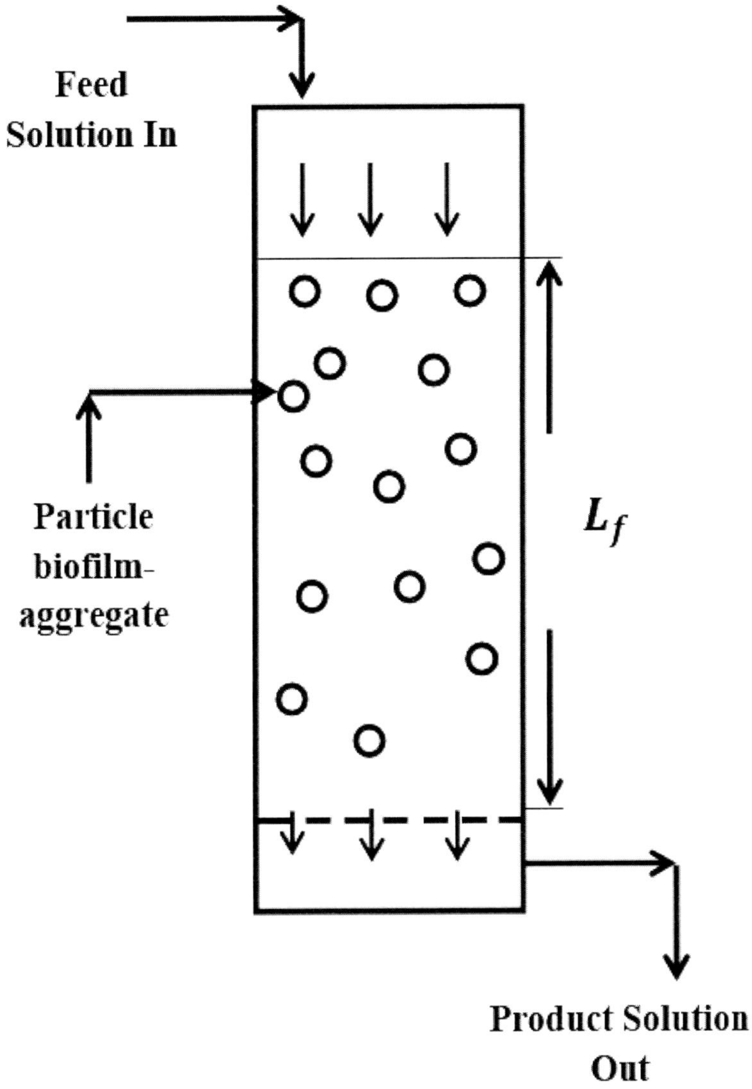

Figure 3.10: Schematic of Inverse Fluidized Bed Biofilm Reactor (two phase)

We all know that if we stack a number of plastic balls at the bottom of a tall glass jar and pour water into it from above, then these plastic balls shall start rising in the liquid. The same principle is used in the above category of bioreactors. The height to which the particle-biofilm aggregate would rise in the down-flowing liquid solution, namely the height of the expanded bed (L_f), is a hydrodynamic parameter and its magnitude depends on the liquid flow rate employed. So is the voidage of the expanded bed ($\in_f$). Also, it is essential that the liquid velocity employed exceeds the minim inverse fluidization velocity, $U_{Lm}(inv)$, which is the liquid velocity at which all the particle-biofilm aggregates just rise into the liquid stream.

The values of all of these hydrodynamic parameters, such as $U_{Lm}(inv)$, L_f and $\in_f$, are to be estimated from available experimental correlations. This is discussed subsequently in this chapter.

Further, as the feed solution is executing downflow under gravity, large liquid flow rates are difficult to be maintained in these types of bioreactors and that restricts the reactor capacity. Also, at high liquid flow rates, some of the particle- biofilm aggregates could get pushed down to the bottom porous plate (due to the large hydrodynamic forces developed) and they tend to accumulate and form a packed bed there. This could disrupt the performance of the bioreactor and the operating cost of the reactor would invariably increase, since the compressed packed bed at the bottom would offer additional resistance to fluid flow.

(b) In the second. Category (Category – II) of bioreactors, inverse fluidization is achieved by using very fine solid particles (for example, nanoparticles) as the discrete phase. Examples are bioreactors employing silica nanoparticles within which the enzyme (biocatalyst) has been immobilized. Here also, though the liquid solution executes downflow, the nanoparticles carrying the immobilized enzyme remain suspended in it (by virtue of their very small size). The bioreactor shall be thus equivalent to a tall column through which a fine suspension of nanoparticles in the feed solution flows down. *Incidentally, inverse fluidized bed bioreactors of this category are the only type of commercial bioreactors that could be used for handling nanoparticles.*

However, handling of nanoparticles in reactors or bioreactors do involve practical difficulties. As stated above, packed bed, fluidized bed (unless circulating fluidized bed is employed) and semifluidized bed reactors are written off and inverse fluidized bed reactors are the only choice in this case. Narayanan, Tejaswi and others [68] have proposed a modified design of inverse fluidized bed bioreactor that is specifically suitable for handling nanoparticles (see figure 3.11). In this design, the nanoparticles (catalyst particles or support particles accommodating the immobilized biocatalyst) are first mixed uniformly with the feed solution in an overhead mixing tank and the resultant suspension is allowed to flow down the reactor column. As the suspension flows down, the substrate diffuses into the nanoparticles, the bioconversion occurs and the products diffuse back into the liquid bulk, From the bottom of the column the product solution with suspended nanoparticles is discharged and then passed through a nanomembrane filter. The nanoparticles retained by the filter membrane are recycled back to the overhead mixing tank (into which the feed solution is also pumped) through a vertical, pneumatic column using compressed air (see figure-3.11). The product solution permeates through the filter membrane and is collected and sent to storage. Since the nanoparticles are extremely fine in size, they can be conveniently transported upward through the pneumatic column without using too high air pressure. As the feed solution containing the suspended nanoparticles is executing downflow under gravity (which eliminates the pumping cost of the feed solution), the only major operating cost associated with the bioreactor operation is that of the nanomembrane filter.

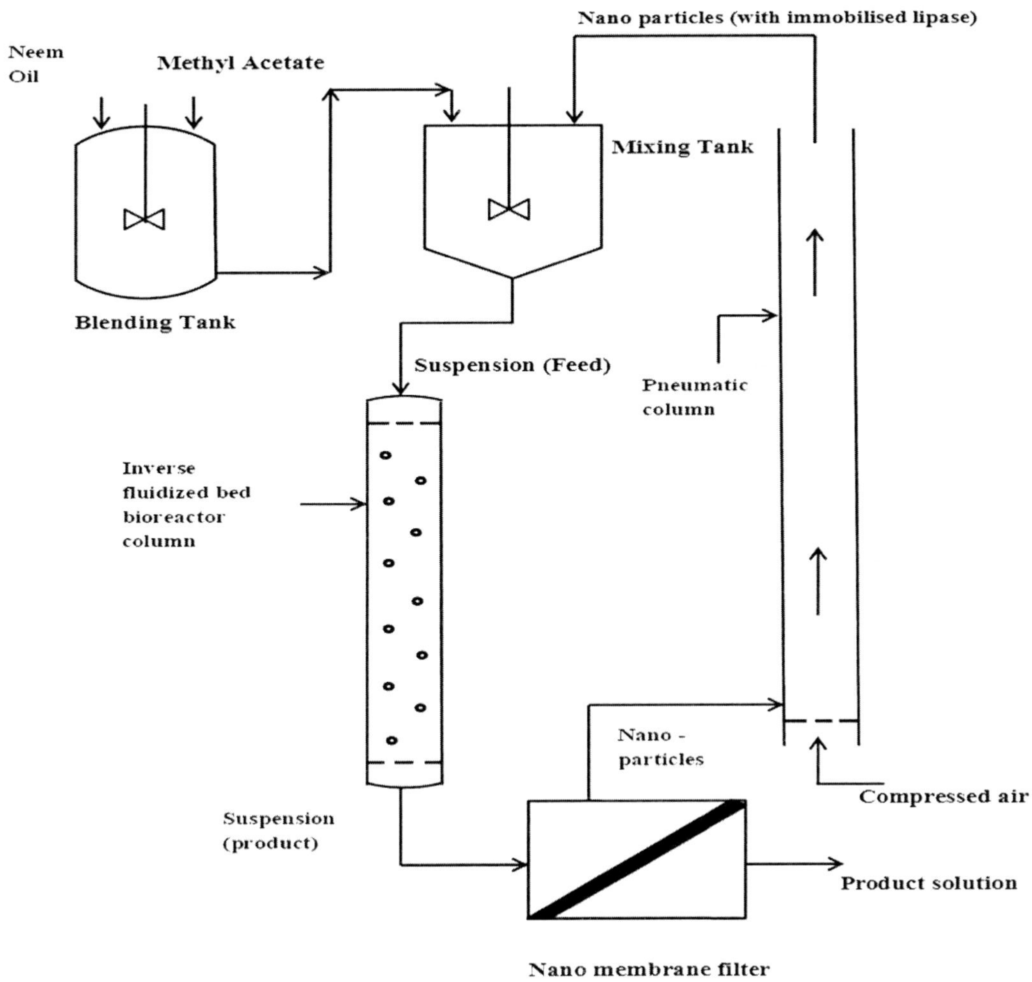

Figure 3.11: Schematic of Two Phase Inverse Fluidized Bed Immobilized Enzyme Bioreactor, employing Nanoparticles as Support Media.

Bioreactors of this category could be constructed using columns of relatively lower diameter (D = 25.4 mm, 38 mm). Also, the column height, L ($L = L_f$) is not a hydrodynamic parameter in this case and could be selected as desired. So is the catalyst loading, C_L (which is the mass ratio of nanoparticles to feed solution), which could also be set at any desired value. To note that since C_L represents the ratio of discrete phase to the liquid phase, it is a parameter that decides the voidage of the bed.

Thus, unlike inverse fluidized bed biofilm reactors (category I), there are no hydrodynamic parameters (that are required to be estimated from selected experimental correlations) associated with the design of bioreactors of this category.

The requirement of maintaining large residence time of the feed solution within the reactor column (so as to obtain an acceptably large substrate conversion) restricts the magnitude of feed flow rate that could be employed in these types of bioreactors and that restricts the reactor capacity. However, the capacity can be increased by using multitube or multicolumn construction, composed of 4 to 6 tubes or columns. In other words, in high capacity installations, four to six bioreactors could be operated in parallel. Nanoparticles, by virtue of their enormously large specific surface, would offer negligible resistance to substrate transfer and consequently, the value of effectiveness factor (η) shall be very close to unity. In other words, in this category of bioreactors, the overall rate (global rate) of reaction shall be more or less equal to the intrinsic rate itself. This enhances the performance of the bioreactor.

Incidentally, bioreactors of this category are not well – suitable for three phase operation, due to the simple reason that if compressed air is admitted from the bottom, a part of the nanoparticles would get carried off in the exit air (entrainment loss). Entrainment loss of nanoparticles could also occur in the gases produced during the process (when they flow up and get discharged from the top of the reactor column).

3.7.1. Estimation of Hydrodynamic parameters

Performance analysis of inverse fluidized bed bioreactors is best performed using the PFDR approach. In the case of inverse fluidized bed biofilm reactors (category-1), we have to firstestimate the hydrodynamic parameters such as the minimum inverse fluidisation velocity of the liquid, $U_{Lm}(inv)$, the expanded bed height attained (L_f) and the voidage of the expanded bed ($\in_f$). We have to invariably resort to experimental correlations reported in the literature in this regard. For the estimation of $U_{Lm}(inv)$ in two-phase, liquid-solid, inverse fluidized beds, Ulaganathan and Krishnaiah [69] have proposed the following *dimensional* correlation:

$$U_{Lm}(inv) = \left(2.93\times10^{-3}\right)\left(Ar_m\right)^{0.202}\left[\left(\rho_L - \rho_{sm}\right)/\rho_L\right]^{0.38} \tag{3.7.1}$$

where Ar_m = modified Archemedes number

$$= \left(\rho_L - \rho_{sm}\right) g\, d_{Pm}^3\, \rho_L / \mu_L^2 \tag{3.7.2}$$

The authors have conducted experiments using water and LDPE granules of size 12.5 to 20 mm. The above correlation is reportedly valid at larger values of Ar_m such as $10^6 < Ar_m < 7 \times 10^7$ and also for $0.4 < \left[\left(\rho_L - \rho_{sm}\right)/\rho_L\right] < 0.9$. What is intriguing about the above correlation is that the *dimensional* parameter, $U_{Lm}(inv)$ has been correlated with the two *dimensionless* parameters such as Ar_m and the density ratio. Accordingly, it is not clear how the dimensional consistency has been maintained.

Vajayalakshmi et al. [70] have conducted experiments using water and polymer beads made of PP and LDPE (size = 4 mm to 8 mm) and have come forward with the following dimensionless correlation:

$$Re_{bm} = 0.233\left(Ar_m\right)^{0.475} \tag{3.7.3}$$

where $Re_{bm} = \left[d_{Pm} U_{Lm}(inv)\rho_L / \mu_L\right]$. (3.7.4)

The above correlation is valid for $3.7 \times 10^4 \le Ar_m \le 8.5 \times 10^5$.

An alternate dimensionless correlation has been proposed by Banerjee et.al [71], developed at relatively lower values of Ar_m such as $17.74 \times 10^3 \le Ar_m \le 26.51 \times 10^3$. The correlation is given below:

$$Re_{bm} = 7.68\times10^{-6}\left(Ar_m\right)^{1.513} \tag{3.7.5}$$

The operating superficial velocity of substrate solution (U_L) could be selected at 12 to 20% higher than the minimum. Thus,

$$U_L = (1.12 - 1.20)\ U_{Lm}(inv) \tag{3.7.6}$$

As stated earlier, too high liquid flow rates are difficult to be maintained in these bioreactors.

Ulaganathan and Krishnaiah [69] have proposed experimental correlation for the estimation of the expanded bed height (L_f) as well. The same is reproduced below :

$$\left(L_f / L_S\right) = (0.572)\left(Re_b\right)^{0.92}\left(Ar_m\right)^{-0.345}\left[\left(\rho_L - \rho_{sm}\right)/\rho_L\right]^{-0.21} \tag{3.7.7}$$

where L_S = height of initial static bed (prior to fluidization)

Re_b = bed Reynolds number

$$= \left[d_{Pm} U_L \rho_L / \mu_L\right] \tag{3.7.8}$$

Once the expanded bed height (L_f) has been estimated, the voidage of the expanded bed ($\in_f$) could be computed from a solid balance as,

$$L_S\left(1-\in_P\right) = L_f\left(1-\in_f\right) \tag{3.7.9}$$

or

$$\in_f = 1-\left(L_S / L_f\right)\left(1-\in_P\right. \tag{3.7.10}$$

where $\in_P$ = voidage of the static bed = 0.35 – 0.40.

Benedict et al. [72] have also studied bed expansion in inverse fluidized beds consisting of 6.0 mm beads made of PP and LDPE (water being the fluidizing medium) and have come up the following correlation :

$$\left(L_S / L_f\right) = 2.656\ \left(Re_{Pm}\right)^{1.258}\ \left(Ar_m\right)^{-0.88} \tag{3.7.11}$$

where Re_{Pm} = modified bed Reynolds number

$$= d_{Pm} U_L \rho_L / [\mu_L\left(1-\in_f\right)] \tag{3.7.12}$$

The above correlation is to be solved simultaneously with equation (3.7.9) to obtain the values of L_f and $\in_f$. This shall invariably demand a trial and error procedure.

In the case of three phase, inverse fluidized bed biofilm reactors (used for conducting aerobic processes), we need separate correlations for the estimation of fractioned liquid holdup ($\in_{fL}$) and fractional gas holdup ($\in_{fg}$) in the bed. We shall have thus four hydrodynamic parameters at hand such as the minimum inverse fluidization velocity of the liquid, $U_{Lm}(inv)$, height of the expanded bed (L_f), fractional liquid holdup ($\in_{fL}$) and fractional gas holdup ($\in_{fg}$) in the bed, which are to be estimated from selected experimental correlations. To note that in most three phases inverse fluidized bioreactors, the liquid moves down as the continuous phase while the gas, either supplied from outside such as compressed air or formed during the process, moves up counter-current to the liquid, in the form of tiny, discrete bubbles. All the correlations presented below consider systems of this kind.

The correlation proposed by Sivasubramanian and Velan [73] for the estimation of minimum inverse fluidization velocity of the liquid, $U_{Lm}(inv)$, in counter-current fluidized beds (in which the down-flowing liquid forms the continuous phase and the gas moves up in the form of tiny, discrete bubbles) is reproduced below :

$$Re_{bm} = 0.07033\left(Ar_m\right)^{0.579} - 16.619\left(Ar_m\right)^{0.113}\left(Re_g\right)^{0.781} \tag{3.7.13}$$

where $Re_g = \left[d_{Pm}U_g\rho_g / \mu_g\right]$ (3.7.14)

U_g = average superficial gas velocity through the bed

The modified Archemedes number (Ar_m) and the bed Reynolds number (Re_{bm}) are those defined earlier in equations (3.7.2) and (3.7.4) respectively. The above correlation is valid for $Re_g \leq 3.0$ and $2.0 \times 10^4 \leq Ar_m \leq 1.18 \times 10^6$.

Fan et.al [74] have reported correlations for the estimation of fluid holdups in three phase, counter-current inverse fluidized beds, as given below :

$$\in_{fL} = -0.26 + 27\ U_L - 2300\ \left(U_L\right)^2 + 9.5\ U_g \tag{3.7.15}$$

$$\in_{fg} = 0.322\ \left(\in_f\right)^{1.35} + \left(U_g / U_L\right)^{0.18} \tag{3.7.16}$$

$$\in_{fL} = \left(\in_f - \in_{fg}\right) \tag{3.7.17}$$

The expanded bed height (L_f) can be now estimated from the solid balance (equation – 3.7.9).

3.7.2 Performance Analysis (Mathematical modeling)

Based on the PFDR approach, the performance equation for the inverse fluidized bed biofilm reactor (Category I) shall be

$$-U_L\left(dC_S / dz\right) + D_{Lf}\left(d^2C_S / dz^2\right) = \eta\left(-r_S\right)\left(int\right) \tag{3.7.18}$$

The above equation, nevertheless, assumes that the mass transfer resistance to substrate transfer in the liquid bulk is negligible and consequently, $C_{SP} \approx C_S$ and $\left(-r_{SP}\right)\left(int\right) \approx \left(-r_S\right)\left(int\right)$.

The governing boundary conditions are,

BC – 1: At z = 0 (at the bottom of the reactor column),

$$C_S = C_{Se} \tag{3.7.19}$$

BC – 2: At z = L_f (at the top of the inverse fluidized bed),

$$C_S = C_{S0} \tag{3.7.20}$$

Readers can easily notice that the above performance equation is exactly analogous to equation (3.2.18) of Section (3.2). As a result, the numerical algorithm for the solution of the above equation shall also be the same as that described in Section (3.2). The same algorithm has been used for the performance analysis of fluidized bed bioreactors (Section 3.5) and that of semifluidized bed bioreactors (Section 3.6) as well. The only difference is that in the present case, computations are started from the bottom of the reactor column (where z = 0 and $C_S = C_{Se}$) and continued upward until z = L_f and $C_S = C_{S0}$. We have, therefore, avoided elaborations on the numerical solution here.

In the case of inverse fluidized bed bioreactors of Category – II that employ nanoparticles also, the performance equation shall be same as equation (3.7.18), except for a few differences:

The first difference is that η = 1.0, in the case of nanoparticles. This, as stated earlier, is due to the enormously large specific surface possessed by the nanoparticles.

Secondly, since the system involves downflow of a suspension of nanoparticles in substrate solution, the term U_L would have to be replaced by U_m, where U_m represents the average downflow velocity of substrate- particle Suspension. These two are related to each other as,

$$U_m = \left(U_L / \in_{fL}\right) \tag{3.7.21}$$

where U_L = average superficial velocity of liquid (substrate solution) down the column

$\in_{fL}$ = fractional liquid holdup (= volume fraction of liquid or substrate solution in the suspension).

The fractional liquid holdup ($\in_{fL}$) is related to the catalyst loading (C_L) as,

$$\in_{fL} = \in_f = \left(1/\rho_L\right) / \left[\left(C_L/\rho_S\right) + \left(1/\rho_L\right)\right]$$

$$= 1 / \left[C_L\left(\rho_L/\rho_S\right) + 1\right] \tag{3.7.22}$$

To note that catalyst loading, as stated earlier, is the mass ratio of nanoparticles to substrate solution (= mass of nanoparticles / mass of substrate solution) and its value could be selected in advance as per requirement. In many cases, C_L = 0.3 to 0.7. Also, the bulk density of the suspension (ρ_{Lm}) shall be

$$\rho_{Lm} = \left(1 + C_L\right) / \left[\left(1/\rho_L\right) + \left(C_L/\rho_L\right)\right] \tag{3.7.23}$$

The bulk viscosity of the suspension (μ_{Lm}) maybe taken equal to that of the substrate solution (μ_L), without serious error.

The performance equation for bioreactors of category II shall be, therefore,

$$-U_m\left(dC_S/dz\right) + D_{Lf}\left(d^2C_S/dz^2\right) = \left(-r_{SP}\right)(int) \tag{3.7.24}$$

If it is assumed that $C_{SP} \approx C_S$, then $\left(-r_{SP}\right)(int) \approx \left(-r_S\right)(int)$.

The boundary conditions shall remain the same as those given in equations (3.7.19) and (3.7.20). The numerical procedure of solution of performance equation shall also be the same. Computations shall be simpler in this case, since evaluation of η shall not be required.

Also, as stated earlier, there are no hydrodynamic parameters associated with the design of these bioreactors (those are otherwise required to be determined from selected experimental correlations). The column height (L or L_f) could be selected as desired and so is the column diameter (D). The feed flow rate (Q_0) could also chosen as per requirement, keeping in mind that large feed flow rates would lower the residence of time of substrate solution in the reactor column and thereby would tend to diminish the substrate conversion. The catalyst loading could also be selected in advance as per requirement or could be chosen by trial. For example, the performance equation may be solved at different value of C_L and the optimal value or the most acceptable value of C_L (at which the substrate conversion attained is the largest) could be arrived at.

As stated earlier, the processes involving three phase operations are difficult to be conducted in these bioreactors, since sizeable entertainment loss of nanoparticles could occur. They are best operated as two - phase, liquid - fluidized bioreactors.

The numerical algorithm in this case could be executed in two ways:

(a) To compute the fractional substrate conversion (α) attained at the chosen column height, L (or L_f). A value of C_{Se} is first assumed (based on an assumed value of α). Computations are performed upward until $z = L$ (or L_f). It is checked whether the value of C_S at $z = L$ is equal to or slightly larger than C_{S0}. If not, the computations are repeated with an alternate value of C_{Se}. The value of C_{Se} is thus finalized by trial and the fractioned substrate conversion attained (α) computed therefrom.

(b) To compute the required height of reactor column (L or L_f), for achieving the desired substate conversion (α). In this case, the value of C_{Se} is known at the outset. Starting from $z = 0$ ($C_S = C_{Se}$), computations are performed upward until $C_S = C_{S0}$. The value of z at $C_S = C_{S0}$ is the required height (L or L_f) of reactor column. Here, no trials are required.

A numerical example in this connection has been presented at the end of this Section.

3.7.3 Advantages and Industrial Applications

The specific advantages of inverse fluidized bad reactors may be summarized as follows:

(i) Due to the downflow mode of operation, the feed solution need not have to be pumped and this brings down the operating cost of the bioreactor significantly.

(ii) When used as biofilm reactors, relatively larger size support particles ($d_P = 20$ to 25 mm) could be used.

(iii) This is the only type of reactor configuration that is adequately suitable for handling nanoparticles (except for circulating fluidized bed bioreactors).

(iv) Nanoparticles, by virtue of their enormously large specific surface, offer negligible resistance to substrate transfer into the particle or into the biofilm (the effectiveness factor (η) shall be close to 1.0) and this enhances the bioreactor performance.

(v) Within the permissible range of feed flow rates, these bioreactors do provide reasonably good substrate conversion and product yield (though relatively lower than that could be achieved in a fluidized bed or semifluidized bed bioreactor of same volume).

Adaptation of inverse fluidization technology to industrial practices has been successful as well as encouraging. Synthesis of polymer grade lactic acid from de-proteinised cheese whey and clarified molasses in inverse fluidized bed biofilm reactors has been investigated by Narayanan and Das [75,76]. The performance of the bioreactor has been analyzed mathematically (using the PFDR approach) and subsequently verified using lab-scale and pilot plant-scale experimental data. The bioreactor has been observed to provide good performance such as around 75-76% conversion of lactose/sucrose in a reactor column of diameter = 0.5 m and expanded bed height = 2.4 m, at a feed flow rate of 59,000 L/hr.

Synthesis of biodiesel from neem oil and methyl acetate in inverse fluidized bed bioreactor of Category-II (figure-3.11) has been studied by Narayanan and Aditi [77]. Nanosilica particles accommodating lipase enzyme were used as the discrete phase. The bioreactor performance has been studied both mathematically and experimentally. It is reported that more than 88% conversion of neem oil to biodiesel is attained in a 25.4 mm diameter reactor column of height = 6.0 m at a catalyst loading of 0.7 and feed flow rate of 360 L/hr, when the molar ratio of oil to acetate in the feed = 1:3. As stated earlier, the capacity of the reactor could be increased from 360 L/hr to 1800-2160 L/hr by using a multi-tube or multi-column construction (composed of 5-6 tubes or columns).

Three phase operation of inverse fluidized bed biofilm reactors for aerobic treatment of industrial effluents (wastewater) has been studied experimentally by many authors and they have reported encouraging results. For example, successful experimental studies on aerobic dephenolisation of refinery wastewater have been reported by Sokol and Korpal [78,79], aerobic digestion of starch wastewater by Rajasimhan and Karthikeyan [80,81] and that of domestic wastewater by Haribabu and Sivasubramanian [82]. Studies on removal of oil from contaminated water in an inverse fluidized column composed of aerogel granules, have been reported by Quevedo et al [83]. The oil recovery reported is around 95%.

Example 3.9

Synthesis of biodiesel (neem oil methyl ester) from neem oil and methyl acetate is to be conducted in an immobilized enzyme inverse fluidized bed bioreactor that handles silica nanoparticles (density = 2100 kg/m^3) containing immobilized lipase enzyme as the catalyst. The oil and the acetate are mixed in the molar ratio of 1:3, then blended with silica nanoparticles (catalyst loading = 0.7) and the resultant suspension is made to flow down a 25.4 mm ID column at the rate of 360 L/hr. Compute the required height of the reactor column, if the substrate conversion desired is 85% and the kinetics of trans-esterification follows equation (2.3.90) of Chapter-2, with $r_{max} = 4.78\times10^{-4}$ mol/(g.s), $K_{m1} = 15.837$ mol/L, $K_{m2} = 1.12$ mol/L and $K_{Si} = 0.053$ mol/L. Take axial dispersion coefficient = 0.03 m^2/s. The property values of oil and acetate are given below:

Property	Neem oil	Methyl acetate
Molecular weight, g/mol	720.0	74.0
Density, kg/m^3	909.0	932.0

Solution :

Since the oil and acetate are mixed in the stoichiometric molar ratio of 1:3, as discussed in Example (3.7), molar concentration of oil in the feed solution shall be,

$$C_{S0} = 1/\left[\left(M_A / \rho_A\right) + \left(3M_B / \rho_B\right)\right]$$

where M_A = molecular weight of neem oil = 720.0 g/mol

M_B = molecular weight of methyl acetate = 74.0 g/mol

ρ_A = density of neem oil = 909 kg/m^3

ρ_B = density of methyl acetate = 932 kg/m^3

Therefore, C_{S0} = 0.9706 mol/L

Now, the density of the feed solution is,

$$\rho_L = C_{S0}\left(M_A + 3M_B\right) = 914.305 \text{ kg/m}^3$$

Since the kinetics of the reaction follows equation (2.3.90) of Chapter – 2,

$$\left(-r_s\right)\left(int\right) = [r_{max}\ C_S C_B] \ / \ [K_{m1} C_S + K_{m2} C_B + K_{m2}\left(C_B^2 / K_{Si}\right) + C_S C_B] \qquad \text{(i)}$$

where C_S = molar concentration of neem oil in substrate solution

C_S = molar concentration of methyl acetate = (3 C_S), as per stoichiometry.

Given that, r_{max} = 4.78×10^{-4} mol/(g.s)

= $(4.78\times10^{-4})\ \left(C_L \rho_L\right)$ mol/(L.s)

= 0.3059 mol/(L.s).

To note that $C_L = 0.7$, $\rho_L = 914.305$ kg/m^3.

Substituting the above value of r_{max} and those of other three kinetic constants in equation (i), we get

$$\left(-r_s\right)\left(int\right) = C_S \ / \ [\ 210.5132\ C_S + 20.9186] \qquad \text{(ii)}$$

The performance equation is that given in equation (3.7.24). Thus,

$$U_L = Q_0 / \left(\pi D^2 / 4\right) = 0.19735 \text{ m/s}$$

To note that Q_0 = (0.360/3600) = 0.0001 m^3/s, D= 0.0254 m. Now, from equation (3.7.22), the fractional liquid holdup ($\in_{fL}$) in the column is

$$\in_{fL} = 1 / \left[C_L\left(\rho_L / \rho_S\right) + 1\right] = 0.7664$$

Therefore, $U_m = \left(U_L / \in_{fL}\right) = 0.2575$ m/s

Given that, $D_{Lf} = 0.03$ m^2/s.

Therefore, after substituting $W\left(z\right) = \left(dC_S / dz\right)$ and $F\left(z\right) = \left(d^2 C_S / dz^2\right) = \left(dW / dz\right)$ in equation (3.7.24), it gets reduced to

$$F\left(z\right) = C_S / [6.3154 C_S + 0.62756] + (8.5833) W\left(z\right) \qquad \text{(iii)}$$

As stated earlier, we shall use the same numerical algorithm presented in Section – 3.2 (equations – 3.2.21 to 3.2.39) for the numerical solution of this performance equation as well. The only difference is that we start the computations from the bottom of the column, where $C_S = C_{Se}$ and proceed upward until the computed value of C_S agrees closely with C_{S0} (feed concentration). The steps involved are summarized below:

Step – 1: $C_{Se} = C_{S0}$ (1 – 0.85) = 0.1456 mol/L

Step – 2: Δz = step size = 0.1 m

Step – 3: $i = 1$

$$C_S(i) = C_S(1) = C_{Se} = 0.1456 \text{ mol/L}$$

$$W(i) = W(1) = 0.0$$

Step – 4: $j = 1$

$$C_S(j) = C_S(i) = 0.1456 \text{ mol/L}$$

$$W(j) = W(i) = 0.0$$

Step – 5: From equation (iii), $F(j) = F(1) = 0.0941$ mol/(L.m^2)

$$m(1) = m(j) = (0.1)\ (0.0941) = 0.0.00941 \text{ mol/(L.m)}.$$

Step – 6: $j = 2$

$$C_S(j) = C_S(2) = C_S(i) + (\Delta z/2)W(i) = C_S(i) = 0.1456 \text{ mol/L}$$

$$W(j) = W(2) = W(i) + (1/2)m(1) = 0.004705 \text{ mol/(L.m)}.$$

Step – 7: From equation (iii),

$$F(j) = F(2) = 0.1345 \text{ mol/(L.m}^2\text{)}$$

$$m(2) = m(j) = (0.1)\ (0.1345) = 0.01345 \text{ mol/(L.m)}.$$

Step – 8: $j = 3$

$$C_S(j) = C_S(3) = C_S(i) + (\Delta z/2)W(i) + (\Delta z/4)m(1)$$

$$= 0.1456 + (0.1/4)\ (0.00941) = 0.145835 \text{ mol/L}$$

$$W(j) = W(3) = W(i) + (1/2)m(2) = 0.006725 \text{ mol/(L.m)}$$

Step – 9: $F(j) = F(3) = 0.151897$ mol/(L.m^2)

$$m(3) = m(j) = 0.0151897 \text{ mol/(L.m)}.$$

Step – 10: $j = 4$

$$C_S(j) = C_S(4) = C_S(i) + (\Delta z)W(i) + (\Delta z/2)m(2)$$

$$= (0.1456) + (0.1/2)\ (0.01345) = 0.1462725 \text{ mol/L}$$

$$W(j) = W(4) = W(i) + m(3) = 0.0151897 \text{ mol/(L.m)}.$$

Step – 11: $F(j) = F(4) = 0.22467$ mol/(L.m²)

$$m(4) = 0.022467 \text{ mol/(L.m)}$$

Step – 12: The substrate concentration at 0.1 m above the bottom of the bed is now given by,

$$C_S(i+1) = C_S(2) = C_S(i) + (\Delta z)W(i) + (\Delta z/6)\left[m(1) + m(2) + m(3)\right]$$

$$= 0.14623 \text{ mol/L}$$

$$W(i+1) = W(2) = W(i) + (1/6)\left[m(1) + 2\,m(2) + 2\,m(3) + m(4)\right]$$

$$= 0.01739 \text{ mol/(L.m)}$$

It can be seen that the computed value of $C_S(i+1)$ is slightly higher than C_{Se}, but much lower than C_{S0}. Computations are to be, therefore, repeated upward starting from Step – 4, after putting $i = i + 1 = 2$. This procedure is to be continued until the computed value of $C_S(i+1)$ becomes equal to or slightly exceeds the feed concentration, C_{S0} (namely, 0.9706 mol/L).

Computations have been performed using a step size of $\Delta z = 0.001$ m and it has been observed that convergence is achieved at $i = 6000$. Therefore,

$$L_f = ((0.001)\ (6000) = \mathbf{6.0\ m}$$

This is the height of the reactor column required to affect 85 % conversion of neem oil to biodiesel (neem oil methyl ester) at a feed flow rate of 360 L/hr. To note that the diameter of the reactor column is only 25.4 mm. By using a multitube (multicolumn) construction composed of say, 5 tubes (channels), the capacity of the reactor could be increased to 1800 L/hr. The low operating cost due to downflow mode of operation remains.

3.8 Design and Analysis of Downflow Stationary Fixed Film (DSFF) Bioreactors

Downflow stationary fixed film (DSFF) bioreactors, relatively recent, promising entries to the family of commercial bioreactors, possess two distinct differences in their construction and operation from other biofilm reactors discussed earlier (such as fluidized bed and semifluidized bed biofilm reactors):

(a) Firstly, unlike fluidized bed and semifluidized bed biofilm reactors, DSFF bioreactors do not use any support particles. The microbial solution forms a biofilm on the inner surfaces of a number of vertical flow channels or pipes, through which the feed solution descends (flows down) and this vertical stack of channels or pipes is enclosed in a large diameter vertical column.

(b) Secondly, the feed solution executes downflow under gravity through the vertical flow channels (see figure 3.12). In this respect, these bioreactors resemble inverse fluidized bed bioreactors in operation.

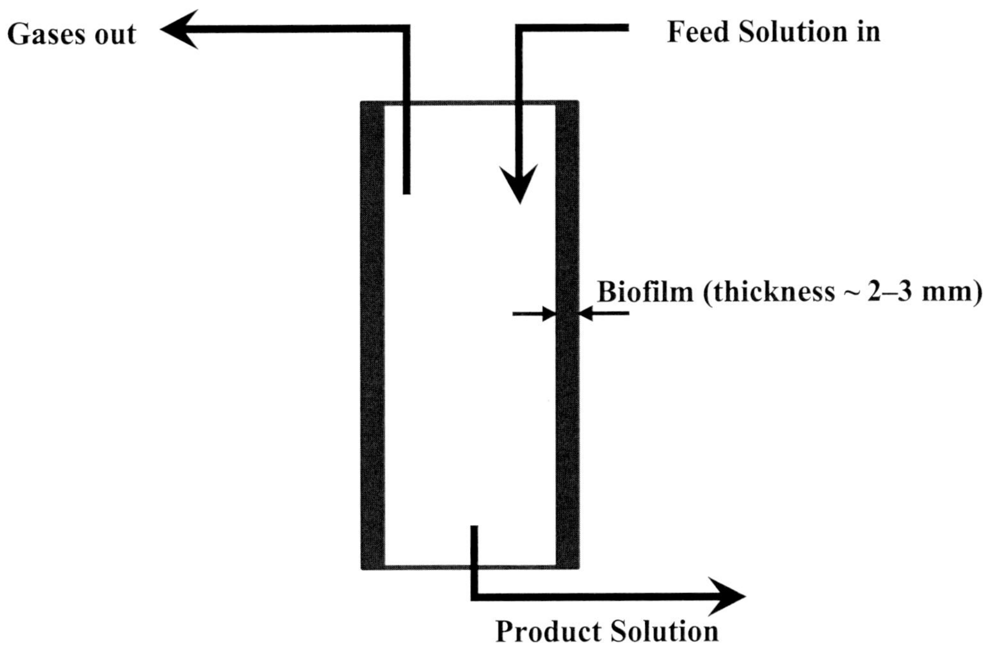

Figure – 3.12

The number of flow channels or pipes could range from 10 to as high as 600 (depending on the capacity of the reactor desired) and they are usually of low diameter (D = 25 to 38 mm), the height of each channel could range from 1.0 to 6.0 m. The channels could be made of metal, metallic alloys, polymer composites or even ceramic (refractory) materials. Biofilms of relatively larger thickness ($\delta = 2 - 3$ mm) could be maintained on the inner surfaces of the channels.

The bioreactor thus consists of a tall, vertical column, accommodating a vertical stack of pipes that are in wall-to-wall contact with each other. Alternately, these vertical flow channels could be fabricated by drilling a number of vertical holes on a solid block (made of clay, redwood or polymer composite), each hole extending over the entire length of the block and thus constituting a flow channel. The entire block is then enclosed in a tall, vertical cylindrical column.

These bioreactors provide reasonably large degree of substrate conversion at reasonably large capacities (feed flow rates), though the volume requirement of these bioreactors is higher than that of fluidized bed or semifluidized bed bioreactors (particularly when large substrate conversion is to be affected at large capacities).

3.8.1 Performance Analysis (Mathematical modeling)

Most of the studies on DSFF bioreactors reported in the literature are those dealing with anaerobic digestion of organic matter. Kennedy and Droste [84] are among the most pioneer investigators who have reported detailed studies on performance characteristics of DSFF bioreactors. They have, however, confined their studies to anaerobic treatment of industrial and domestic wastewater. Both mathematical analysis and experimental investigations have been reported by them. However, the mathematical analysis reported by them does suffer from two serious deficiencies, such as the bioreactor has been assumed equivalent to an ideal CSTR (completely backmixed reactor) and the effectiveness factor that accounts for the resistance to substrate transport into the biofilm has been assumed equal to unity. The influence of gases produced (methane, carbon dioxide) on the rate of bioconversion of organic matter has been neglected and a more or less two phase operation of the reactor has been assumed.

On the other hand, bioconversion has been assumed to occur both in the liquid bulk as well as in the biofilm, the latter being more predominant. The rate of detachment of biofilm from the channel surface and the rate of its attachment (replenishment) have been incorporated in the analysis through appropriately defined attachment and detachment coefficients (k_A, k_D), such that

$$r\,(\text{attachment}) = k_A x_L \tag{3.8.1}$$

$$r\,(\text{detachment}) = k_D\left(f x_f\right) \tag{3.8.2}$$

where k_A, k_D = cell mass attachment coefficient and detachment coefficient respectively, d[1]

x_L, x_f = cell mass concentration in the liquid bulk and that in biofilm respectively, g/L

f = ratio of volume of biofilm to free volume of flow channel

= (volume of biofilm) / (volume of substrate solution)

$= \left[\left(\pi D^2/4\right)-\left(\pi(D-2\delta)^2/4\right)\right] / \left[\pi(D-2\delta)^2/4\right]$

$$= \left[D/(D-2\delta)\right]^2 - 1 \tag{3.8.3}$$

It has been assumed that both the rate of attachment of biofilm and the rate of its detachment are first order in cell mass concentration in the liquid bulk and that in the biofilm respectively.

Since a complex culture of microbes is used in anaerobic wastewater treatment (anaerobic digestion of organic matter), the process does occur in a number of consecutive steps. In the present case, it is assumed that the process occurs predominantly in two steps or stages, such as

$$\text{Organic matter } (S_1) \longrightarrow \text{organic acids } (S_2) \tag{3.8.4}$$

$$\text{Organic acids } (S_2) \longrightarrow CH_4 + CO_2 \tag{3.8.5}$$

The first step is catalyzed by *acidogens* and the second step by *methanogens*. Both of the above steps are assumed to follow Monod's kinetics (with due allowance for endogenous decay), such that

$$\left(-dC_{S1}/dt\right) = \left(\mu_1/Y_1\right)\left[\left(x_1\right)_L + f\left(x_1\right)_f\right] \tag{3.8.6}$$

where $\mu_1 = \left(\mu_{m1} C_{S1}\right)/\left(K_{S1} + C_{S1}\right)$ (3.8.7)

Y_1 = *true* yield coefficient for microbe – 1 (acidogens)

= (mass of microbe – 1 produced) / (mass of S_1 consumed) (3.8.8)

And $\left(-dC_{S2}/dt\right) = \left(\mu_2/Y_2\right)\left[\left(x_2\right)_L + f\left(x_2\right)_f\right] - Y_A\left(\mu_1/Y_1\right)\left[\left(x_1\right)_L + f\left(x_1\right)_f\right]$ (3.8.9) where $\mu_2 = (\mu_{m2} C_{S2}) / (K_{S2} + C_{S2})$ (3.8.10)

Y_2 = *true* yield coefficient for microbe – 2 (methanogens)

= (mass of microbe – 2 produced) / (mass of S_2 consumed) (3.8.11)

Y_2 = yield coefficient for the production of S_2, g. S_2/ g. S_1 (3.8.12)

To note that substrate (S_2) is produced in the first step and is consumed in the second step. The expression for the rate of production of microbe – 1 and that of microbe – 2 shall be, therefore, as follows :

$$d(x_1)_L / dt = (\mu_1 - k_{d1})(x_1)_L - k_{A1}(x_1)_L + k_{D1} f(x_1)_f \tag{3.8.13}$$

and $$d(x_2)_L / dt = (\mu_2 - k_{d2})(x_2)_L - k_{A2}(x_2)_L + k_{D2} f(x_2)_f \tag{3.8.14}$$

where k_{d1}, k_{d2} = endogenous decay coefficient for microbe – 1 and that for microbe – 2 respectively, d[1].

The performance equation of the bioreactor in terms of the concentration of S_1 shall be

$$\tau = \left[(C_{S1})_0 - (C_{S1})_e\right] / (-dC_{S1}/dt)_e \tag{3.8.15}$$

The expression for $(-dC_{S1}/dt)_e$ is that given in equations (3.8.6) and (3.8.7), with C_{S1} having been replaced by $(C_{S1})_e$ and $(x_1)_L$ and $(x_1)_f$ being treated as *steady state concentration* of microbe – 1 in the liquid bulk and in the biofilm respectively.

The performance equation may also be written in terms of the liquid phase concentration of microbe – 1, as

$$\tau = \left[(x_1)_L - (x_1)_0\right] / \left[d(x_1)_L / dt\right]_e \tag{3.8.16}$$

In most cases, $(x_1)_0 = 0$. Also, a mass balance for microbe – 1 in the biofilm yields,

$$k_{A1}(x_1)_L - k_{D1} f(x_1)_f + (\mu_1 - k_{d1}) f(x_1)_f = 0 \tag{3.8.17}$$

At a specified value of reactor space time (τ), the above three equations can be solved simultaneously to obtain the value of $(C_{S1})_e$. The fractional conversion of S_1 attained (α) is then given by

$$\alpha = \left[(C_{S1})_0 - (C_{S1})_e\right] / (C_{S1})_0 \tag{3.8.18}$$

The steady state concentrations of substrate-2 (S_2) and that of microbe-2 can also be estimated from the performance equations of the bioreactor written in terms of $(C_{S2})_e$ and $(x_2)_L$. Thus

$$\tau = \left[(C_{S2})_0 - (C_{S2})_e\right] / (-dC_{S2}/dt)_e \tag{3.8.19}$$

$$= \left[-(C_{S2})_e\right] / (-dC_{S2}/dt)_e \tag{3.8.20}$$

To note that $(C_{S2})_0 = 0$. The expression for $(-dC_{S2}/dt)_e$ is that given through equations (3.8.9) and (3.8.10). Similarly,

$$\tau = \left[(x_2)_L - (x_2)_0\right] / \left[d(x_2)_L / dt\right]_e \tag{3.8.21}$$

In most cases, $(x_2)_0 = 0$. The expression for $\left[d(x_2)_L / dt\right]_e$ is that given through equations (3.8.14) and (3.8.10). Now, a mass balance for microbe-2 in the biofilm yields,

$$k_{A2}(x_2)_L - k_{D2} f(x_2)_f + (\mu_2 - k_{d2}) f(x_2)_f = 0 \tag{3.8.22}$$

By solving the above three equations simultaneously, we obtain the values of steady state concentrations of substrate-2 and microbe-2, namely $(C_{S2})_e$, $(x_2)_L$ and $(x_2)_f$. To note that the values of steady state cell mass concentrations in the biofilm, such as $(x_1)_f$ and $(x_2)_f$, in this case shall depend on the magnitudes of the attachment and detachment coefficients (k_A, k_D) and these coefficients are experimentally determined. They could also be functions of reactor space time (τ) and the feed concentration (see the numerical example discussed at the end of this section).

The above-presented performance analysis of DSFF bioreactor does involve approximations. However, it is mathematically simpler and on occasions, it does predict dependable values of design parameters.

A more rigorous analysis based on the PFDR approach has been reported by Narayanan [85] and Narayanan and Pandey [86]. An appropriately defined effectiveness factor (η) has also been incorporated in the analysis. If we assume the bioconversion under consideration follows Haldane – Andrews kinetics, then the performance equation of the bioreactor shall be

$$-U_L(dC_S / dz) + D_{Ld}\,(d^2C_S / dz^2) = \eta(-r_S)(int)$$

$$= \eta\mu_m(app)C_S / \left(K_S + C_S + C_S^2 / K_{Si}\right) \tag{3.8.23}$$

where $\mu_m(app) = (\mu_m / Y)(fx_f)$ (3.8.24)

$$U_L = Q_O / \left(\pi(D - 2\delta)^2 / 4\right) \tag{3.8.25}$$

f = defined in equation (3.8.3)

The boundary conditions are

BC – 1: At $z = 0$ (at the bottom of the flow channel), $C_S = C_{Se}$ (3.8.26)

BC – 2: At $z = L$ (at the top of the flow channel), $C_S = C_{S0}$ (3.8.27)

The axial dispersion coefficient (D_{Ld}) may be estimated from any of the reported correlations such as that proposed by Levenspiel (see equation – 3.2.3 of Section 3.2) or determined experimentally through RTD studies (tracer experiments).

The effectiveness factor (η) may be computed from any of the correlations discussed in Chapter 2. In this connection, the characteristic dimension (L^*) is to be defined as follows :

$$L^* = \text{(volume of biofilm) / (surface area of biofilm)}$$

$$= \left[(D^2) - (D - 2\delta)^2\right] / \left[4(D - 2\delta)\right] \tag{3.8.28}$$

The performance equation can be seen to be identical with equation (3.2.18) of Section (3.2). It is also analogous to the performance equation for fluidized bed biofilm reactors (Section 3.5) and that for inverse fluidized bed biofilm reactors (equation – 3.7.18). The numerical algorithm for the solution of the performance equation shall also be , therefore, analogous (same as that described through equations – 3.2.21 to 3.2.39 in Section 3.2). However, as in the case of inverse fluidized bed bioreactors (Section – 3.7), here also the computations are to be started from the bottom of the reactor column, where $C_S = C_{Se}$ and proceeded upward until the computed value of C_S becomes equal to or slightly larger than C_{S0} .

For the convenience of the readers, the scheme is briefly summarized below:

The scheme involves computation of fractional substrate conversion (α) attained in a single flow channel of height, L. As stated earlier, computations are started from the bottom of the reactor (where $z = 0$, $C_S = C_{Se}$) and continued upward until the computed value of C_S is equal to C_{S0} (or slightly exceeds C_{S0}).

Step 1: Select the step size (Δz). For example, let Δz = 1.0 mm = 0.001 m.

Step 2: Compute the value of $\mu_m(app)$ from equations (3.8.24).

Step 3: Select the fractional substrate conversion desired (α). For example, let $\alpha = 0.80$.

Step 4: Compute substrate concentration in the exit stream as, $C_{Se} = C_{S0}(1-\alpha)$.

Step 5: Now, solve the performance equation for the bioreactor (equation – 3.8.23) using the same procedure as that employed for the design of fluidized bed / inverse fluidized bed biofilm reactor. Thus,

$$\text{Put } i = 1$$

$$C_S(1) = C_{Se}$$

$$W(1) = 0.0$$

Step 6:

$$\text{Put } j = 1$$

$$C_S(j) = C_S(i)$$

$$W(j) = W(i)$$

Step 7: Compute the characteristic dimension (L^*) of the flow channel from equation (3.8.28) and then compute the effectiveness factor, $\eta(j)$, from any of the correlations discussed in Chapter 2.

Step 8: Compute $F(j)$ from equations (3.2.20). However, since the bioconversion has been assumed to follow Haldane – Andrews kinetics (and not Monod's kinetics), the expression for $F(j)$ or $F(z)$ shall be slightly different. Thus, after substituting $W(z) = (dC_S/dz)$ and $F(z) = (dW/dz) = (d^2C_S/dz^2)$ in equation (3.8.23), we get

$$F(z) = \left[\eta(z)\mu_m(app)/D_{Ld}\right]C_S/\left(K_S + C_S + C_S^2/K_{Si}\right) + \left(U_L/D_{Ld}\right)W(z) \qquad (3.8.29)$$

Now, compute $F(j)$ from the above equation, keeping in mind that $F(j)$ is nothing but the value of $F(z)$ at $C_S = C_S(j)$.

Step 9: Compute $m(j)$ as

$$m(j)=(\Delta z)F(j)$$

If j = 4, GO TO Step – 11.

Step 10: Put $j = j+1$.

Compute $C_S(j)$ and $W(j)$ from equations (3.2.32) to (3.2.36). Then, GO TO Step – 7 (repeat computations starting from Step – 7).

Step 11: Compute $C_S(i+1)$ and $W(i+1)$ from equations (3.2.37) and (3.2.38).

Step 12: Compute $L(cal)$ as,

$$L(cal)= (i)(\Delta z)$$

Step 13: If $L(cal) \geq L$, GO TO Step – 14.

Otherwise, put $i = i+1$ and repeat the computations starting from Step – 6.

Step 14: Compute $C_{S0}(cal)$ as,

$$C_{S0}(cal)=C_S(i+1)$$

Step 15: If $C_{S0}(cal) \geq C_{S0}$, proceed to Step – 16. Otherwise, re-choose α and repeat the computations starting from Step – 4.

Step 16: Print the values of $Q_0, \alpha, L, C_{S0}, C_{Se}$.

3.8.2 Advantages and Industrial Applications

As stated earlier, anaerobic digestion of organic wastes is the one that has been maximum experimented in DSFF bioreactors. The reported results are, no doubt, encouraging [87].

Hall et.al [88] have reported studies on anaerobic digestion of pharmaceutical wastewater in a DSFF bioreactor and also in a fluidized bed biofilm reactor. The have observed that both of them provide comparable performance (around 70% destruction of BOD) under the same operating conditions, but the operation of DSFF bioreactor was more stable and smooth, in spite of the fact that the feed solution had a high concentration of suspended solids.

Narayanan and Pandey [86] have successfully investigated synthesis of lactic acid from molasses and de-proteinized cheese whey in DSFF bioreactors. They have reported that more than 70% conversion of sucrose (present in molasses) to lactic acid could be obtained in a DSFF bioreactor composed of a 25.4 mm ID flow channel, 6.0 m in height, coated with a biofilm of *Enterococcus faecalis*, at a feed flow rate of 520 to 550 L/hr. Also, by using a multi-column (multi-tube) construction composed of 20 vertical flow channels, the capacity of the bioreactor could be increased to 10,400 to 11,000 L/hr.

The simplicity of construction and low operating cost (due to downflow mode of operation) are also to be kept in mind.

Example 3.10:

A DSFF bioreactor composed of two hundred vertical flow channels or pipes each of 25 mm in diameter and 3.0 m in length is to be used for the anaerobic treatment of wastewater from a food processing unit. The inside surface of each channel is coated with a 3 mm thick biofilm and the feed solution with a BOD of 100 mg /L is admitted from the top at the rate of 250 L/d.

If the bioconversion follows the scheme represented by equations (3.8.4) and (3.8.5) and the kinetic parameters associated with are those listed in the table below, estimate the fractional destruction of BOD attained. Assume the reactor to be equivalent to an ideal CSTR and neglect resistance to substrate transfer into the biofilm. What shall be the cell mass concentration in the biofilm?

Parameter	Microbe – 1 (acidogens)	Microbe – 2 (methanogens)
μ_m, d^{-1}	0.453	0.45
K_S, g /L	0.275	0.0397
k_d, d^{-1}	0.022	0.075
k_A, d^{-1}	0.033	0.033
k_D, d^{-1}	0.030	0.030
Y, g /g	0.2 g.microbe – 1 /g. S_1	0.13 g.microbe – 2 /g. S_2

Take yield coefficient for production of $S_2 = 0.75$ g. S_2 / g. S_1. Assume constant density system.

Solution:

The total active volume of the bioreactor (that is occupied by the substrate solution) is

$$V = (\pi/4)(D-2\delta)^2 L(N)$$

where $D = 25$ mm $= 0.025$ m

$\delta = 3.0$ mm $= 0.003$ m

$L = 3.0$ m

$N =$ number of channels $= 200$

Therefore, $V = 0.170117$ m^3 = 170.117 L

And $\tau = (170.117 / 250) = 0.68$ d

Now, from equation (3.8.3),

$$f = \left[D/(D-2\delta)\right]^2 - 1 = (25/19)^2 - 1 = 0.7313$$

From equation (3.8.17),

$$k_{A1}(x_1)_L - k_{D1} f (x_1)_f + (\mu_1 - k_{d1}) f (x_1)_f = 0 \quad \text{(i)}$$

Given that,

$k_{A1} = 0.033$ d^{-1}

$k_{D1} = 0.03$ d^{-1}

$k_{d1} = 0.022 \text{ d}^{-1}$

Also, $f = 0.7313$ (computed above)

Substituting the above values in equation (i) and rearranging, we get

$$(\mu_1\beta) - 0.052\beta + 0.045125 = 0 \quad \text{(ii)}$$

where $\beta = (x_1)_f / (x_1)_L$.

Now, from equation (3.8.13),

$$\left[d(x_1)_L / dt\right]_e = (\mu_1 - k_{d1})(x_1)_L - k_{A1}(x_1)_L + (k_{D1}f)(x_1)_f$$

$$= (\mu_1 - 0.022)(x_1)_L - (0.033)(x_1)_L + (0.021939)(x_1)_f \quad \text{(iii)}$$

And equation (3.8.16) predicts,

$$(1/\tau) = \left[d(x_1)_L / dt\right]_e / (x_1)_L$$

Substituting equation (iii) and the value of τ in the above equation, we get, on re-arrangement,

$$\mu_1 + (0.021939)\beta - 1.5256 = 0 \quad \text{(iv)}$$

Solving equations (ii) and (iv) simultaneously, we get

$$\beta = 67.1517$$

$$\mu_1 = 0.05133 \text{ d}^{-1}$$

Now, from equation (3.8.7),

$$\mu_1 = \mu_{m1}(C_{S1})_e / \left[K_{S1} + (C_{S1})_e\right]$$

where $\mu_{m1} = 0.453 \text{ d}^{-1}$

$K_{S1} = 0.275$ g /L

and $\mu_1 = 0.05133 \text{ d}^{-1}$ (computed above).

Therefore, $(C_{S1})_e = 0.035$ g /L

The fractional destruction of BOD (α) achieved is,

$$\alpha = (0.1 - 0.035) / (0.1) = \mathbf{65\ \%}$$

Let us now proceed to compute the cell mass concentration in the biofilm. Thus, from equation (3.8.15),

$$\tau = \left[(C_{S1})_0 - (C_{S1})_e\right] / (-dC_{S1} / dt)_e$$

or, $(-dC_{S1} / dt)_e = (0.1 - 0.035) / (0.68) = 0.0956$ g /(L.d)

Substituting the above value of rate of conversion of S_1 in equation (3.8.6), we get

$$0.0956 = (\mu_1 / Y_1)\left[(x_1)_L + f\beta(x_1)_L\right]$$

Substituting $\mu_1 = 0.05133 \text{ d}^{-1}$, $Y_1 = 0.2$, $f = 0.7313$ and $\beta = 67.1517$ and solving, we get

$$\left(x_1\right)_L = 7.4337 \text{ mg /L}$$

$$\left(x_1\right)_f = \beta\left(x_1\right)_L = 500 \text{ mg /L}$$

This is the concentration of microbe – 1 in the biofilm. The value of $\left(x_2\right)_f$ can be computed in a similar way. Thus, substituting the known values of kinetic constants in equation (3.8.22) and re- arranging,

$$\left(\mu_2 \beta\right) - 0.105\beta + 0.045125 = 0 \qquad \text{(v)}$$

Similarly, from equations (3.8.21) and (3.8.14),

$$\mu_2 + \left(0.021939\right)\beta - 1.5786 = 0 \qquad \text{(vi)}$$

To note that here, $\beta = \left(x_2\right)_f / \left(x_2\right)_L$. On solving simultaneously,

$$\beta = 67.1517$$

$$\mu_2 = 0.1043 \text{ d}^{1}$$

The value of β is thus the same as that for microbe – 1. From equation (3.8.10), therefore,

$$\left(C_{S2}\right)_e = 0.012 \text{ g /L}$$

And from equation (3.8.20),

$$\left(-dC_{S2} / dt\right)_e = -0.017647\, g / \left(L.d\right)$$

The minus sign indicates that it is the rate of consumption of S_2. Substituting in equation (3.8.9) and solving, we get,

$$\left(x_2\right)_f = 90.478 \text{ mg /L}$$

$$\left(x_2\right)_L = 1.3474 \text{ mg /L}$$

3.9 Design and Analysis of UASB (Upflow Anaerobic Sludge Blanket) Bioreactors

UASB (Upflow Anaerobic Sludge Blanket) bioreactors constitute a novel design of bioreactors, which entered the world market relatively late, but attained popularity quite fast.

As the name implies, they are restricted to anaerobic processes (presently to anaerobic wastewater treatment processes) and could be classified under *sludge bed reactors*. These bioreactors differ from conventional packed bed, fluidized bed or DSFF bioreactors in the sense that they do not use any support particles and there is no biofilm attached to any solid surface. The reactor principally consists of a tall, vertical column with a liquid distributor at the bottom and a gas cap and disengagement space at the top (there are hardly much internals). The column is fed with a specific amount of complex microbial culture or microbial sludge, cultured within a specific volume of proposed feed solution (could be made nutrient – rich in advance) and this is kept undisturbed for a couple of months (2 to 6 months). Microbial cells form tiny granules (sludge granules) of around 0.5 – 2.0 mm in size and these form a sludge bed of height L_P at the bottom of the column (see figure 3.13).

As the feed solution is admitted from the bottom and it rises through this sludge bed, anaerobic digestion of dissolved organic matter present in the feed solution occurs, the gases formed during the process (methane and carbon dioxide) move up in the form of bubbles, each bubble carrying a part of sludge granules with it in the form of a wake or tail. A *sludge blanket* of height L_f thus gets developed above the sludge bed, which shall consist of sludge granules suspended in the ascending stream of feed solution (substrate solution) and a number of gas bubbles each carrying some sludge granules with it in the form of a wake (wake formation does occur in many conventional gas-fluidized beds as well). The sludge blanket thus assembles a three phase fluidized bed, with the difference that the gas bubbles present carry a part of the sludge granules in the wakes attached to them.

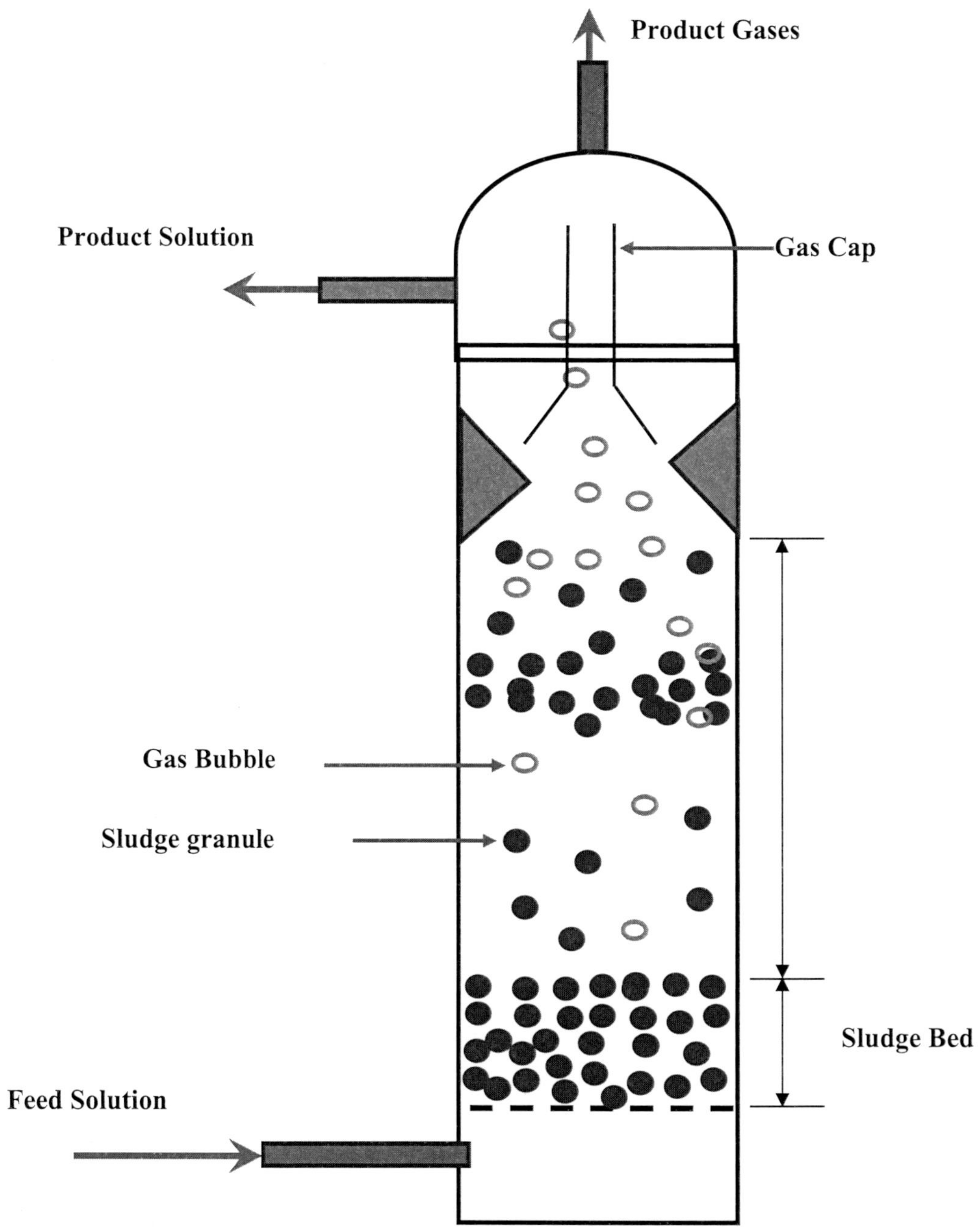

Figure 3.13: Schematic of UASB bioreactor

Once the sludge bed of height L_P and the sludge blanket of height L_f have been formed (which could take 2 to 8 months), the bioreactor operates steadily with feed solution admitted continuously from the bottom at the specified flow rate (Q_0), the gases leaving from the top through the top outlet and the treated effluent through a side peripheral outlet at the top (see figure – 3.13). The gas cap at the top helps in inhibiting entrainment loss of granules in the outgoing fluid or gas stream.

As stated earlier, this bioreactor is presently restricted to anaerobic processes that handle a complex culture of microbes. Aerobic microbes have not been observed to exhibit this property of granule formation so far. Also, a single culture of microbes is found incapable of forming such sludge granules (due to reasons discussed subsequently).

The bioreactor, by itself, is simple to construct and install. It can be operated at high capacities (high feed flow rates) and it affects significantly high BOD destruction even with high strength feedstocks.

The main drawback of the UASB bioreactor is that it demands enormously large startup time (2 to 8 months). This is due to the fact that the formation of microbial sludge granules is a slow, evolutionary, biological process (and not a physical agglomeration or flocculation of microbial cells).

3.9.1 Mechanism of Sludge granule formation

The concept of sludge bed or sludge blanket reactors was first introduced by Lettinga and coworkers [89] during the late seventies. They observed that anaerobic microbial sludges have a tendency to agglomerate and form compact sludge granules (though the process is slow and takes months to complete). Subsequently, a large number of researchers investigated this phenomenon and came forward with diversified theories and possible mechanisms. Ultimately, it got more or less established that these sludge granules are micro-colonies or consortia of different varieties of microbes and sludge granulation is a spontaneous, biological process [90]. To put it more clearly, we know that anaerobic digestion of organic matter is a complex process that involves different types of microbial species such as hydrolytic microbes, acidogens, acetogens and methanogens. These different groups of microbes conduct sequential processes such as decomposition (hydrolysis) of organic matter into simple sugars or monomers (hydrolytic microbes), conversion of these monomers into organic acids, alcohols (acidogens), then to acetates, hydrogen and carbon dioxide (acetogens) and finally to methane and carbon dioxide (methanogens). To ensure that the substrate handled is being subjected to these transformations efficiently in the proper sequence, it is essential that these different types of microbial cells be spatially organized in an optimum fashion such that there shall be a close, synergistic relationship among them. As a result, the microbial cells shall have a natural tendency to gather together so as to form stable, contiguous multicellular associations (microcolonies or consortia), which we call *the sludge granules*. The driving force behind the sludge granulation is thus the need for developing efficient bacterial balance and an optimum combination of multiple microbial species so that their biological functions get performed in the most optimum manner, in the precise sequence. This theory is substantiated by the fact that these sludge granules are much more stable than individual suspended cells and are more resistant to presence of toxins such as heavy metals and sulfur compounds.

Being such a slow, evolutionary process, formation of stable sludge granules would cost a lot of time and that explains the tremendously long startup time demanded by UASB reactors. This theory also explains why this phenomenon is exhibited by complex culture of microbes only, since such spatial arrangement or organization of different classes of microbes becomes necessary only in a complex culture.

It must be kept in mind that suspended growth of microbes occurs in UASB bioreactors. As there are no support particles, no attached growth of microbes occurs. As dead cells fall off from granules, they are immediately replaced by new cells. The size and organization of sludge granules thus remain more or less constant and uniform throughout the operation of the bioreactor.

As stated earlier, physiochemical factors such as surface charge and hydrophobicity of bacterial surface, addition of flocculating or coagulating agents, secretion of extracellular polymers (ECPs, which act as coagulants) etc. would contribute to speeding up of the granulation process, but are not deciding or controlling parameters. For example, since microbial cells have negatively charged surfaces under usual pH conditions, addition of multivalent metallic ions such as those of calcium, magnesium and iron (no doubt, in controlled amounts) would bring down the electrostatic repulsion between bacterial cells and would thereby increase their tendency to agglomerate. Similarly, addition of a popular flocculating agent or coagulant has been observed to augment the rate of microbial granulation. An example is the addition of polymer Chitosan, which has been reported to make formation of sludge granules 2 to 2.5 times faster. The extracellular polymers (ECPs) secreted by microbes also aid in the cohesion and adhesion of microbial cells and thereby improve the structural strength and stability of microbial granules.

Hydrophobicity of bacterial surface also plays a role in sludge granulation. High surface hydrophobicity means a decrease in Gibbs energy and this promotes cell to cell interaction. Acetogens and methanogens exhibit high surface hydrophobicity, while acidogens display more or less hydrophilic characteristics. This explains why acidogenic cells are usually found in the outer layer of sludge granules.

Filamentous bacteria, if present, do initiate and accelerate granule formation. These bacteria have a tendency to attach to precursors and form a three-dimensional, branched network and other bacteria get entrapped in this network. These structural aggregates soon grow in size due to multiplication of entrapped bacteria and become denser and get spherically shaped due to the hydrodynamic shear forces generated by the upflowing liquid solution and product gases.

3.9.2 Performance Analysis

In spite of their simplicity in construction, installation and operation, the performance analysis of these bioreactors is relatively complex. Though laboratory scale investigations on the performance of these bioreactors have been reported by many authors [92-94], no attempt has been made to study the dependence of reaction zone dimensions (L_P, L_f) on the process / system parameters such as the feed flow rate, feed composition and sludge granule characteristics and to develop correlations for the same.

Since the processes handled in these bioreactors are mainly those dealing with anaerobic digestion of organic matter, the type of microbial sludge used does not vary much from one reactor to another and so are the kinetics of bioconversion. It is possible to fix the height of the sludge bed at a prescribed value by adjusting the amount of microbial sludge per unit volume of feed solution used at the startup. This height usually varies from 1.5 to 2.5 m, height of sludge blanket developed (L_f) being (1.5 L_P) to (2 L_P). The sludge bed thus occupies one-third to 40% of reactor volume, the rest 60% to two-third being occupied by the sludge blanket. Larger share of bioconversion occurs in the sludge blanket. This has been found to be true in many commercial bioreactors under operation.

Once the values of L_P and L_f are thus more or less finalized, we can proceed to analyze the performance of the bioreactor. There are, in essence, two approaches possible:

Model – I (PFR – CSTR Combination model):

This, in fact, is an approximate model but is mathematically simpler (due to obvious reasons). If we approximate the sludge bed to a PFR, then,

$$\tau_P = \int_{C_{Sb}}^{C_{SO}} \left[1/\eta(-r_S)(int)\right] dC_S \tag{3.9.1}$$

where, τ_P = space time of PFR (sludge bed) = $(V_P \in_{PL})/Q_0$ (3.9.2)

V_P = volume of sludge bed = $(\pi D^2/4)L_P$ (3.9.3)

Q_0 = volumetric flow rate of feed solution

$\in_L$ = fractional liquid holdup in the sludge bed

C_{Sb} = substrate concentration at the top of the sludge bed (entrance to sludge blanket).

It shall not be too incorrect to assume that accumulation of product gases within the sludge bed is negligibly small and accordingly,

$$\in_{PL} \cong \in_P = 0.35 - 0.40 \tag{3.9.4}$$

It has also been assumed that the mass transfer resistance to substrate transfer in the liquid bulk (from liquid bulk to granule surface) is negligible and consequently, $C_{SP} \approx C_S$.

The above performance equation is to be solved through numerical integration (using Simpson's rule or trapezoidal rule) after substituting the expression for intrinsic rate ($-r_S$)(int) and the value of c_{sb} is to be determined by trial.

For example, let the bioconversion follow Monod's kinetics. Then,

$$(-r_S)(int) = \mu_m(app) C_S/(K_S + C_S) \tag{3.9.5}$$

where $\mu_m(app) = (\mu_m/Y)x_f(1-\in_P)/\in_{PL}$

$$= (\mu_m/Y)x_f(1-\in_{PL})/\in_{PL} \tag{3.9.6}$$

Here, x_f is the concentration of microbial cells in each sludge granule and shall be equal to the apparent density of the granule. For computing the value of the effectiveness factor (η), we may utilize any of the correlations discussed in Chapter – 2.

To note that the characteristic dimension of sludge granule (L^*) is given by

$$L^* = (d_P/6) \tag{3.9.7}$$

Equation (3.9.1) may be, therefore, rearranged to the following form,

$$\left[\mu_m(app)(A \in_{PL})/Q_0\right]L_P = \int_{C_{Sb}}^{C_{S0}} F(C_S)\,dC_S \qquad (3.9.8)$$

where $F(C_S)=(K_S+C_S)/(\eta C_S)$ (3.9.9)

A = column cross-sectional area $=(\pi D^2/4)$ (3.9.10)

The value of C_{Sb} is to be estimated from equation (3.9.8), by evaluating the integral numerically using Simpson's rule or trapezoidal rule. This invariably demands a trial and error procedure. A value of C_{Sb} is first assumed, the integral evaluated numerically and then L_P computed from equation (3.9.8). This value of L_P is now checked against the prescribed value of L_P and if the two differ significantly, computations are repeated with an alternate value of C_{Sb}.

If the sludge blanket is assumed equivalent to a three phase, ideal CSTR, then the performance equation for the same shall be

$$\tau_m = (C_{Sb} - C_{Se})/[\eta(-r_{Se})(int)] \qquad (3.9.11)$$

where $\tau_m = (V_m \in_{fL}/Q_O)$ (3.9.12)

V_m = total volume of the sludge blanket $=(\pi D^2/4)L_f$ (3.9.13)

η_e = effectiveness factor at $C_S = C_{Se}$

Based on Monod's kinetics,

$$(-r_{Se})(int) = \mu_m(app)C_{Se}/(K_S + C_{Se}) \qquad (3.9.14)$$

where $\mu_m(app)=(\mu_m/Y)x_f(1-\in_f)/\in_{fL}$ (3.9.15)

$\in_f = (\in_{fL} + \in_{fg})$ (3.9.16)

$\in_{fL}, \in_{fg}$ = fractional liquid holdup and gas holdup respectively in sludge blanket.

The values of $\in_{fL}$ and $\in_{fg}$ may be estimated from any of the experimental correlations presented in Section (3.5), under three phase fluidized bed bioreactors. Here also, a degree of approximation exists. The sludge blanket has been assumed to be a three phase fluidized bed and the fraction of sludge granules carried off in the wakes of gas bubbles is not separately accounted for.

A numerical example based on the above analysis has been illustrated at the end of this section.

Model – II: PFDRs - in- Series Model:

This, no doubt, is more accurate and more rigorous approach to analyze the performance of UASB bioreactors. In this case, it is being assumed that the sludge blanket is equivalent to a PFDR (PFDR – 1) with relatively high degree of axial dispersion (axial dispersion coefficient = D_{Lf} = 0.03 to 0.033 m^2/s) and the sludge bed below is PFDR – 2 with D_{LP} = 0.007 to 0.0075 m^2/s. The performance equations shall be then similar to those for the semifluidized bed bioreactor (discussed in Section – 3.6). Thus, for PFDR – 1(sludge blanket), the performance equation is

$$-U_L\left(dC_S / dz\right)+D_{Lf}\left(d^2C_S / dz^2\right)= \eta\left(-r_S\right)\left(int\right) \tag{3.9.17}$$

where U_L = superficial velocity of feed solution through the reactor column

$$=\left(Q_0 / A\right) \tag{3.9.18}$$

D_{Lf} = axial dispersion coefficient in sludge blanket = 0.03 – 0.033 m^2/s

The governing boundary conditions are,

BC – 1: At z = 0 (at the top of the sludge blanket), $C_S = C_{Se}$ (3.9.19)

BC – 2: At $z = L_f$ (at the top of the sludge bed or inlet to sludge blanket), $C_S = C_{Sb}$ (3.9.20)

Similarly, the performance equation for PFDR – 2 (sludge bed) is

$$-U_L\left(dC_S / dz\right)+D_{LP}\left(d^2C_S / dz^2\right)= \eta\left(-r_S\right)\left(int\right) \tag{3.9.21}$$

where D_{LP} = axial dispersion coefficient in sludge bed = 0.007 – 0.0075 m^2/s

The boundary conditions are,

BC – 1: At z = 0 (at the top of the sludge bed), $C_S = C_{Sb}$ (3.9.22)

BC – 2: At $z = L_P$ (at the feed inlet), $C_S = C_{S0}$ (3.9.23)

Mass transfer resistance to substrate transfer in the liquid bulk has been neglected in this case also, such that $C_{SP} \approx C_S$.

The numerical algorithm for the solution of the above two performance equations shall also be exactly similar to that employed in the case of semifluidized bed bioreactors (Section – 3.6). A brief summary of the same is presented below :

Let us assume that the bioconversion follows Monod's kinetics. Also, the height of the sludge bed (L_P) and that of the sludge blanket formed (L_f) have already been specified.

Step 1: Select the step size (Δz). For example, let $\Delta z = 1.0$ mm = 0.001 m.

Step 2: Assume a value of the fractional substrate conversion (α). For example, let $\alpha = 0.80$.

Step 3: Compute the substrate concentration in the product stream as

$$C_{Se} = C_{S0}\left(1-\alpha\right) \tag{3.9.24}$$

Step 4: Compute the value of $\mu_m\left(app\right)$ for the sludge blanket from equation (3.9.15).

Step 5: Now, proceed to solve the performance equation for the sludge blanket (PFDR – 1) using the same numerical algorithm described in Section (3.2). Thus,

$$\text{Put } i = 1$$

$$C_S\left(1\right) = C_{Se}$$

$$W\left(1\right) = 0.0$$

Step 6: Put j = 1

$$C_S\left(j\right) = C_S\left(i\right)$$

$$W(j) = W(i)$$

Step 7: Compute the effectiveness factor, $\eta(j)$ from any of the correlations discussed in Chapter – 2 and $F(j)$ from equation (3.2.20), after replacing D_L by D_{Lf}.

Step 8: Compute $m(j)$ as

$$m(j) = (\Delta z) F(j) \tag{3.9.25}$$

If j = 4, GO TO Step 10.

Step 9: Put $j = j + 1$

Compute $C_S(j)$ and $W(j)$ from equations (3.2.32) to (3.2.36).

GO TO Step – 7 (repeat the computations starting from Step 7).

Step 10: Compute $C_S(i+1)$ and $W(i+1)$ from equations (3.2.37) and (3.2.38) respectively.

Step 11: Compute $L_f(cal)$ as,

$$L_f(cal) = (i)(\Delta z) \tag{3.9.26}$$

Step 12: If $L_f(cal) \geq L_f$, GO TO Step – 13. Otherwise, put $i = i + 1$ and repeat the computations starting from Step – 6.

Step 13: Compute C_{Sb} as

$$C_{Sb} = C_S(i+1) \tag{3.9.27}$$

Print C_{Sb}.

Step 14: Compute the value of $\mu_m(app)$ for the sludge bed from equation (3.9.6).

Step 15: Now, proceed to solve the performance equation for the sludge bed (PFDR – 2) similarly. Thus,

$$\text{Put } i = 1$$

$$C_S(1) = C_{Sb}$$

$$W(1) = 0.0$$

Step 16: Put $j = 1$

$$C_S(j) = C_S(i)$$

$$W(j) = W(i)$$

Step 17: Compute $\eta(j)$ from any of the correlations discussed in Chapter – 2 and $F(j)$ from equation (3.2.20), after replacing D_L by D_{LP}.

Step 18: Compute $m(j)$ as, $m(j) = (\Delta z) F(j)$.

If $j = 4$, GO TO Step – 20.

Step 19: Put $j = j+1$.

Compute $C_S(j)$ and $W(j)$ from equations (3.2.32) to (3.2.36).

GO TO Step – 17 (repeat the computations starting from Step 17).

Step 20: Compute $C_S(i+1)$ and $W(i+1)$ from equations (3.2.37) and (3.2.38) respectively.

Step 21: Compute $L_P(cal)$ as,

$$L_P(cal) = (i)(\Delta z) \tag{3.9.28}$$

Step 22: If $L_P(cal) \geq L_P$, GO TO Step – 23.

Otherwise, put $i = i+1$ and repeat the computations starting from Step – 16.

Step 23: Compute $C_{S0}(cal)$ as

$$C_{S0}(cal) = C_S(i+1) \tag{3.9.29}$$

Step 24: If $C_{S0}(cal) \geq C_{S0}$, proceed to Step – 25. Otherwise, re-choose α and repeat the computations starting from Step – 3.

Step 25: Print α. Print also the values of Q_0, U_L, L_f, L_P, C_{Sb}, C_{S0}, C_{Se}.

Kalyuzhnyi and Fedorovieh [93] have proposed a dispersed flow model for UASB bioreactors. They have, however, assumed the entire reactor to be equivalent to a PFDR and have neglected resistance to substrate transfer into the sludge granule (the effectiveness factor (η) has been assumed to be equal to unity). The study has also been restricted to processes that follow Monod's kinetics with little endogenous decay. Narayanan and Vikas [92] have proposed an improved analysis, though equating the entire reactor to a PFDR, but including the resistance to substrate transport into sludge granules through an approximately defined effectiveness factor (η). The variation of axial dispersion coefficient (D_L) and gas velocity (U_g) with axial distance (z) has also been accounted for. Lately, Narayanan and coworkers [87,94] have come forwarded with a well-tested, rigorous software package that predicts the performance of UASB bioreactors based on the above - described PFDRs – in – series model. Both Monod's and Haldane – Andrews kinetics have been considered.

3.9.3 Advantages and Industrial Applications

The specific advantages / merits of UASB bioreactors may be summarized as given below:

(i) The construction and installation of these bioreactors are relatively simple,

(ii) they provide substantially high degree of BOD destruction (more than 85%), even with high strength feed stocks and also at high feed flow rates (high reactor capacities),

(iii) They are efficient to handle large volumes of high strength feed stocks, with reasonably low reactor volume requirement.

The chief disadvantage of these bioreactors is the excessively large startup time demanded by them, which, as described above, is invariable and cannot be avoided. These reactors are also restricted to anaerobic bioprocesses handling complex culture of microbes.

Today more than 800 UASB bioreactors are in successful operation all over the world, out of which about 80 are in India. Most of them handle anaerobic digestion of wastewater (industrial effluents) discharged from distilleries, paper and pulp industries, tanneries, pharmaceutical and textile industries, dairy farms and food processing industries. An elaborate survey on the performance characteristics of UASB bioreactors operating in different parts of the world has been presented by Seghezzo et. al [91]. The data reported by them demonstrate that all of these bioreactors are in stable, continuous operation and are consistently providing high degree of BOD / COD removal (more than 85 – 90 %) at elevated capacities.

UASB bioreactors have a promising future. Extending the application of these bioreactors to alternate anaerobic processes (other than wastewater treatment) such as to hydrogen production etc. are worthwhile areas for investigation (many research projects in these areas are in progress in different parts of the world). Experimental studies are also very much worthwhile to develop correlations for predicting the sludge bed height and the height of the sludge blanket formed under specified operating conditions.

Expanded granular sludge bed (EGSB) bioreactors from a close analogy to UASB bioreactors. These bioreactors are relatively taller (larger in height) and they operate with larger fluid velocities. In other words, the feed solution is admitted from the bottom at a flow rate / velocity that is much higher than that employed in a typical UASB bioreactor and consequently, all the sludge granules get completely fluidized and thereby form an expanded bed. There shall hardly be any sludge blanket (see figure 3.14).

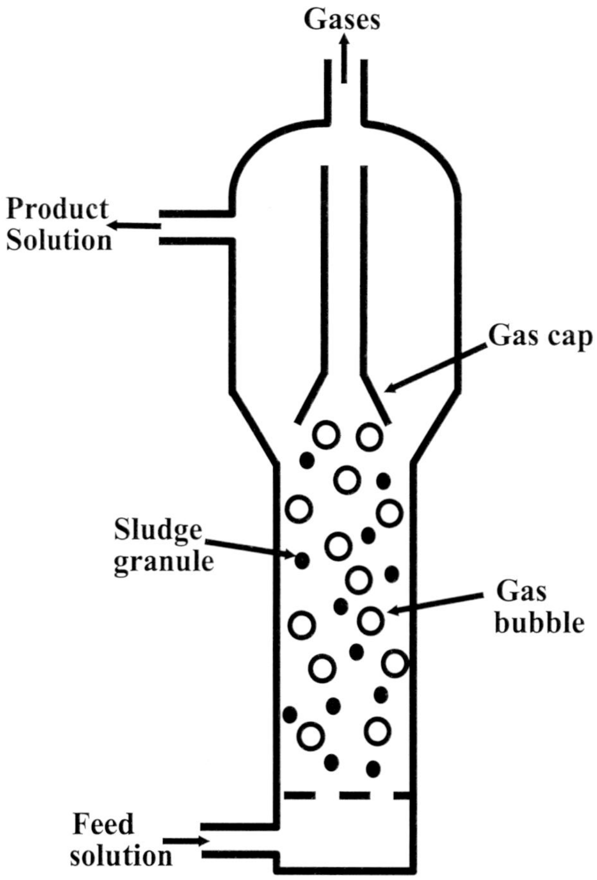

Figure 3.14: Schematic of EGSB bioreactor

The entire reactor column thus resembles a fully fluidized, three phase bioreactor, with the sludge granules remaining suspended in the ascending stream of feed solution (substrate solution), the product gases moving up in the form of tiny, discrete bubbles (figure 3.14). Narayanan and Vikas [95] have reported a rigorous performance analysis of these bioreactors based on the PFDR – approach. These reactors are best suitable for handling low strength feedstocks and in such cases, they provide enhanced BOD destruction and larger biogas generation.

Example 3.11: Anaerobic digestion of wastewater from a pulp and paper industry (BOD = 12.5 g/L) is to be carried out at the rate of 1900 m^3/din a 2.0 m diameter UASB bioreactor. The bioconversion intrinsically follows Monod – type kinetics with $\mu_m = 3.17\ d^{-1}$, $K_S = 0.277$ g /L and overall yield coefficient for cell mass production = 0.60. Estimate the fractional BOD destruction achieved if the height of sludge bed formed is 1.35 m (total voidage = 0.4) and that of sludge blanket = 2.5 m. Average size of each sludge granule = 2.0 mm and its apparent density = 1050 g/L. Take effective diffusivity of substrate (organic matter) into sludge granules = 2.75×10^{-10} m^2/s, interfacial tension = 0.1 N/m and average biogas production = 0.7 m^3 per kg of BOD destroyed. The fractional gas holdup in the sludge bed maybe assumed negligible and assume that the bioreactor is equivalent to a PFR – CSTR combination.

Solution: If the sludge bed is assumed equivalent to a PFR, then its performance equation is that given in equation (3.9.8). However, since the value of C_{Sb}, which is the substrate concentration at the top of the sludge bed, is not known at the outset, we have to resort to a trial and error procedure. Thus, let

$$C_{Sb} = 5.5 \text{ g/L}$$

Equation (3.9.8), therefore, becomes,

$$L_P = \left\{U_L / \left[\in_{PL} \mu_m(app)\right]\right\} \int_{5.5}^{12.5} F(C_S)dC_S \quad \text{(i)}$$

where $F(C_S) = (K_S + C_S)/(\eta C_S) = (0.277 + C_S)/(\eta C_S)$ (ii)

Since Q_0 = 1900 m^3/d = 0.022 m^3/s and $A = (\pi D^2/4) = (\pi(2.0)^2/4) = 3.1416\ m^2$,

$$U_L = (Q_0 / A) = 0.007 \text{ m/s}$$

Since fractional gas holdup in the sludge bed has been assumed negligible,

$$\in_{PL} = \in_P = 0.40$$

And $\mu_m(app) = (\mu_m / Y)x_f(1 - \in_P)/\in_{PL}$

= (3.17/0.60) (1050) (0.60)/(0.40)

= 8321.25 g /(L.d)

Equation (i) thereby reduces to

$$L_P = (0.1817) \int_{5.5}^{12.5} F(C_S)dC_S \quad \text{(iii)}$$

To evaluate the integral of above equation numerically, we need to estimate the value of the effectiveness factor (η) at each value of C_S. Let us use equations (2.4.28) to (2.4.29) for the computation of η. Thus,

$$L^* = \left(d_P / 6\right) = (2.0 \times 10^3)/6 = 0.333 \times 10^3$$

Therefore, from equation (2.4.24), $\varnothing' = 11.8525$

Using the above value of $\varnothing'$ in equation (2.4.22), we get

$$\eta_d = (0.11932)\left[(1+\beta)/\beta\right]\left[\beta - \ln(1+\beta)\right]^{1/2} \quad \text{(iv)}$$

where $$\beta = \left(C_S / 0.277\right) \quad \text{(v)}$$

Since $\tanh(\varnothing') = \tanh(11.8525) = 1.0$, from equation (2.4.28),

$$\eta = 1 - (0.08437)\left[\eta_d\, coth(\eta_d) - 1\right], \text{ for } \eta_d \le 1.0 \quad \text{(vi)}$$

In the present case, all η_d - values shall all be less than 1.0 and we can use the above equation with confidence.

Let us now proceed to evaluate the integral of equation (iii) numerically using Simpson's rule. For this, at each value of C_S ranging from $C_S = 5.5$ g /L to $C_S = 12.5$ g/L, the value of η is computed from equations (iv) to (vi) and that of $F(C_S)$ from equation (ii). The results are tabulated in Table (3.12.1). The step size chosen is 0.35 g/L.

Table 3.12.1

Substrate concentration (C_S), g/L	**Effectiveness factor (η)**	**$F(C_S)$**
5.50	0.9927	1.058
5.85	0.9922	1.0555
6.20	0.99176	1.05336
6.55	0.9913	1.05145
6.90	0.9908	1.04978
7.25	0.99035	1.0483
7.60	0.98988	1.047
7.95	0.9894	1.04593
8.30	0.9889	1.0449
8.65	0.9885	1.044
9.00	0.9880	1.0433
9.35	0.98755	1.0426
9.70	0.987	1.0421
10.05	0.9866	1.04152
10.40	0.9860	1.04121
10.75	0.9857	1.04065
11.10	0.9852	1.04035
11.45	0.98477	1.04

Substrate concentration (C_S), g/L	**Effectiveness factor (η)**	**$F(C_S)$**
11.80	0.9843	1.0398
12.15	0.9838	1.0396
12.50	0.9834	1.0394

Now, from Simpson's rule,

$$I=(h/3)\left[f_0+4(f_1+f_3+.....+f_{n-1})+2(f_2+f_4+.....f_{n-2})+f_n\right]$$

where $h=$ step size $= 0.35$ g/L, $f_0 = 1.058$, $f_1 = 1.0555$ and so on. Finally, $f_n = 1.0394$.

Therefore, $I = 7.3149$ g/L

And, from equation (iii) $L_P = (0.1817)\ (7.3149) = 1.329 \approx 1.33$ m

The above – computed value of L_P is very close to the value specified (namely, L_P = 1.35 m). We may, therefore, take, with allowable error, C_{Sb} = 5.5 g/L as final and discontinue further trials.

Now, since the sludge blanket has been assumed equivalent to an ideal CSTR, the performance equation shall be that given in equation (3.9.11). However, in this case also, we have to resort to a trial and error procedure, since the value of C_{Se} is not known at the outlet. Thus, let

$$C_{Se} = 1.0 \text{ g/L}$$

Since then average gas production is 0.7 m^3 per kg of BOD destroyed, the volume rate of gas production (Q_g) is,

$$Q_g = Q_0(C_{S0}-C_{Se})(0.7)$$

or $U_g = U_L(C_{S0}-C_{Se})(0.7)$

$= (0.007)\ (12.5 - 1.0)\ (0.7) = 0.05635$ m/s

This is the average superficial velocity of gases through the reactor column. We have to now estimate the total fluid holdup $(\in_f)$ and the fractional liquid holdup $(\in_{fL})$ in the sludge blanket from selected experimental correlations. Let us thus use Dakshinamorthy et al's correlation (equation - 3.5.10) for the computation of total fluid holdup (..) and Kato et al's correlation (equation - 3.5.27) for the computation of fractional liquid holdup $(\in_{fL})$. For this, we have to first, estimate the terminal free settling velocity (V_t) of each sludge granule, which once again demands a trial and error procedure. Thus, let

$$V_t = 0.02 \text{ m/s}$$

$$Re_P = (d_P V_t \rho_L / \mu_L)$$

$$= (0.002)\ (0.02)\ (1000)\ /\ (0.001) = 40.0$$

$$\ln(Re_P) = 3.6888$$

From table (2.1) of Chapter – 2,

$$a_0 = 3.32,\ a_1 = (-0.929),\ a_2 = 0.0429,\ a_3 = 0.0014$$

Therefore, from equation (2.5.30) of Chapter – 2,

$$\ln(f_D) = 0.5471278$$

or $f_D = 1.72828$

Now, from equation (2.5.27),

$V_t = [(4/3)\ (1029 - 1000)\ (9.8)\ (2.0 \times 10^{3})\ /(1000)\ (1.72828)]^{1/2}$

$= 0.0209$ m/s

To note that the apparent density of sludge granule could get lowered due to gas entrapment. It shall be, therefore, more accurate to take $\rho_S = (\varphi x_f)$, where φ is the correction factor to account for gas entrapment. The value of φ usually varies from 0.97 to 0.98. Taking $\varphi = 0.98$,

$$\rho_S = (0.98)\ (1050) = 1029 \text{ g /L or kg/ m}^3$$

We have substituted the above value of ρ_S in the computation of V_t. Since the above computed value of V_t does not differ significantly from that assumed at the outset, we may discontinue further trials and take an average value of V_t as the final value. Thus,

$$V_t = (0.02 + 0.0209)/2 = 0.02045 \text{ m/s}$$

$$Re_P = (0.002)\ (0.02045)\ (1000)\ /\ (0.001) = 40.9$$

Now, from equation (3.5.10),

$$\in_f = k(U_L / V_t)^m (U_g \mu_L / \sigma)^{0.08}$$

Since Re_P is less than 500, $k = 2.12$, $m = 0.41$. Therefore,

$$\in_f = 0.75$$

Let us now compute the fractional liquid holdup $(\in_{fL})$ in the sludge blanket from equations (3.5.27) to (3.5.32). Thus,

$$N = (U_g)^4 \rho_L / (g\sigma) = 0.0103$$

$$a = 5.1 + 86.2\ (N)^{0.28} = 28.49$$

$$b = (Re_P)^{0.9}\left[0.1 + 0.443(N)^{0.165}\right] = 8.7$$

$$m = (b+1)/(a+2.7b) = 0.1866$$

$$k = 1 - 9.7(N)^{0.092} / \sqrt{350 + (Re_P)^{1.1}} = 0.685266$$

Therefore. From equation (3.5.27),

$$\in_{fL} = (0.685266)[0.007/0.02045]^{0.1866} = 0.56$$

The fractional gas holdup in the sludge blanket is, therefore,

$$\in_{fg} = (\in_f - \in_{fL}) = 0.19$$

The space time (τ_m) for the sludge blanket is, therefore,

$$\tau_m = (V_m \in_{fL} / Q_0) = (L_f \in_{fL}) / U_L$$

$$= (2.5)\ (0.56)\ /\ (0.007) = 200.0\ \text{s}$$

And from equation (3.9.20),

$$\mu_m(app) = (\mu_m / Y)\, x_f (1 - \in_f) / \in_{fL}$$

$$= (3.7\ /\ 0.60)\ (1050)\ (0.25)\ /\ (0.56) = 2472.148\ \text{g}\ /\ (\text{L.d})$$

The value of effectiveness factor (η_e) at $C_S = C_{Se} = 1.0$ g/L can be now estimated from equations (iv), (v) and (vi). Thus,

$$\beta = (C_{Se} / 0.277) = (1.0\ /\ 0.277) = 3.61$$

$$\eta_d \text{ (from equation – iv)} = 0.21985$$

$$\eta_e \text{ (from equation – v)} = 0.9986$$

Substituting the above-computed values of η_e, τ_m, $\mu_m(app)$ in equation (3.9.11) and re – arranging,

we get

$$C_{Se}^2 + (0.49155) C_{Se} - 1.5235 = 0$$

On solving, $C_{Se} = 1.012$ g/L

Since the above – computed value of C_{Se} is very close is that assumed at the outset, we can safely discontinue further trials. The fractional BOD destruction achieved (α) is,

$$\alpha = (C_{S0} - C_{Se}) / C_{S0} = (12.5 - 1.0)\ /\ (12.5) = 92\ \%$$

The bioreactor thus affects substantially large degree of BOD destruction. The total active height of bioreactor (the total height of reaction zone) is,

$$L = (L_P + L_f) = (1.35 + 2.5) = 3.85\ \text{m}$$

The sludge bed thus occupies 35% of the reaction zone and the rest 65% is occupied by the sludge blanket. If we include a disengagement zone of height 2.0 m above the sludge blanket, then the total height of the reactor column shall be, (3.85 + 2.0) = 5.85 m ≈ 6.0 m.

As anticipated, the major share of bioconversion occurs in the sludge blanket (from 5.5 g/L BOD to 1.0 g/L BOD, namely 82 %). The sludge bed contributes BOD destruction from 12.5 g/L to 5.5 g/L (namely, 56%).

3.10 Comparison among Bioreactors

All classes of bioreactors discussed above do possess their own merits and demerits. All the pros and cons associated with their construction, operation and maintenance are to be kept in mind, while making a choice among them. High reactor capacity, enhanced fractional substrate conversion, reasonably low reactor volume requirement, low operating cost, ease of catalyst handling are all factors affecting the choice of the bioreactor for a particular industrial application, few among them could also be complimentary to each other.

A brief comparison among the popular bioreactors (discussed in this Chapter) is summarized below in tables (3.10.1A) and (3.10. 1B).

Table – 3.10.1A

Bioreactor	Merits	Limitations
Stirred tank	Simple in construction and operation. Uses suspended growth of microbes. Suitable for aerobic and anaerobic processes.	Restricted to low capacities.
Moving bed biofilm reactor (slurry reactor)	Heterogeneous version of stirred tank. High cell concentration in biofilm promotes rate of bioconversion.	Capacity-wise inferior to column reactors. Biofilm could get disturbed due to high rate of agitation.
Packed bed bioreactors	Simple in construction. Provides reasonably large degree of bioconversion within the range of flow rates handled. Lower operating cost than fluidized bed bioreactors.	Restricted to low capacities. Continuous operation with continuous regeneration or reactivation of biocatalyst not possible.
Trickle bed biofilm reactor	Uses attached growth of microbes. Low operating cost due to downflow mode of operation. High cell mass concentration in biofilm promotes rate of bioconversion.	Mainly for aerobic BOD removal. Low capacity due to low feed flow rate maintained.
Fluidized bed bioreactor	Operates at high capacities, provides high degree of bioconversion. Degree of bioconversion increases with increase in feed flow rate due to bed expansion. Once fully fluidized, pressured drop across the bed remains constant and does not increase with increase in feed flow rate. Suitable for handling deactivating enzymes, since could be operated in the circulating mode (as circulating fluidized bed).	Entrainment loss of particles / particle-biofilm aggregates possible. Operating cost higher than trickle bed (packed bed).

Bioreactor	Merits	Limitations
Semifluidized bed bioreactor	Higher degree of bioconversion (than fluidized beds) at higher capacities and low reactor volume requirement. Degree of bioconversion increases with increase in feed flow rate, even when the reactor volume is kept constant. No problems of entrainment loss or carry over, due to presence of top restraint.	Higher operating cost than fluidized beds. Continuous, circulating mode of operation not possible.

Table – 3.10.1B

Inverse fluidized biofilm reactor	Low operating cost due to down flow mode of operation. Larger size support particles could be used. Reasonably large degree of bioconversion. No problem of entrainment loss.	Lower capacity than fluidized bed and semifluidized bed bioreactors. Larger reactor volume requirement.
Inverse fluidized bed immobilized enzyme bioreactor (employing nanoparticles)	Exclusive design for handling nanoparticles. Low operating cost. Reasonably high degree of bioconversion.	Restricted to relatively lower feed flow rates. Not suitable for three phase (G-L-S) operation with upflow of gases.
DSFF bioreactor	Simple in construction and operation. No support particles required. Low operating cost due to downflow mode of operation. Multiple tubes / columns could be used to increase capacity.	Presently restricted to anaerobic operation. Large reactor volume requirement at high capacities.
UASB reactor	Simple in construction. No support particles used. Provides significantly high degree of bioconversion at distinctly high capacities and even with high strength feedstock.	Restricted to anaerobic processes, employing complex culture of microbes. Enormously large start-up time.

Nomenclature

a specific interfacial area for mass transfer, m^2/m^3

a_b specific surface of the bed, m^2/m^3 *of bed*

A area of cross-section of bioreactor / column, m^2

Ar_m modified Archemedes number, dimensionless

C_L mass ratio of nanoparticles to substrate solution (catalyst loading)

C_S substrate concentration in liquid bulk, g/L or mol/L

$C_S(z)$ substrate concentration at any z, g/L or mol/L

C_{Sb} substrate concentration at interface between packed section and fluidized section, g/L, mol/L

C_{Se} substrate concentration in product solution, g/L or mol/L

C_{S0} substrate concentration in feed solution, g/L or mol/L

C_{SP} substrate concentration at biofilm – liquid interface, g/L or at particle surface, mol/L

d_p diameter of support particle, m

d_{pm} diameter of particle-biofilm aggregate, m

D diameter of reactor column, m

D_1, D_2 minimum and maximum diameter respectively of diverging-converging column, m

D_a impeller diameter, m

D_e effective diffusivity of substrate into biofilm / into particle, m^2/s

D_L axial dispersion coefficient, m^2/s

D_{Ld} axial dispersion coefficient in DSFF bioreactor column, m^2/s

D_{Lf} axial dispersion coefficient in fluidized bed /section, m^2/s

D_{LP} axial dispersion coefficient in packed bed / packed section, m^2/s

D_t tank diameter, m

Eo Eötvös number, dimensionless

f volume fraction of biofilm in particle-biofilm aggregate ; volume ratio of biofilm to free volume of flow channel, dimensionless

f_b bed friction factor, dimensionless

f_D drag coefficient, dimensionless

Fr_g gas phase Froude number, dimensionless

Fr_L liquid phase Froude number, dimensionless

G_g mass velocity of gas, kg/(m^2.s)

G_L mass velocity of liquid, kg/(m^2.s)

G_t mass velocity of fluid through tube/pipe, kg/(m^2.s)

k_A cell mass attachment coefficient, s^{-1}

k_d endogenous decay coefficient, s^{-1}

k_D cell mass detachment coefficient, s 1

k_{LP} liquid to particle mass transfer coefficient, m/s

K_C Contois kinetic constant, g/g

K_m Michaelis-Menten kinetic constant, mol/L

K_S Monod's kinetic constant, g/L

K_{Si} substrate inhibition coefficient, g/L

K_T turbulent constant, dimensionless

L height of each flow passage of DSFF bioreactor, m

L^* characteristic dimension of particle or biofilm, m

L_f height of fluidized bed (expanded bed), m

L_{mf} height of minimum fluidized bed, m

L_P height of packed bed or packed section, m

L_S height of static bed (prior to fluidization); segment length of diverging-converging column

L_{Sf} height of semifluidized bed, m

$\dot{m}$ mass flow rate of fluid, kg/s

Mo Morton number, dimensionless

n impeller speed, rps

N_d dispersion number (see equation-3.2.4), dimensionless

P_{gL} power consumption for agitation of gassed liquid, W

P_{gLS} power consumption for agitation of gassed liquid-solid suspension, W

P_L power consumptionfor agitation of ungassed liquid, W

Q_0 substrate flow rate, m^3/s

Q_g gas flow rate, m^3/s

r_{max} Michaelis-Menten kinetic constant, mol/(L.s)

$(-r_S)$ rate (global rate) of bioconversion, g/(L.s) or mol/(L.s)

$(-r_S)(int)$ intrinsic rate of bioconversion, g/$(L.s)$ or mol/(L.s)

$(-r_{Se})$ rate (global rate) of bioconversion at $C_S = C_{Se}$, g/(L.s) or mol/(L.s)

$(-r_{Se})(int)$ intrinsic rate of bioconversion at $C_S = C_{Se}$, g/$(L.s)$ or mol/(L.s)

$(-r_{SP})(int)$ intrinsic rate of bioconversion at $C_S = C_{SP}$, g/$(L.s)$ or mol/(L.s)

R bed expansion ratio = L_{Sf} / L_S

Re Reynolds number, dimensionless

Re_b bed Reynolds number, dimensionless

Re_m mixing Reynolds number (impeller Reynolds number), dimensionless

Re_{mf} Reynolds number at minimum fluidization, dimensionless

Re_P particle Reynolds number

Re_T modified Reynolds number (see equation – 3.5.9), dimensionless

t batch time, s

U_g average superficial velocity of gas, m/s

U_L average superficial velocity of liquid (substrate solution), m/s

$U_{Lm}(inv)$ minimum inverse fluidization velocity of liquid, m/s

U_{Lmf} minimum fluidization velocity of liquid in three phase fluidized bed, m/s

U_m downflow velocity of suspension through inverse fluidized bed column, m/s

U_{mf} minimum fluidization velocity, m/s

U_{SL} operating superficial velocity of liquid through semifluidized bed, m/s

$U_{SL}(min)$ minimum semifluidization velocity of liquid, m/s

V volume of reactor (reaction mixture), m^3

V_t terminal free settling velocity of particle or particle-biofilm aggregate, m/s

We_L Weber number, dimensionless

x cell mass concentration, g/L

x_e cell mass concentration in product solution, g/L

x_f cell mass concentration in biofilm, g/L

x_0 cell mass concentration at $t = 0$; that in feed solution, g/L

X mass fraction of catalyst particles (particle-biofilm aggregates) in reaction mixture, g/g

Y overall yield coefficient for cell mass synthesis , mg/mg

z axial coordinate

α fractional substrate conversion, dimensionless

β parameter defined in equation (2.2.12) of Chapter-2 ; pressure drop ratio (see eq. 3.4.1)

biofilm thickness, m

$\in$ voidage / void fraction of the bed, dimensionless

$\in_f$ total voidage of fluidized bed or fluidized section or sludge blanket, dimensionless

$\in_{fL}$ fractional liquid holdup in fluidized bed / section, dimensionless

$\in_{fg}$ fractional gas holdup in fluidized bed / section, dimensionless

$\in_g$ fractional gas holdup in the bed / reaction mixture, dimensionless

$\in_L$ fractional liquid holdup in the bed / reaction mixture, dimensionless

$\in_P$ total voidage of packed bed / section, dimensionless

$\in_{Pg}$ fractional gas holdup in the packed bed / section, dimensionless

$\in_{PL}$ fractional liquid holdup in the packed bed / section, dimensionless

$\in_S$ fractional solid holdup, dimensionless

η effectiveness factor, dimensionless

$\eta(z)$ effectiveness factor at any z, dimensionless

η_d parameter defined in equation (27), dimensionless

η_e effectiveness factor at $C_S = C_{Se}$, dimensionless

θ angle of convergence / divergence, degrees

μ_f fluid viscosity, kg/(m.s)

μ_g viscosity of gas, kg/(m.s)

μ_L viscosity of liquid (substrate solution), kg/(m.s)

μ_m maximum specific growth rate of microbes, s^{-1}

ρ_g density of gas, kg/m^3

ρ_L density of liquid / substrate solution, kg/m^3

ρ_{Lm} bulk density of suspension, kg/m^3

ρ_m density of microbial solution, kg/m^3

ρ_S density of support particle, kg/m^3

ρ_{Sm} density of particle-biofilm aggregate, kg/m^3

σ_L interfacial tension, N/m

τ space time, s

$\varnothing$ Thiele - type modulus, dimensionless

$\varnothing'$ parameter defined in equation (2.4.24) and in (2.4.26) of Chapter - 2, dimensionless

ψ parameter to account for gas entrapment in sludge granule, dimensionless

Exercises

3.A: Discuss briefly what do you understand by,

(a) Tanks – in – series model,

(b) PFDR approach,

(c) Axial dispersion coefficient,

(d) Microbial sludge granule formation.

3.B: Distinguish between (use sketches, wherever possible, to elucidate your answers):

(a) Biofilm reactors and immobilized cell bioreactors

(b) Moving bed and fixed bed biofilm reactors

(c) Semi-fluidized and inverse fluidized biofilm reactors,

(d) UASB and EGSB bioreactors

(e) Inverse fluidized biofilm reactors and immobilized enzyme bioreactors (employing nanoparticles)

(f) Immobilized enzyme fluidized bed and circulating fluidized bed bioreactors.

3.C: Comment on the following statements, giving reasons. You may agree with or contradict against any of the statements, but do justify your answers:

(a) In a packed bed bioreactor, the fractional substrate conversion decreases with increase in feed flow rate, but the conversion increases with increase in feed flow rate in fluidized bed bioreactors,

(b) In a semifluidized bed bioreactor, fractional substrate conversion increases with increase in feed flow rate, though the total volume of the reactor remains fixed,

(c) UASB bioreactors demand enormously large startup time,

(d) In biofilm reactors, the biofilm thickness and the cell mass concentration in the biofilm remain more or less constant throughout the operation of the reactor,

(e) For handling nanoparticles accommodating immobilized enzyme, inverse fluidized bed bioreactors are the only option.

3.D: A batch moving bed biofilm reactor, 0.8 m in diameter, fitted with four baffles and agitated with a 500 rpm Rushton impeller is being used for the synthesis of polymer – grade lactic acid from de- proteinized cheese whey (lactose content = 9.0 g/L). The reactor tank is composed of 1.0 mm silica granules, each surrounded by a 0.3 mm thick biofilm of *Lactobacillus helveticus* culture, suspended in the substrate solution (mass fraction of particle – biofilm aggregates in suspension = 0.03). Cell mass concentration in biofilm = 100 g/L. Determine the batch time required to affect 70% conversion of dissolved lactose to lactic acid, if the bioconversion follows Monod – type kinetics with μ_m = 0.7 hr^{1}, K_S = 0.22 g/L and overall yield coefficient for cell mass production = 0.65. Assume that mass transfer resistance in liquid bulk is negligible and take effective diffusivity of lactose into biofilm = 3.0×10^{10} m^2/s.

3.E: Xanthum gum is being synthesized continuously through aerobic fermentation of de - proteinized cheese whey in a 1.7 m^3 moving bed biofilm reactor (tank diameter = 0.8 m), the feed solution (lactose content = 9.0 g/L) is being admitted at 175 L/hr and air sparged from below at 6800 L/hr. The discrete phase is composed of 1.0 mm silica granules (density = 2100 kg/m^3), each surrounded by a 0.3 mm thick biofilm of *Xanthomonas campestris* (density of microbial solution = 1025 kg/m^3, mass fraction of particle – biofilm aggregates in suspension = 0.03, cell mass concentration in biofilm = 95 g/L). The tank is fitted with four baffles and is being agitated with a 400 rpm, six – bladed turbine impeller ($D_a = D_t / 3$). Compute the fractional substrate conversion achieved, if aerobic bioconversion of lactose to Xanthan gum follows Contois – type kinetic equation with μ_m = 0.7 hr^{1}, K_S = 0.22 g/L and overall yield coefficient for cell mass production = 0.5. Assume that the bioreactor is equivalent to three ideal CSTRs of equal volume in series (tanks – in – series model) and neglect mass transfer resistance in liquid bulk. Take effective diffusivity of substrate into the biofilm = 2.9×10^{10} m^2/s.

3.F: Hydrolysis of benzoylarginine ethyl ester is to be conducted in a 0.1 m ID packed bed bioreactor composed of 0.79 mm iron oxide particles that accommodate immobilized papain enzyme. Papain content of each particle = 10 mg per 30 g of particle, particle density = 1900 kg /m^3, bulk density of the bed = 1100 kg /m^3. The ester solution with an ester content of 0.014 mol/L is fed from the bottom at the rate of 85.0 L /hr. The process follows Michaelis – Menten kinetics with r_{max} = 19×10^{6} mole of ester / (min. mg catalyst), K_m = 0.005 mol/L. The fractional substrate conversion desired is 50 %. Estimate the depth of the catalyst bed required, if

(a) true plug flow may be assumed through the catalyst bed,

(b) the reported value of axial dispersion coefficient is 0.007 m^2/s.

Liquid to particle mass transfer coefficient = 3.2 m /hr. Take effective diffusivity of substrate into particle = 2.097×10^{10} m^2/s.

3.G: Anaerobic fermentation of a starchy substrate is to be conducted in a fluidized bed biofilm reactor that is composed of 2.5 mm support particles (particle density = 1300 kg/m^3) each surrounded by a microbial film of thickness 0.3 mm. The feed solution with a substrate concentration of 420 mg/L is admitted from the bottom at the rate of 1000 m^3/hr and the fractional substrate conversion desired is 60%. Estimate the required diameter and height of the bioreactor, if

(a) axial dispersion and resistance to substrate transport in the liquid bulk are negligible,

(b) reported value of axial dispersion coefficient is 0.0315 m^2/s and the liquid to particle mass transfer coefficient is predicted by the correlation proposed Koloini and coworkers (see Section 2.5 of Chapter – 2).

The process follows Monod's kinetic equation, with μ_m = 77.8 d^1, K_S = 6.3 mg/L and overall yield coefficient for cell mass production = 0.60. Neglect endogenous decay. Cell mas concentration in the biofilm = 3750 mg/L. Take interfacial tension = 0.08 N/m, density of microbial solution = 1030 kg/m^3, liquid phase diffusivity of substrate in solution = 1.0×10^9 m^2/s and effective diffusivity of substrate into biofilm = 4.5×10^{10} m^2/s. Assume that formation of product gases amounts to 0.9 mol per gram of cell mass synthesized. For the estimation phase holdups in the bed, the correlation proposed by Dakshinamurthy and coworkers and that proposed by Kato and coworkers may be employed.

3.H: Aerobic treatment of wastewater (industrial effluent) from a paper and pulp unit is to be performed in a semifluidized bed biofilm reactor (column diameter = 0.5 m), that uses 2.5 mm silica granules (density = 2200 kg/m^3), each surrounded by a 0.3 mm thick biofilm (cell mass concentration in biofilm = 4000 mg/L, density of microbial solution = 1030 kg/m^3). The feed solution with a BOD of 250 mg/L is admitted from the bottom at the rate of 82000 L/hr and air is sparged from below at 10,600 L/hr. The initial static bed height was 0.5 m and bed expansion ration (R) maintained is 2.0. the bioconversion (BOD destruction) reportedly follows Monod's kinetics with μ_m = 0.187 hr^1, K_S = 60.0 mg/L and overall yield coefficient for cell mass production (assumed constant) = 0.60. Assume that mass transfer resistance (for substrate transport) in liquid bulk is negligible and take effective diffusivity of substrate (dissolved organic matter) into biofilm = 3.0×10^{10} m^2/s. Determine the fractional BOD removal attained,

(a) if the bioreactor is assumed equivalent to a CSTR – PFR combination

(b) if the bioreactor is assumed equivalent to two PFDRs in series, with axial dispersion coefficient in the packed section = 0.0072 m^2/s and that in fluidized section = 0.032 m^2/s.

The correlations proposed by Chern et.al (equations – 3.6.8 and 3.6.13) and those by Saberian – Broudjenni et.al (see table – 3.6.2A, ref. No. 27) are recommended for the estimation of hydrodynamic parameters. You may very well use alternate correlations and compare the results.

3.I: Synthesis of biodiesel from jatropha oil and butyl acetate (discussed in Example – 3.7) is to be carried out in a semifluidised bed immobilized enzyme bioreactor, 0.5 m in diameter and 1.2 m in height (height of semifluidized bed). Bed expansion ratio used = 2.0. Lipase enzyme is immobilized in 1.0 mm silica particles (density of silica = 2100 kg /m^3) and oil – acetate mixture is 1:3 molar ratio is admitted from the bottom of the reactor column at 32,000 L/hr. Compute the degree of trans – esterification of jatopha oil attained,

(a) if the bioreactor is assumed equivalent to a PFR – CSTR combination

(b) if the bioreactor is assumed equivalent to two PFDRs in series, the value of axial dispersion coefficient in PFDR – 1 (packed section) being 0.0075 m^2/s and that in PFDR – 2 (fluidized section) 0.0315 m^2/s.

The kinetics of trans – esterification may be assumed to follow equation (2.3.90) of Chapter – 2, with values of kinetic constants being same as those mentioned in Example (3.7). The mass transfer resistance to substrate transport in the liquid bulk may be assumed negligible and take effective diffusivity of substrate into particle = 2.95×10^{10} m^2/s. The property values of jatropha oil and butyl acetate are those given in Example (3.7).

3.J: An inverse fluidized bed biofilm reactor is to be used for the synthesis of lactic acid from de-proteinized cheese whey using a culture of *Lactobacillus helveticus*. The reactor consists of a 0.5 m ID column, containing 12.5 mm polymer (polypropylene) beads, each surrounded by a biofilm of 0.3 mm thickness (cell mass concentration in biofilm = 150 g/L, density of microbial solution = 1030 kg/m^3). Density of polypropylene = 910.0 kg/m^3. The feed solution with a lactose content 9.0 g/L is admitted from the top at the rate of 59,000 L/hr. If the bioconversion follows Monod - type kinetics with μ_m = 0.7 hr^1, K_S = 0.22 g /L and overall yield coefficient for cell mass production = 0.65, determine

(a) the height of the inverse fluidized bed formed,

(b) the fractional conversion of lactose to lactic acid attained in the bed, if the bioreactor could be assumed equivalent to a plug flow dispersion reactor (PFDR) with an average value of axial dispersion coefficient = 0.0315 m^2/s. Neglect mass transfer resistance in liquid bulk and take effective diffusivity of substrate (lactose) into the biofilm = 2.9×10^{10} m^2/s. The original static bed of polymer beads were 0.5 m in height, with an average voidage of 0.35.

3.K: Molasses (after clarification and adequate dilution) is to be fermented to convert its sucrose content to lactic acid using a culture of *Enterococcus faecalis* in a downflow stationary fixed film (DSFF) bioreactor which consists often 25.4 mm ID vertical flow channels, each 6.0 m in height, the inside surface of each having been coated with a 1.5 mm biofilm (cell mass concentration in biofilm = 450 g/L). Clarified molasses diluted to a substrate concentration of 150 g/L is fed from the top at 5200 L/hr and is equally distributed among the flow channels. If the bioconversion follows Monod - type kinetics, with μ_m = 2.564 hr^1, K_S = 0.1 g /L and overall yield coefficient for cell mass production (assumed constant) = 0.77, determine the fractional conversion of sucrose attained.

Assume that the bioreactor is equivalent to a plug flow dispersion reactor (PFDR) and use Levenspiel's correlation (equations – 3.2.3 to 3.2.7) for the estimation of axial dispersion coefficient. Neglect mass transfer resistance in liquid bulk and take effective diffusivity of substrate into biofilm = $2.9 \times 10^{10}\ m^2/s$.

3.L: Rework Example (3.12), by assuming the bioreactor to be equivalent to two PFDRs in series, the axial dispersion coefficient in the sludge blanket (PFDR – 1) being equal to 0.0315 m^2/s and that in the sludge bed (PFDR – 2) 0.0075 m^2/s and compare the results.

References

1. Narayanan, C. M. and Bhattacharya, B. C., Unit Operations and Unit Processes, Vol.2, CBS Publishers, New Delhi, 2005.
2. Narayanan, C.M. and Vikas Narayan, Studies on Synthesis of Lactic Acid and Xanthan Gum from Cheese Whey Permeate in Two Phase and Three Phase Moving Bed Biofilm Reactors, Korean J. Chem. Eng., **38** (9), 1888 – 902, 2021. **DOI: 10.1007/s11814-021-0821-5.**
3. Levenspiel, O., Axial Dispersion in Straight Pipes, Ind. Eng. Chem., **50**, 343-46, 1958.
4. Narayanan, C.M., Biotechnology and Bioprocess Engineering, Galgotia Publishers, New Delhi, 2011.
5. Bischoff, K. B. and Levenspiel, O., Axial Dispersion in Packed Beds, Chem. Eng. Sci., **17**, 257-60, 1962.
6. Wilhelm, R. H., Dispersed flow through packed beds, Pure Appl. Chem., **5**, 403-07, 1962.
7. Carberry, J. J., Chemical and Catalytic Reaction Engineering, McGraw Hill, New York, 1976.
8. Godlewska, M. M., and Karcz, J., Agitation of a Gas–Solid–Liquid System in a Vessel with High Speed Impeller and Vertical Tubular Coil, Chem. Papers, **66**(6) , 566 – 73, 2012.
9. Patrick AJ, Kennedy MJ., Power Consumption of Agitated Gas – Liquid – Solid Systems with Particular Relevance to Solid Substrate Submerged Culture Fermentations, Biotech. Letters, **17**(5), 487 – 92, 1995.
10. Sato, Y., Hirose, T., Takahashi, F. and Toda, M., Pressure Drop and Liquid Holdup in Packed Bed Reactor with Concurrent Gas – Liquid Downflow, J. Chem. Eng. Japan, **6**(2), 147-52, 1973.
11. Midoux, N., Favier, M. and Charpentier, J. C., Flow Pattern, Pressure Loss and Liquid Holdup Data in Gas-Liquid Downflow Packed Beds with Foaming and Nonfoaming Hydrocarbons, J. Chem. Eng. Japan, **9**(5), 350-55, 1976.
12. Bird, R. B., Stewart, W. E. and Lightfoot, E. N., Transport Phenomena, Wiley, New York, 1960.
13. Narayanan, C. M. and Bhattacharya, B. C., Mechanical Operations for Chemical Engineers, 4th edition, Khanna Publishers, New Delhi, 2004.
14. Kumar, R. K., Rao, A. V. R., Sankarshana, T. and Khan, A., Liquid Holdup in Concurrent Gas-Liquid Upflow through Packed Column with Random and Corrugated Structured

Packing, Proc. World Conf. on Eng. and Computer Sci., 792-97, San Francisco, October 24-26, 2012.
15. Wen C. Y. and Yu Y. H., A Generalized Method for Predicting the Minimum Fluidization Velocity, AIChE J., **12**, 610-612, 1966.
16. Begovich, J.M. and Watson, J.S., Hydrodynamic Characteristics of Three-Phase Fluidized Beds, Fluidization, University Press, Cambridge, 190–195, 1978.
17. Fortin, Y., Réacteurs à lit Fluidisé Triphasique: Caractéristiques Hydrodynamiques et Mélange des Particules Solides, Ph.D. Thesis, Institut National Polytechnique de Lorraine, Nancy, France, 1984.
18. Costa, E., Lucas, D. and Garcia, P., Fluid Dynamics of Gas – Liquid – Solid Fluidized Beds, Ind. Eng. Chem. Proc. Des. Dev., 25, 849 – 54, 1986.
19. Song, G. H., Bavarian, F. and Fan, L. S., Hydrodynamics of Three-Phase Fluidized Bed Containing Cylindrical Hydrotreating Catalysts, Can. J. Chem. Eng., 67, 265-275, 1989.
20. Nacef, S., Hydrodynamique des Lits Fluidises Gaz – Liquide – Solide, PhD Thesis, Institut National Polytechnique de Lorraine, Nancy, France, 1991.
21. Zhang, J.P., Epstine, R. and Grace, J.R., Minimum Fluidization Velocities for Gas – Liquid–Solid Three Phase Systems, Powder Technol.,100, 113-18, 1998.
22. Ruiz, R.S., Alonso, F. and Ancheyta, J., Minimum Fluidization Velocity and Bed Expansion Characteristics of Hydrotreating Catalysts in Ebullated Bed Systems, Energy & Fuels, 18, 1149-55, 2004.
23. Jena, H. M., Sahho, B. K., Roy, G. K. and Meikap, B. C., Characterization of Hydrodynamic Properties of a G-L-S Three Phase Fluidized bed with Glass Bead Particles, Chem. Eng. J., 145, 50-56, 2008.
24. Richardson, J. F. and Zaki, W. N., Sedimentation and Fluidization, Part – I, Trans. Inst. Chem. Engrs., 32, 35-38, 1954.
25. Muroyama, K. and Fan, L., Fundamentals of Gas-Liquid-Solid Fluidized Beds, AIChE J., 31(1), 1-34, 1985.
26. Dakshinamurthy, P., Subramanyam, V. and Rao, J.N., Bed Porosities in Gas-Liquid Fluidization, Ind. Eng. Chem. Proc. Des. Dev., **10**, 322-28, 1971; **11**, 318-19, 1972.
27. Saberian-Broudjenni, M., Wild, G., Charpentier, J.C., Fortin, Y., Euzen, J.P. and Patoux, R., 1987. Contribution to the hydrodynamic study of gas-liquid-solid fluidized-bed reactors, Int. Chem. Eng., **27**, 423-40, 1987.
28. Jena, H. M. and others, Statistical Analysis of Phase Holdup Characteristics of a Gas – Liquid – Solid Fluidized Bed, Can. J. Chem. Eng., 87, 1- 10, 2009.
29. Ramesh, K. and Murugesan, T., Minimum Fluidization Velocity and Gas Holdup in Gas – Liquid – Solid Fluidized Reactors, J. Chem. Tech. Biotech., 77, 129-36, 2002.
30. Kim, S.P., Baker, C.G.J. and Bergougnou, M.A., Phase Holdup Characteristics of Three Phase Fluidized Beds, Can. J. Chem. Eng., 53, 134-139, 1975.
31. Kato, Y., Uchido, K., Kago, T. and Morooka, S., Liquid Holdup and Heat Transfer Coefficient between Bed and Wall in Liquid-Solid and Gas-Liquid-Solid Fluidized Beds, Powder Technol., 28(2), 173-79, 1981.
32. Kim, S. D. and Kim, C. H., Axial Dispersion Characteristics of Three Phase Fluidized Beds, J. Chem. Eng. Japan, 16, 172-77, 1983.

33. De, T., Sikder, J. and Narayanan, C. M., Studies on Synthesis of Biodiesel by Trans-esterification of Neem Oil with Immobilized Lipase in Cylindrical and Diverging – Converging Fluidized Bed Bioreactors, Int. J. Chem. Environ. Eng., 6(5), 322 - 336, 2015.
34. De, T., Sikder, J. and Narayanan, C. M., Biodiesel Synthesis and Bioreactor Design – An Overview, Indian J. of Chemical Technology, **24**, 575 – 92, 2017.
35. De ,T., Sikder. J. and Narayanan, C. M., Characteristics and Applications of Biodiesels and Design of Bioreactors for their Industrial Manufacture, Chapter – 13, 285 – 320, in N. K. Rathinam and R. K. Sani (eds.), Next Generation Bio-manufacturing Technologies, ACS Symposium Series, Washington, 2019. **DOI : 10.1021/bk-2019-1329-ch013.**
36. Biswas, S. and Narayanan, C. M., Computer Aided Design and Analysis of Three Phase Fluidized Bed Biofilm Reactors for Waste Water Treatment, Asian J. Biochem. Pharmaceutical Res., **5**(2), 224 - 49, 2015.
37. Narayanan, C. M., Case Studies on Synthesis of PLLA Bioplastic starting from Food and Agricultural Wastes, Int. J. Chem. Eng. Processing, **1**(1) ,1-13, 2015.
38. Narayanan, C.M., Design of Fluidized Bed Bioreactors for Bio-desulfurization of Petroleum Oils – Development of CAD (Software) Packages, Int. J. Chem. Eng. Res., **2**(2), 97 - 108, 2010.
39. Narayanan, C.M., Modeling and Simulation of Fluidized Bed Bioreactors with Liquid Phase Oxygen Utilization, Int. J. Transport Phenomena, **11**(2), 127 - 132, 2009.
40. Narayanan, C.M., Process Intensification of Heat and Mass Transfer Equipment / Processes Using Variable Area Design – A Review, Annual Technical Volume (interdisciplinary coordination committee), The Institution of Engineers (I), **2**, 128 - 139, December, 2017.
41. Narayanan, C.M., Studies on Performance Characteristics of Variable Area Heat Exchangers, J. Chem. Eng. Japan, **31** (6), 903 - 09, 1998.
42. Narayanan, C.M., Studies on Transport Phenomena in Polymer Solutions and Suspensions flowing through Tubes of Tortuous Wall Geometry, Heat Mass Transfer (Wärme und Stoffübertragung), 50(2), 161 - 168, 2014. DOI : **10.1007/s00231 – 013 – 1238 – 4.**
43. Narayanan, C.M. and Bhattacharya, B.C., Mathematical Simulation of Condensation inside Diverging Converging Tubes, Proc. Int. Conf. on Advances in Chem. Eng., Kanpur, 381 - 387, 1989.
44. Chakraborty, T.K., Narayanan, C.M. and Bhattacharya, B.C., Experimental Studies on Heat Transfer with Phase Change on Diverging Converging Surfaces, Can. J. Chem. Eng., **69**, 281 - 86, 1991.
45. Narayanan C.M., Transport Phenomena in Falling films on Diverging – Converging Surfaces, Ind. J. Technology, **25** (4), 176 - 84, 1987.
46. Narayanan, C.M., Studies on Gas Absorption with Chemical Reaction in Falling films on Wavy Surfaces, J. Energy Heat Mass Transfer, **15** , 305 - 310, 1993.
47. Goel, S., Sahay, R.R., Bhattacharya, M. and Narayanan, C.M., Enhanced Collection Efficiency for Water Heating Solar Collectors, Ind. Chem. Engr., **30** (3), 27 - 32, 1988.
48. Narayanan, C. M., Studies on Performance Characteristics of Solar Parabolic Trough Concentrator with Variable Area Absorber, Int. J. Renewable Energy Technol., **5**(1), 1 - 11, 2014.

49. Narayanan, C.M., Performance Analysis of Constricted Tube Ultrafiltration Unit for the Production of Food – grade Protein Concentrates from Food Wastes, J. Chem. Eng. Material Sci., **2** (9), 149 - 55, 2011.
50. Basak, A., Saha, A., Jha, S. and Narayanan, C.M., Studies on Performance Characteristics of Diverging – Converging Fluidized Bed Biofilm Reactor with Special Reference to Lactic Acid Synthesis from Molasses and Cheese Whey, Int. Rev. Chem. Eng., **6**(4), 142 - 52, 2014.
51. Narayanan, C.M., Performance and Applications of Semifluidized Bioreactors – A Review, Int. J. Chem. Reactor Eng., July, 2018. DOI: **10.1515/ijcre-2017-0178**.
52. Kurian J. and Raja Rao M., Hydrodynamics of Semi – Fluidised Bed, Ind. J. Technol., **8**, 275 - 294, 1970.
53. Roy G. K. and Sharma K. J. R., Dynamics of Liquid-Solid Semifluidisation III., Chem. Eng. J., **4**, 294 - 296, 1972.
54. Roy G. K. and Sharat Chandra H.N., Liquid – Solid Semifluidization of Homogeneous Mixtures – Prediction of Semifluidization Velocity, Chem. Eng. J., **12**, 77 - 80, 1972.
55. Jena H.M., Hydrodynamics of Gas-Liquid-Solid Fluidized and Semi-fluidized Beds, PhD Thesis, National Institute of Technology, Rourkela, India, 2009.
56. Lee, S.L.P. and de Lasa, H.L., Phase Holdups in Three Phase Fluidized Beds, AIChE J., **33**, 1359-70, 1987.
57. Sivakumar,V., Kannadasan, T. and Senthilkumar, K., Prediction of Gas Holdup in Three Phase Fluidized Bed, Polish. J. Chem. Tech., **12**(4), 64 - 71, 2010.
58. Chern, S.H., Fan, L.S. and Muroyama, K., Hydrodynamics of Concurrent Gas- Liquid-Solid Semifluidization with Liquid as the Continuous Phase, AIChE J., **30**, 288-94, 1984.
59. Singh A. N., Takhalate H. K., Storck A. and Sengupta P., A Correlation for the Prediction of Packed Bed Height in Liquid-Solid Semifluidisation, Chem. Eng. J., **20**, 69 -73, 1980.
60. Mydlarz, J., Prediction of the Packed Bed Height in Liquid-Solid Semifluidisation of Homogeneous Mixtures, Chem. Eng. J., **34**, 155 - 58, 1987.
61. Das, Shrijita and Narayanan, C.M., Studies on Synthesis of Lactic Acid from Molasses and Cheese Whey in Semifluidized Bed Biofilm Reactors, Int. J. Environ. Waste Management, **19**(1), 1-20, 2017.
62. Biswas, S. and Narayanan, C. M., Studies on Waste Water Treatment in Three Phase Semifluidized Bed Bioreactors – Computer Aided Analysis and Software Development, J. Modern Chemistry Chem. Technol., Special Issue, **7**(1), 1-21, 2016.
63. Narayanan, C. M., Studies on Computer Aided Design and Analysis of Three Phase Semifluidized Bed Bioreactors , Chem. Product Process Modelling, **10**(1), 55-70, 2015. DOI : **10.1515/cppm.2014 – 0029**.
64. De, T., Sikder, J. and Narayanan, C. M., Biodiesel Synthesis Using Immobilized Lipase Enzyme in Semifluidized Bed Bioreactors – Bioreactor Design and Performance Analysis, AIChE – Environ. Progr. Sustainable Energy, **36**(5), 1537-45, 2017. DOI : **10.1002/ep.12602**.
65. Narayanan, C.M., Studies on Performance Characteristics of Circulating Fluidized Bed Bioreactors employing Immobilized Enzyme Nanoparticles , IUP J. Chem. Eng., Vol.III, No.4, 17-26, 2011.

66. Jena, H.M., Roy, G.K. and Meikap, B. C., Development of Comparative Study of a Semifluidized Bed Bioreactor for Treatment of Waste Water from Process Industries, Proc. Plant Eng. Environ. Management, **23**, 73-75, 2005.
67. Narayanan, C.M., Performance Analysis of Semifluidized Bed Biofilm Reactors with Liquid Phase Oxygen Utilization, ECI Symp. Series, Proc. 14th Int. Conf. on Fluidization: From Fundamentals to Products, Noordwijkerhout, Netherlands, May 26-31, 2013. http://dc.engconfintl.org/fluidization_xiv/12
68. Narayanan, C.M., Tejaswi, M., Harshika Prasad and Neelanjana Chakraborty, Studies on Performance Analysis and Computer Aided Design of Inverse Fluidized Bed Bioreactors with Nano-support Particles, Int. J. Chem. Reactor Eng., 12(1), 303 - 315, 2014. DOI: **10.1515 / ijcre.2004 – 0027**.
69. Ulaganathan, N. and Krishnaiah, K., Hydrodynamics of Two Phase Inverse Fluidized Beds, Bioproc. Eng., **15**, 159 - 64, 1996.
70. Vijayalakshmi, A. C., Balamurugan, M., Sivakumar, M., Newton, J. S. and Velan, M., Minimum Fluidization Velocity and Friction Factor in a Liquid-Solid Inverse Fluidized Bed Reactor, Bioproc. Eng., **22**, 461-66, 2000.
71. Banerjee, J., Basu, J. K. and Ganguli, U. P., Some Studies on the Hydrodynamics of Reverse Fluidization Velocities, Ind. Chem. Engr., **41**, 35-38, 1999.
72. Benedict, R. J. F., Kumaresan, G. and Velan, M., Bed Expansion and Pressure Drop Studies in Liquid-Solid Inverse Fluidized Bed Reactor, Bioproc. Eng., **19**, 137-42, 1998.
73. Sivasubramanian, V. and Velan, M., Minimum Fluidization Velocity in a Three Phase Inverse Fluidized Bed Reactor with Newtonian and Non-Newtonian Fluids, J. Chem. Eng. Japan, **37**(12), 1436 - 44, 2004.
74. Fan, L.S., Muroyama, K. and Chern, S.H., Hydrodynamic Characteristics of Inverse Fluidization in Liquid-Solid and Gas-Liquid-Solid Systems, Chem. Eng. J., **24**, 143 - 50, 1982.
75. Narayanan, C.M. and Das, Shrijita, Computer Aided Design and Performance Analysis of Inverse Fluidized Bed Biofilm Reactors With Special Reference to Bioplastic Synthesis, Adv. Chem. Eng. Sci., **6**, 130-39, 2016. http://dx.doi.org/10.4236/aces.2016.62015
76. Narayanan, C.M., Das, Shrijita and Pandey, Aditi, Green Technologies for Manufacture of Valuable Products from Food Wastes and Agricultural Wastes, Chapter 1, Volume 2, 1-54, Handbook of Food Bioengineering, Elsevier, 2017.
77. Narayanan, C.M. and Pandey, Aditi, Studies on Biodiesel Synthesis Using Nanosilica Immobilized Lipase in Inverse Fluidized Bed Bioreactors, J. Advances Chem., **15**(1), 6072 – 86, 2018.
78. Sokol, W., Treatment of Refinery Wastewater in a Three Phase Fluidized Bed Bioreactor with a Low Density Biomass Support, Biochem. Eng. J., **15**, 1-10, 2003.
79. Sokol, W. and Korpal, W., Phenolic wastewater treatment in a three-phase fluidised bed bioreactor containing low density particles, J. Chem. Technol. Biotech., 80, 884 - 91, 2005.
80. Rajasimman, M. and Karthikeyan, C., Aerobic Digestion of Starch Wastewater in a Fluidized Bed Bioreactor with Low Density Biomass Support, J. Hazard. Mater., **143**, 82-86, 2007.

81. Rajasimman, M. and Karthikeyan, C., Optimization Studies in an Inverse Fluidized Bed Bioreactor for Starch Wastewater Treatment, Int. J. Environ. Res., **3**, 569 - 74, 2009.
82. Haribabu, K. and Sivasubramanian, V., Biodegradation of Organic Content in Wastewater in Fluidized Bed Bioreactor Using Low Density Biosupport, Desalin. Water Treat., **57**, 4322 - 27, 2016.
83. Quevedo, J.A., Patel, G. and Pfeffer, R., Removal of Oil from Water by Inverse Fluidization of Aerogels, Ind. Eng. Chem. Res., **48**, 191 - 201, 2009.
84. Kennedy, K. J. and Droste, R. L., Anaerobic Wastewater Treatment in Downflow Stationary Fixed Film Reactors, Water Sci Technol., **24**, 157-77, 1991.
85. Narayanan, C. M., Performance Analysis of Anaerobic Downflow Stationary Fixed Film (DSFF) Bioreactors and Software Development, Int. J. Chem. React. Eng., **7**, A61, 2009.
86. Pandey, Aditi and Narayanan, C. M., Studies on Synthesis of Lactic Acid from Agricultural and Food Wastes in Downflow Stationary Fixed Film (DSFF) Bioreactors, Int. J. Trans. Phenom., **14**, 241–54, 2017.
87. Narayanan, C. M. and Vikas Narayan, Advances in Biogas Manufacture : Bioreactor Design and Biogas Utilization, Lambert Academic Publishing, Mauritius, 2020. **ISBN : 978 – 620 – 3 – 19653 – 5.**
88. Hall, E.R., Jovanovic, M. and Pejic, M., Pilot Studies of Methane Production in Fixed Film and Sludge Blanket Anaerobic Reactors, Proc. Fourth Bioenergy Res. Dev. Seminar, Winnipeg, March 29-31, 1982.
89. Lettinga, G., Roersma, R. and Grin, P., Anaerobic Treatment of Raw Domestic Sewage at Ambient Temperatures Using a Granular Bed UASB Reactor, Biotechnol. Bioeng., **25**, 1701 - 23, 1983.
90. Liu, Y., Xu, H. L., Yang, S. F. and Tay, J. H., Mechanisms and Models for Anaerobic Granulation in Upflow Anaerobic Sludge Blanket Reactor, Water Res., **37**, 661 - 73, 2003.
91. Seghezzo, L., Zeeman, G., van Lier J. B., Hamelers, H. V. M., Lettinga, G., A Review: The Anaerobic Treatment of Sewage in UASB and EGSB Reactors, Bioresour. Technol., **65**, 175 - 90, 1998.
92. Narayanan, C. M. and Vikas Narayan, Multiparameter Models for Performance Analysis of Upflow Anaerobic Sludge Blanket (UASB) Bioreactors, J. Chem. Tech. Biotech., **83**, 1170-76, 2008.
93. Kalyuzhnyi, S.V., Fedorovieh, V. V. and Lens, P., Dispersed Plug Flow Model for Upflow Anaerobic Sludge Bed Reactors with focus on Granular Sludge Dynamics, J. Ind. Microbiol. Biot., **33**, 221 - 37, 2006.
94. Narayanan, C. M., Samui, E., Chatterjee, A. and Das, S., A Parametric Study of Performance Characteristics of Upflow Anaerobic Sludge Blanket Bioreactors, Res. J. Chem. Sciences, **2**(6), 12-20, 2012.
95. Narayanan, C. M. and Vikas Narayan, Software Package for Process Simulation and Computer Aided Design of Expanded Granular Sludge Bed Reactors, Ind. Chem. Engr., **50**(2), 122-28 , 2008.

Chapter - 4

Medium Sterilization Systems and Processes

In many biotechnological processes, it is often desirable to sterilize the feed solution before feeding it to the bioreactor. In industries such as drugs and pharmaceutical industries, it would be essential to use a pre-sterilized medium or substrate and in the case of aerobic processes, the air admitted would also have to be sterilized in advance. This is particularly true in processes handling synthetic substrates. While handling complex substrates such as wastewater or industrial effluents, pre-sterilization of feed solution shall not be required, neither shall be required to use sterile air.

Sterilization means destruction of all microbes present in the feed solution and thus making it microbe-free, so that when the inoculum of the specific microbe is seeded to it, only that particular microbial culture shall flourish in the medium, leading to the formation of the desired product(s) only (formation of side products gets avoided). We, therefore, define a parameter called del factor (∇) to depict degree of sterilization attained (or to predict the efficacy of the sterilization process) as

$$\nabla = \text{del factor} = ln\left(n_0 / n\right) \tag{4.1}$$

where n_0 = microbial population (number of microbes per unit volume) at t = 0 (at the start of sterilisation process)

n = microbial population at the end of the sterilisation process.

The higher the value of ∇, the larger is the degree of sterilisation achieved. For example, if $n_0 = 10 \times 10^{10}$ cells/m^3 and $n = 1.0 \times 10^{-3}$ cells/m^3, then

$$\nabla = ln\left[\left(10 \times 10^{10}\right) / \left(1.0 \times 10^{-3}\right)\right] = ln\left(10^{14}\right) = 32.236$$

Thermal sterilization is the one that is most popularly employed in the case of liquid substrates. Sterilization by filtration is used for sterilizing gases such as air. Thermal sterilization is conducted in three stages such as the heating, holding and cooling stages. During the heating cycle, the substrate solution is heated to the sterilization temperature, T_{st} (by steam sparging, by jacketed heating or by using a heat exchanger), it is then held at that temperature for a specific period of time during the holding stage and finally, cooled to the operating temperature of the bioreactor (usually, ambient temperature) during the cooling cycle. The ultimate del factor (ultimate degree of sterilization) attained shall be, therefore, the sum of those attained during the heating, holding and cooling stages. Thus,

$$\nabla\left(overall\right) = \nabla\left(heating\right) + \nabla\left(holding\right) + \nabla\left(cooling\right) \tag{4.2}$$

Experimental studies have demonstrated that the microbial death during thermal sterilization more or less follows first order kinetics (the rate of microbial destruction or the rate of sterilization is directly proportional to the microbial concentration). Thus,

$$(-dn/dt) = k_d n \tag{4.3}$$

where k_d = specific death rate, s^{-1}

n = microbial population at any time t

It has also been observed that the temperature - dependence of k_d could be expressed using an Arrhenius-type correlation:

$$k_d = k_{d0}\, exp(-E_d / RT) \tag{4.4}$$

The values of correlation constants, k_{d0} (s^{-1}) and E_d (in J/kmol) shall depend on the type of microbial culture handled.

Equation (4.3) can be now integrated after substituting the above correlation into it. This yields an expression for del factor as,

$$\nabla = ln(n_0 / n) = k_{do} \int_0^t exp(-E_d / RT)\, dt \tag{4.5}$$

Once we know *T* as a function of time, *t* (how does the temperature of the medium varies with time during the sterilization process), then that expression could be substituted in the above equation and the integral evaluated numerically, to obtain the value of del factor attained.

4.1 Batch Thermal Sterilization

Thermal sterilization maybe conducted as a batch process, the batch time being decided based on the extent of microbial destruction desired. Batch sterilization is simpler to conduct and control, as compared to continuous processes. However, in large capacity installations, continuous sterilization processes could be more desirable.

4.1.1 Heating Cycle

During the heating stage, the feed solution may be heated through any of the following means:

(a) by sparging steam through the substrate solution (this may be conducted in the bioreactor itself, if a tank reactor is being employed),

(b) by passing an isothermal heating fluid such as dry, saturated steam through the jacket or through a submerged heating coil (could be practiced in the reactor itself, in case of tank reactors),

(c) bypassing an non-isothermal heating fluid through the jacket or through the submerged heating coil,

(d) by using an external heat exchanger (such as a plate heat exchanger) which utilizes an isothermal heating fluid (such as dry, saturated steam) or a non-isothermal heating fluid.

Steam Sparging

Sparging of steam into the feed solution present in the sterilization tank is the most straight- forward process. In this case, all the steam condenses inside the tank, rejecting its latent heat to the feedstock and thereby heating it from temperature T_0 to temperature T_{St}. The process occurs quite fast, but all the condensate formed mixes with the feed solution and thereby dilutes it. This mode of heating can be, therefore, recommended only if such dilution of fermentation medium is not unacceptable. Also, since the process is fast, little microbial death occurs during this heating cycle. Microbial death (and thereby sterilization of the medium) chiefly occurs during the holding and cooling periods that follow.

If dry, saturated steam at temperature T_S (with enthalpy H_S J/kg) is sparged at the rate of $\dot{m}_S$ kg/s through M kg of substrate solution in the tank, then at the end of time (t), we shall have ($M + \dot{m}_S t$) kg of diluted sterilised medium (substrate solution) at temperature T in the tank. A heat balance (taking reference temperature = T_0) thus yields,

$$(\dot{m}_S t)(H_S - h_L) = MC_{Pm}(T - t_0) + (\dot{m}_S t)C_{PL}(T - T_0) \tag{4.1.1}$$

or

$$T = T_0 + \left[(\dot{m}_S t)(H_S - h_L) / \left(MC_{Pm} + (\dot{m}_S t)C_{PL}\right)\right] \tag{4.1.2}$$

where H_S = specific enthalpy of dry, saturated steam at T_S, J/kg

h_L = specific enthalpy of liquid water at the reference temperature (T_0), J/kg

C_{Pm}, C_{PL} = specific heat of substrate solution and that of liquid water, J/(kg.K)

T = temperature of fermentation medium (substrate solution) at any time t

To note that since the value of H_S is usually retrieved from the standard steam table (which is based on a reference temperature of $0°$ C or 273 K), its value is required to be corrected for the chosen reference temperature (such as T_0) by subtracting the specific enthalpy of liquid water at T_0 (namely, h_L) from it. This is what has been indicated in the above equation. The value of h_L could also be retrieved from the steam table or may be computed as

$$h_L = C_{PL} T_0 \tag{4.1.3}$$

Taking C_{PL} = 4.18 kJ/(kg.K),

$$h_L = (4.18)\, T_0 \text{ kJ/kg} \tag{4.1.4}$$

After substituting the expression for T from equation (4.1.2) in equation (4.5), it could be integrated numerically to obtain the value of del factor (∇) attained.

The main drawback of the above process is that it consumes a large amount of steam and all the condensate gets mixed with the feedstock, thereby dilutes it significantly.

Circulation of Isothermal Heating Fluid (Steam) through Jacket or Heating Coil

We may pass an isothermal heating fluid (such as dry, saturated steam) through the jacket of the vessel or through a submerged helical coil (the coil remains submerged in the substrate solution) and thereby accomplish heating of the feedstock to the sterilization temperature. In such a case, the substrate solution in the tank must be agitated vigorously using a mechanical impeller (turbine, paddle, propeller) to distribute the heat uniformly. For such a system, a heat balance yields,

$$M\, C_{Pm}\, dT = UA(T_S - T)\, dt \tag{4.1.5}$$

On integration and re-arrangement, we get

$$T = T_S - (T_S - T_0)exp\left[-(UA)t/(MC_{Pm})\right] \tag{4.1.6}$$

where U = overall heat transfer coefficient, $W/(m^2.K)$

A = heat transfer area, m^2

In this case, heat transfer from the heating fluid to the feed solution occurs in three consecutive steps. As a result, we can evaluate the overall heat transfer coefficient (U) by summing up the resistances to all these three steps. For example, heat is first transferred from steam in the jacket to the outer wall of the vessel or reactor. This involves condensation heat transfer and since the magnitude of condensation heat transfer coefficient (h_0) is substantially large, resistance to this step shall not be controlling. Condensation heat transfer coefficient for steam is around 10,000 W/(m^2. K) and with allowable error, we may assume that this steam-side heat transfer coefficient (h_0) remains constant at this value. In other words, in most design calculations on this system, we may safely assume,

$$h_0 = 10{,}000 \text{ W/(m}^2\text{. K)} \tag{4.1.7}$$

From the outer surface of the tank, heat is then transferred to the inner surface by conduction. The resistance to this step is controlled by the thermal conductivity (k_P) of the material of construction of the tank wall and the tank wall thickness. However, in most situations, the resistance to this conduction heat transfer shall be of very low magnitude and could be neglected without any serious error.

Finally, heat is transferred from the inner surface of the tank to the feed solution. The magnitude of the heat transfer coefficient (h_i) governing this step shall depend on the impeller characteristics (type of impeller used to agitate the medium, impeller speed, impeller diameter). Experimental correlations have been reported by different researchers for the estimation of h_i and most of them are of the following form:

$$Nu = C_1 (Re_m)^a (Pr)^b (\mu_f / \mu_W)^c \tag{4.1.8}$$

where Nu = Nusselt number = $(h_i D_t / k_f)$ (4.1.9)

Re_m = mixing Reynolds number (also called impeller Reynolds number)

$= (nD_a^2 \rho_f / \mu_f)$ (4.1.10)

Pr = Prandtl number of substrate solution = $(C_{Pm}\mu_f / k_f)$ (4.1.11)

D_t = ID (inside diameter) of the tank

n, D_a = speed (rps) and diameter of impeller

$C_{Pm}, \rho_f, \mu_f, k_f$ = specific heat, density, viscosity and thermal conductivity respectively of substrate solution at the average medium temperature.

μ_W = viscosity of substrate solution at the inner wall temperature of the tank, kg/(m.s)

The values of correlation constants (C_1, a, b, c) are listed in the table (4.1).

Table 4.1: Values of correlation constants of equation (4.1.8)

Type of Impeller	C_1	a	b	c	Re_m	Ref. no.
Disk, flat blade turbine	0.54	(⅔)	(⅓)	0.14	$40 - 3\times 10^5$	[1]
Pitched blade turbine	0.53	(⅔)	(⅓)	0.24	80 – 200	[2]
Paddle	0.36	(⅔)	(⅓)	0.21	$300 - 3 \times 10^5$	[2], [3]
Propeller	0.54	(⅔)	(⅓)	0.14	2000	[4]
Anchor	0.36	(⅔)	(⅓)	0.18	$300 - 4 \times 10^4$	[2]

Once we have estimated the values of h_0 and h_i, the value of the overall heat transfer coefficient (U_0 or U_i) could be evaluated as

$$\left(1/U_0\right)=\left(1/h_0\right)+\left(1/h_i\right)\left(A_0/A_i\right)+R_d \quad (4.1.12)$$

$$\left(1/U_i\right)=\left(1/h_i\right)+\left(1/h_0\right)\left(A_i/A_0\right)+R_d \quad (4.1.13)$$

where R_d = overall combined dirt factor, $(m^2.K)/W$

U_0 = overall heat transfer coefficient based on heat transfer area A_0, W/($m^2.K$)

U_i = overall heat transfer coefficient based on heat transfer area A_i, W/($m^2.K$)

To note that,

$$\left(U_0A_0\right)=\left(U_iA_i\right)=\left(UA\right) \quad (4.1.14)$$

The dirt factor (R_d) accounts for the additional resistance to heat transfer that comes into play due to fouling of the heat transfer surfaces (the inner and outer walls/surfaces of the tank/reactor). Typical values of minimum dirt factor specified by TEMA (tubular exchangers manufacturers' association) are listed in table (4.2). To note that the dirt factor is nothing but the reciprocal of fouling coefficient (h_d).

Table 4.2: Typical values of overall combined dirt factor (R_d)

Process Fluid	Dirt factor (R_d), (m^2. K)/w
Vegetable oils	6.0×10^{-4}
Brine, organic liquids, refrigerants	2.0×10^{-4}
Sea water	0.9×10^{-4}
River water	2.0×10^{-4}
Boiler feed water	0.9×10^{-4}
Fuel oil	9.0×10^{-4}
Quenching oil	7.0×10^{-4}
Machine and transformer oils	2.0×10^{-4}

Source: TEMA

If medium heating is performed by passing an isothermal heating fluid (dry, saturated steam) through a submerged heating coil (a helical oil that is kept submerged in the feed solution), then also the heat balance equation shall be same as that given in equation (4.1.6). The steam- side heat transfer coefficient (in this case, it is h_i) can be taken equal to around 10,000 W/(m^2. K), as it is not controlling. Heat is transferred from steam to the inner surface of the helical coil, then it is conducted to the outer surface (the resistance to this conduction heat transfer can be treated as negligible, as mentioned earlier) and finally, heat is transferred from the outer surface of the coil to the substrate solution in the tank (which is well-agitated using a mechanical impeller). The heat transfer coefficient (h_0) that decides the rate of heat transfer from coil outer surface to substrate solution shall be a function of the impeller Reynolds number, Re_m (defined in the equation - 4.1.10) and the type of impeller used. Chilton and coworkers [3] have proposed the following experimental correlation for the estimation of h_0, when the solution is being agitated by a paddle impeller :

$$Nu = 0.87\left(Re_m\right)^{2/3}\left(Pr\right)^{1/3}\phi \tag{4.1.15}$$

where $Nu = \left(h_0 D_t / k_f\right)$ (4.1.16)

ϕ = viscosity correction factor = $\left(\mu_f / \mu_W\right)^{0.14}$ (4.1.17)

The above correlation is valid for $300 \le Re_m \le 4.0\times10^5$.

Oldshue and Gretten [5] have proposed an alternate correlation for the estimation of h_0, for the case when the solution is being agitated by a disc flat blade turbine :

$$\left[h_0 D_0 / k_f\right] = 0.17\left(Re_m\right)^{0.67}\left(Pr\right)^{0.37}\left(D_a / D_0\right)^{0.1}\left(D_0 / D_t\right)^{0.5} \tag{4.1.18}$$

where D_0 = outer diameter (OD) of coil tube (OD of tube used for the construction of the helical coil).

The above correlation is valid for $400 \le Re_m \le 2.0\times10^5$.

Once the value of h_0 has been computed from any one of the above correlations and h_i is taken equal to around 10,000 W/(m^2.K), the value of overall heat transfer coefficient (U_0 or U_i) could be computed from equation (4.1.12) or (4.1.13).

The del factor (∇) attained can be thereafter evaluated through numerical integration of equation (4.5), after substituting the expression for T from equation (4.1.6).

For the heating cycle, the use of an isothermal heating fluid such as dry, saturated steam is most preferred, since the condensation heat transfer coefficient, as stated earlier, is significantly large and this helps in attaining a high rate of heat transfer, though the use of a non-isothermal heating fluid is not unusual. For the cooling operation, on the other hand, the use of an isothermal cooling fluid (such as a refrigerant) shall be difficult on many occasions and the use of a non- isothermal coolant (such as cooling water) would be more handy.

Circulation of Non-isothermal Heating Fluid Through Jacket or Heating Coil

If heating of the substrate solution is performed by passing a non-isothermal heating fluid through the vessel jacket or through the submerged heating coil (the hot liquid entering at temperature T_{h1} and leaving at T_{h2}), then the heat balance equation gets modified to

$$M\,C_{Pm}\,dT = \dot{m}_h\,C_{Ph}\left(T_{h1} - T_h\right)dt \tag{4.1.19}$$

Also,

$$\dot{m}_h\,C_{Ph}\left(T_{h1} - T_h\right) = UA\left(-\Delta T\right)_{ln} \tag{4.1.20}$$

where $\left(-\Delta T\right)_{ln}$ = logarithmic mean temperature difference (LMTD)

$$= \left(T_{h1} - T_h\right) / ln\left[\left(T_{h1} - T\right) / \left(T_h - T\right)\right] \tag{4.1.21}$$

We have, nevertheless, assumed that the effective temperature difference driving force for heat transfer is equal to the logarithmic mean temperature difference, $\left(-\Delta T\right)_{ln}$. This has been observed to be a reasonably accurate assumption in the case of the present systems handled, such as jacketed vessels and submerged helical coils.

Substituting the expression for $\left(-\Delta T\right)_{ln}$ from equation (4.1.21) in equation (4.1.20) and rearranging, we get

$$T_h = T + \left(T_{h1} - T\right) exp\left[-UA / \left(\dot{m}_h C_{Ph}\right)\right] \tag{4.1.22}$$

Equation (4.1.19) can be now integrated after substituting the above expression for T_h in it. The result is,

$$ln\left[1 / \left(1 - \epsilon\right)\right] = C\left[1 - exp\left(-NTU\right)\right]t \tag{4.1.23}$$

where $\epsilon = \left(T - T_0\right) / \left(T_{h1} - T_0\right)$ (4.1.24)

$$C = \left(\dot{m}_h\,C_{Ph}\right) / \left(M\,C_{Pm}\right) \tag{4.1.25}$$

(NTU) = number of transfer units = $\left(UA\right) / \left(\dot{m}_h C_{Ph}\right)$ (4.1.26)

For the computation of the overall heat transfer coefficient (U), we need to estimate the values of individual heat transfer coefficients (h_i and h_0). In the case of heating performed using a submerged heating coil, the value of h_0 (heat transfer coefficient between the outer surface of the helical coil and the substrate solution) shall be the same as that computed from equation (4.1.15) or (4.1.18). For the computation of h_i (heat transfer coefficient between the heating fluid and the inner surface of the heating coil), it has been recommended [6] that the same correlation that has been proposed for forced convection heat transfer in flow through straight tubes or pipes can be employed, except that a correction factor (C_{hc}) must be incorporated into it. In the other words,

$$h_i\left(helical\,coil\right) = h_i\left(straight\,tube\right)C_{hc} \tag{4.1.27}$$

where $C_{hc} = 1.0 + 3.5\left(D_i / D_H\right)$ (4.1.28)

D_i = ID of coil tube (inside diameter of tube used for construction of the helical coil)

D_H = diameter of the helix (see fig. 4.1)

Figure (4.1): Helical Coil

To note that the length of the helical coil = ($\pi D_H n$), where n is the number of turns. The above correlation (4.1.27) incorporating the correction factor is valid for forced convection heat transfer in turbulent flow through the helical oil only ($Re_t \geq 10{,}000$, which is so in many industrial applications, no doubt).

The forced convection heat transfer coefficient in turbulent flow of Newtonian fluids through a straight tube or pipe, namely h_i (st. tube), can be estimated from the Dittus-Boelter equation, modified by Sieder and Tate [7,8] :

$$Nu = 0.023\left(Re_t\right)^{0.8}\left(Pr\right)^{1/3}\phi \tag{4.1.29}$$

where $Nu = \left(h_i D_i / k_f\right)$ (4.1.30)

$Re_t = \left(D_i G_t / \mu_f\right)$ (4.1.31)

G_t = mass velocity of fluid through the tube/pipe = $\dot{m} / \left(\pi D_i^2 / 4\right)$ (4.1.32)

ϕ = viscosity correction factor (defined earlier, see equation - 4.1.17)

It is important to keep in mind that the above correlation is valid for heat transfer in fully developed turbulent flow of the fluid only (for $Re_t \geq 10{,}000$). It is also applicable to all Newtonian fluids, ***except water***.

For fully developed turbulent flow of water through a straight tubes or pipe (at $Re_t \geq 10{,}000$), the forced convection heat transfer coefficient (h_i) may be computed from the *dimensional* correlation reported by Perry [9] and subsequently modified by Narayanan and Bhattacharya [8]. This correlation is based on the graphical data reported by Eagle and Ferguson [10] :

$$h_i\left(st.tube\right) = \left(1057.0\right)C_f\left[1.352 + 0.02t_{am}\right]\left(G_t / \rho_f\right)^{0.8} / \left(D_i\right)^{0.2} \tag{4.1.33}$$

where C_f = correction factor [8] = $0.1864 ln\left(1/D_i\right) + 0.22455$ (4.1.34)

t_{am} = arithmetic mean temperature of water, °C

When heating of the substrate solution is accomplished by circulating a non-isothermal heating fluid through the jacket (the substrate solution inside the tank is being agitated vigorously), then the value of h_i (heat transfer coefficient between the inner surface of the tank/reactor and the substrate solution) shall be the same as that predicted by the correlation given in equation (4.1.8). In the case of h_0 (heat transfer coefficient between the heating fluid in the jacket and the outer surface of the tank), it has been recommended that it can be estimated by assuming that the flow of the heating fluid through the jacket is equivalent to flow through the annular space between two concentric pipes/columns, the outer diameter of inner pipe/column being the OD of the tank, D_{to} and the ID of the outer pipe/column being the diameter of the jacket, D_j . To note that,

$$D_j = \left(D_{to} + 2\delta\right) \tag{4.1.35}$$

where δ = jacket clearance = clearance between the inner surface of the jacket and the outer surface of the tank.

Based on this approximation, the value of h_0 may be computed from equation (4.1.29) or equation (4.1.33), after replacing the ID of the tube (D_i) by the equivalent diameter (D_e) of the annulus. By definition,

$$D_e = 4\left(R_H\right) \tag{4.1.36}$$

where R_H = hydraulic radius = (a_a / l_W) (4.1.37)

a_a = flow area = $(\pi/4)\left[\left(D_j\right)^2 - \left(D_{to}\right)^2\right]$ (4.1.38)

l_W = wetted and heated perimeter = $\left(\pi D_{to}\right)$ (4.1.39)

Therefore, the expression for D_e shall be,

$$D_e = \left[\left(D_j\right)^2 - \left(D_{to}\right)^2\right] / D_{to} \tag{4.1.40}$$

Equation(4.1.29), thus, gets modified to

$$Nu_a = 0.023 (Re_a)^{0.8} (Pr_a)^{1/3} \phi_a \qquad (4.1.41)$$

where Nu_a = annulus Nusselt number = $(h_0 D_e / k_{fa})$ (4.1.42)

Re_a = annulus Reynolds number = $(D_e G_a / \mu_{fa})$ (4.1.43)

$G_a = (\dot{m}_a / a_a)$ (4.1.44)

$\dot{m}_a$ = mass flow rate of annulus fluid (heating fluid)

Pr_a = Prandtl number of annulus fluid (heating fluid) = $(C_{Pa}\mu_{fa} / k_{fa})$ (4.1.45)

ϕ_a = viscosity correction factor for annulus fluid = $(\mu_{fa} / \mu_{Wa})^{0.14}$ (4.1.46)

To note that $[C_{Pa}, \mu_{fa}, k_{fa}]$ are property values (specific heat, viscosity and thermal conductivity respectively) of annulus fluid (heating fluid) at its average bulk temperature and μ_{Wa} is the viscosity of annulus fluid at the outer surface temperature of the tank.

Similarly, equation (4.1.33) gets modified to (this correlation is to be used when the heating fluid employed is hot water),

$$h_0 = (1057.0) C_f [1.352 + 0.02 t_{am}] (G_a / \rho_{fa})^{0.8} / (D_e)^{0.2} \qquad (4.1.47)$$

It must be, nevertheless, kept in mind that both of the above correlations (equations - 4.1.41 and 4.1.47) are valid for fully developed turbulent flow of the heating fluid through the jacket ($Re_a \geq 10{,}000$).

As stated earlier, when a non-isothermal heating fluid is being employed, the process shall be slower and would demand a longer heating period (since liquid-liquid heat transfer coefficients are much lower in magnitude than condensation heat transfer coefficient).

Heating using an external heat exchanger

Heating of substrate solution by passing it through an external heat exchanger adds flexibility to the operation. The cost of installation, maintenance and operation of the heat exchanger shall be, no doubt, additional. Plate exchangers are popularly employed in this connection, though the use of a double-pipe or shell-and-tube heat exchanger is not unusual.

If an isothermal heating fluid (such as dry, saturated steam) is being used in the heat exchanger, then the scheme shall be that sketched in figure (4.2). The substrate solution leaves the tank at temperature T and is circulated through the heat exchanger at $\dot{m}$ kg/s, where it is heated to temperature T and is thereafter recycled back to the tank. Dry, saturated steam enters the exchanger at temperature, T_S and the condensate (saturated liquid) leaves at the same temperature, T_S. The process is continued until the temperature of the substrate solution in the tank becomes equal to the sterilization temperature, T_{Si} (until T becomes equal to T_{Si}).

Since an isothermal heating fluid (saturated steam) is being used, the effective temperature difference driving force, $(-\Delta T)(eff)$, shall be equal to the logarithmic mean temperature difference, $(-\Delta T)_{ln}$. This is irrespective of the type of exchanger used. Accordingly, the heat balance equations shall be,

$$M C_{Pm} dT = \dot{m} C_{Pm} (T' - T) dt \qquad (4.1.48)$$

Also, $\dot{m}C_{Pm}(T'-T)=UA(-\Delta T)_{ln}$ (4.1.49)

where $(-\Delta T)_{ln}=(T'-T)/ln[(T_S-T)/(T_S-T')]$ (4.1.50)

From the above two equations,

$$T'=T_S-(T_S-T)exp[-(UA)/(\dot{m}C_{Pm})] \quad (4.1.51)$$

Substituting the above expression for T' in equation (4.1.48) and integrating, we get

$$ln[(T_S-T_0)/(T_S-T)]=(\dot{m}/M)[1-exp(-UA/\dot{m}C_{Pm})]t \quad (4.1.52)$$

The above equation implicitly predicts how the temperature of the substrate solution (T) varies with time (t). This is to be used in equation (4.5) to evaluate the del factor (∇) attained.

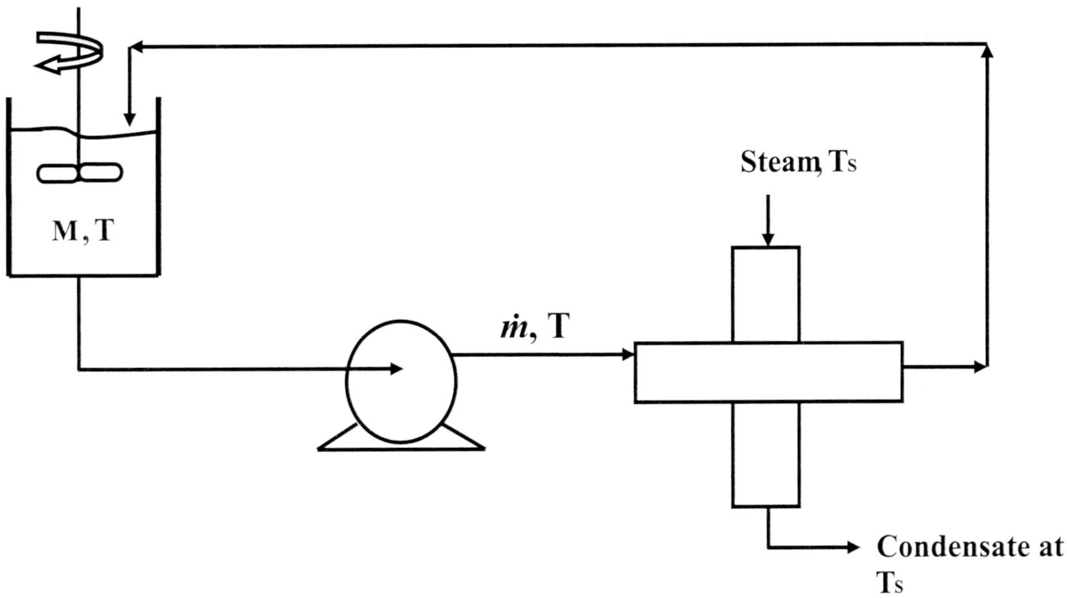

Figure 4.2: Batch sterilization using external heat exchanger (isothermal heating fluid)

As stated earlier, the above heat balance equation is valid for all types of heat exchangers. The value of U (overall heat transfer coefficient), however, shall depend on the type of heat exchanger used. Let us, for example, consider a plate heat exchanger employing a looped flow arrangement (see figure 4.3). The exchanger consists of a number of vertical, rectangular, corrugated plates (called *thermal plates*, usually made of stainless steel) that are stacked one aside the other. The entire stack is pressed between two support plates (called *backing plates*), one fixed and the other movable. The two fluids (the cold fluid and the hot fluid) flow through the narrow channels between the corrugated plates. Each plate has four corner openings called *ports*. Two of the ports are fully gasketed (fully surrounded by a gasket) and the other two are partially gasketed (partially surrounded by the gasket). The former are called *transfer ports* and the latter *flow ports*. The gaskets are usually made of compressed asbestos, synthetic rubber or teflon.

In the looped flow arrangement (figure - 4.3), each liquid stream (the cold fluid stream as well as the hot fluid stream) is divided into a number of parallel streams and each stream is pumped through an alternate flow channel. As a result, when one fluid (say, the cold fluid) flows down on one side of the plate, the other fluid (the hot fluid) flows up on the other side of the plate (see figure - 4.3). The flow pattern is thus very close to true countercurrent flow.

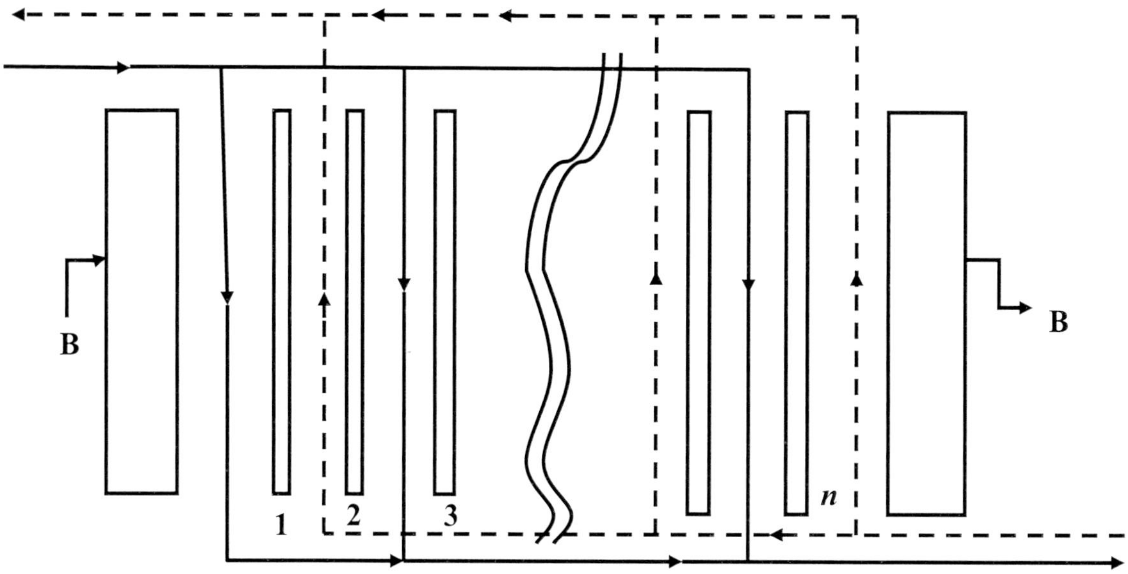

Figure 4.3: Plate heat exchanger (looped flow arrangement). B: Backing plates. 1,2,3,......n: thermal plates.

What is to be kept in mind is that since the hot fluid is divided into N_h streams and the cold fluid divided into N_C streams and each stream traverses only one of the flow channels, the flow rate of the hot fluid through the exchanger shall be ($\dot{m}_h / N_h$) and that of the cold fluid, ($\dot{m}_C / N_C$). If the number of thermal plates (N_P) used in the exchanger is odd, then

$$N_C = N_h = (N_P + 1)/2 \tag{4.1.53}$$

If N_P is even, then there will be ($N_P/2$) number of one stream and $(N_P+1)/2$ number of other stream. For example, if $N_C = (N_P/2)$, then $N_h = (N_P/2)+1$.

For computing the value of the overall heat transfer coefficient (U), we have to first evaluate the individual heat transfer coefficients such as that for the cold fluid (h_C) and that for the hot fluid (h_h). They can be estimated from any of the reliable experimental correlations reported in literature (for forced convection heat transfer in flow between two parallel, corrugated plates). One such correlation [11, 12] is reproduced below:

$$Nu = C_1 (Re)^m (Pr)^{1/3} \phi \tag{4.1.54}$$

where $Nu = (h_C D_e / k_{fC}) or (h_h D_e / k_{fh})$ (4.1.55)

$$Re = (D_e G / \mu_f) \tag{4.1.56}$$

$$G = \dot{m} / (Bz) \tag{4.1.57}$$

B, z = width of each plate and spacing between successive plates respectively.

For $Re \geq 150$, $C_1 = 0.45, m = 0.61$

For $$Re < 150,\ C_1 = 0.60, m = 0.50 \tag{4.1.58}$$

In the case of looped flow arrangement,

$$\dot{m} = \left(\dot{m}_C / N_C\right) or \left(\dot{m}_h / N_h\right) \tag{4.1.59}$$

The equivalent diameter (D_e) of the flow channel can be computed as

D_e = (4.0) (hydraulic radius)

= (4.0) (flow area)/(welted and heated perimeter)

$$= (4.0)\ (Bz)/(2B) = (2z) \tag{4.1.60}$$

The above correlation can be used for both cold fluid as well as the hot fluid. Once the values of h_C and h_h are thus known, the value of overall heat transfer coefficient (U) can be estimated from equation (4.1.12) or (4.1.13). To note that for plate heat exchanger, $A_0 = A_i = A =$ projected area of each plate. Also, the interspacing between plates (z) shall be equal to gasket thickness (when compressed).

Since the plates are usually made of stainless steel (whose thermal conductivity is quite low), we may include conduction resistance offered by the plate also in the computation of U. Thus, for plate exchangers, equation (4.1. 12) or (4.1.13) gets reduced to

$$(1/U) = (1/h_C) + (1/h_h) + (\delta / k_P) + R_d \tag{4.1.60a}$$

where δ is the thickness of each plate and k_P is the thermal conductivity of plate material.

We may very well circulate a non-isothermal heating fluid (such as hot water or a high-boiling thermal liquid like dowtherm) through the heat exchanger. The scheme is sketched in figure (4.4). Here, as the substrate solution gets heated from T to T' in the exchanger at the rate of $\dot{m}$ kg/s, the heating fluid (that is circulated at $\dot{m}_h$ kg/s) gets cooled from temperature T_{h1} to T_h. The process shall be invariably slower and shall demand larger heating period.

The system performance in cases like this, is influenced by the flow pattern in the heat exchanger. If true countercurrent flow exists in the exchanger (fairly true in the case of double pipe heat exchangers), then the effective temperature difference driving force, $(-\Delta T)(eff)$, shall be equal to the logarithmic mean temperature difference, $(-\Delta T)_{ln}$. In cases otherwise, suitable correction factors would have to be incorporated.

Let us assume that a plate heat exchanger with looped flow arrangement is being used. As stated earlier, the flow pattern in such an exchanger is close to true countercurrent flow. However, since the plates are corrugated, the fluids traverse a tortuous path through the flow channels and consequently, the effective temperature difference driving force deviates slightly from logarithmic mean temperature difference. To account for this deviation, a correction factor (F) is to be incorporated, such that

$$(-\Delta T)(eff) = F(-\Delta T)_{ln} \tag{4.1.61}$$

Experimental investigations have demonstrated that $F = 0.942$, if the number of thermal plates (N_P) in the exchanger is odd and $F = 0.967$, if N_P is even. It can be thus seen that the value of F does not deviate significantly from unity.

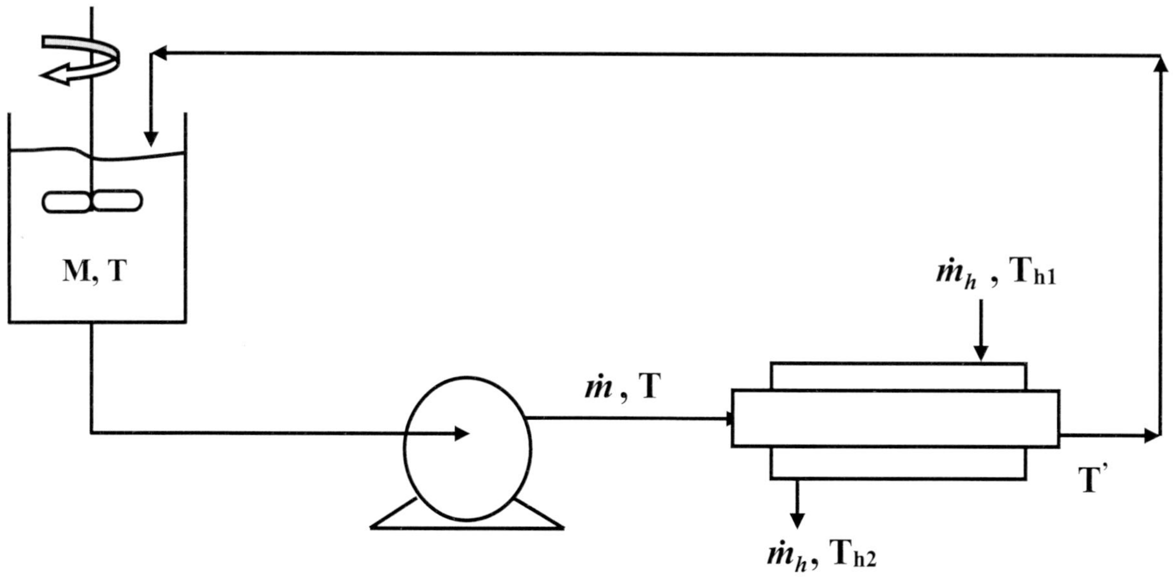

Figure 4.4: Batch sterilization using external heat exchanger (non - isothermal heating fluid).

The heat balance equations for the present system shall be

$$M\,C_{Pm}\,dT = \dot{m}\,C_{Pm}\left(T' - T\right)dt \tag{4.1.62}$$

or,
$$T' = T + \left(M / \dot{m}\right)\left(dT / dt\right) \tag{4.1.63}$$

Also, $M\,C_{Pm}\,dT = \dot{m}_h\,C_{Ph}\left(T_{h1} - T_h\right)dt$

or
$$T_h = T_{h1} - \left[\left(MC_{Pm}\right) / \left(\dot{m}_h C_{Ph}\right)\right]\left(dT / dt\right) \tag{4.1.64}$$

Further,
$$M\,C_{Pm}\,dT = FUA\left(-\Delta T\right)_{ln}\,dt \tag{4.1.65}$$

where
$$(-\Delta T)_{ln} = \left(T_{h1} - T'\right) - \left(T_h - T\right) / ln\left[\left(T_{h1} - T'\right) / \left(T_h - T\right)\right] \tag{4.1.66}$$

If we substitute for T' and T_h from equations (4.1.63) and (4.1.64) respectively in the above equation and then substitute the resultant expression for $(-\Delta T)_{ln}$ in equation (4.1.65) and rearrange it, we get,

$$dT / \left(T_{h1} - T\right) = \left[\left(K_1 - 1\right)\left(\dot{m} / M\right) / \left(K_1 C - 1\right)\right]dt \tag{4.1.67}$$

On integration,

$$T = T_{h1} - \left(T_{h1} - T_0\right)exp\left(-K_2 T\right) \tag{4.1.68}$$

where $K_2 = \left(K_1 - 1\right)\left(\dot{m} / M\right) / \left(K_1 C - 1\right)$ (4.1.69)

$$K_1 = exp\left[FUA\left(C - 1\right) / \left(\dot{m}C_{Pm}\right)\right] \tag{4.1.70}$$

$$C = \left(\dot{m}C_{Pm}\right) / \left(\dot{m}_h C_{Ph}\right) \tag{4.1.71}$$

The value of U can be estimated from equation (4.1.60a) after evaluating the individual heat transfer coefficients (h_C and h_h) from equation (4.1.54).

Plate heat exchangers are extremely compact. For the same heat transfer surface, a plate heat exchanger would occupy 50% less headroom (floor space) than a shell and tube heat exchanger. The plates are usually made of stainless steel, which is an expensive material of construction. Also, the operation of these exchangers is restricted by the gasket characteristics. In the present status, these exchangers are not suitable for operating at temperatures exceeding 150 °C and for operating pressure exceeding 20 atm. For thermal sterilization of feedstocks to bioreactors, these can be conveniently used, since sterilization processes hardly employ too high operating pressure and the operating temperature seldom exceeds 150 °C. Further, it is possible to conduct all the three operations, such as heating, holding and cooling, in a single exchanger. The entire stack of plates is divided into three or four sections by using connecting plates. The first section (the first set of plates) forms the preheating section (where the feed solution is preheated to around 80-85 °C), the second section acts as the heating section (where the feedstock is heated to the sterilization temperature), the third section constitutes the holding section and the last section forms the cooling section. The feed solution flows from one section to the other, until it is completely sterilized and cooled to the operating temperature of the bioreactor (fermentation temperature).

4.1.2 Cooling Cycle

After heating the feed solution to the sterilization temperature (T_{St}), it is held at that temperature (at T_{St}) for a specific period (holding period) and thereafter, it is cooled to the fermentation temperature (operating temperature of the bioreactor), which is usually around 25 − 30 °C.

Cooling can be accomplished by using an isothermal cooling fluid such as a refrigerant (liquid ammonia or freon) or by using a quenching fluid such as chilled brine (*quench cooling*). The refrigerant may be circulated through the vessel jacket or through a submerged cooling coil, it entering as saturated liquid at temperature T_S and leaving as saturated vapor at the same temperature. These vapors are to be condensed and recycled back to the jacket or the cooling coil. The process shall be fast, demanding less cooling time, but the use of such a refrigerant may not be handy or convenient in many industrial applications.

A still further alternative is to inject the substrate solution through an expansion valve into a vacuum chamber (flash drum). Due to the sudden decrease in pressure, a part of the solution vaporizes (flash vaporization) taking heat from the liquid bulk and as a result, the liquid / solution cools down. This is called *flash cooling*. Both quench cooling and flash cooling occur very fast (within a very short duration of time) and consequently, hardly any sterilization (microbial death) occurs during this cooling period.

A more conventional method is to use a non-isothermal cooling fluid such as cooling water. The process shall be, no doubt, slower, demanding much larger cooling time. If cooling of the substrate solution is affected by passing cooling water through the vessel jacket or through a submerged cooling coil, then the heat balance equation shall be the same as that derived for the heating cycle, such as equation (4.1.23), except for a few modifications. Thus,

$$ln\left[1/\left(1-\epsilon\right)\right]=C\left[1-exp\left(-NTU\right)\right]t \tag{4.1.72}$$

where $\epsilon=\left(T_{St}-T\right)/\left(T_{St}-T_{C1}\right)$ (4.1.73)

$$C = (\dot{m}_C C_{PC}) / (M C_{Pm}) \quad (4.1.74)$$

$$(\text{NTU}) = (UA) / (\dot{m}_C C_{PC}) \quad (4.1.75)$$

$\dot{m}_C$ = mass flow rate of cooling water

T_{C1} = inlet temperature of cooling water

To note that cooling is being affected from temperature T_{St} (sterilization temperature) to the fermentation temperature (say, T_0 itself). The computation of overall heat transfer coefficient (U) shall follow the same correlations as those discussed under heating cycle.

Cooling can very well be affected by passing the substrate solution through an external heat exchanger, such as a plate heat exchanger (with looped flow arrangement), with cooling water being used as the coolant in the exchanger. The scheme shall be then similar to that sketched in figure (4.4). The heat balance equation shall be similar to that developed for the heating cycle, such as equation (4.1.68). Thus,

$$T = T_{C1} + (T_{St} - T_{C1}) exp(K_4 t) \quad (4.1.76)$$

where $K_4 = (K_3 - 1)(\dot{m} / M) / (C - K_3)$ (4.1.77)

$$K_3 = exp[FUA(1 - C) / (\dot{m} C_{Pm})] \quad (4.1.78)$$

$$C = (\dot{m} C_{Pm}) / (\dot{m}_C C_{PC}) \quad (4.1.79)$$

In this case also, computation of the overall heat transfer coefficient (U) shall be based on the correlations discussed under heating cycle (in the previous subsection).

4.2 Continuous Thermal Sterilisation Processes

Continuous thermal sterilization of feed solutions is practiced mainly in large capacity installations (when the amount of substrate solution to be sterilized per hour is significantly large). As stated in the earlier section, heating of the substrate solution to the sterilization temperature using an isothermal heating fluid (such as dry, saturated steam) is most preferred, since the process shall be fast and the duration of heating cycle shall be very short. Due to the same reason, there shall not be any significant microbial death during this period. Heating of feed solution by sparging steam into the feedstock is practiced in a good number of industries employing continuous sterilization process.

The hot substrate solution (at the sterilization temperature) is then passed through a long, well- insulated holding tube or tube coil. Since the tube or coil is well-insulated, the temperature of the substrate solution remains more or less constant at T_{St} (sterilization temperature) during this period (holding period). The residence time of the substrate solution in the holding tube or coil (namely, the holding period or the duration of the holding cycle) shall depend on the length of the tube or coil and the flow rate of the substrate solution. Since the volumetric flow rate of the feed solution is usually decided by the capacity of the installation, the tube length becomes the most important design parameter that is to be adjusted so as to obtain the desired holding time. Substantial degree of sterilization (microbial death) occurs during this holding period.

The substrate solution is thereafter cooled down to the fermentation temperature (operating temperature of the bioreactor). Quench cooling (cooling by using a quench fluid such as chilled brine) is utilized in a good number of industrial practices, mainly due to the fact the cooling is affected very fast (the substrate solution gets cooled down to the fermentation temperature within a short duration of time). Since the length of the cooling period is also thus very short, little microbial death occurs during this stage also. In a scheme like this therefore, microbial death and thereby sterilization of the medium occurs more or less during the holding period only.

Such a scheme of continuous thermal sterilization process is sketched in figure (4.5).

The feed solution is admitted to the sterilization tank at the rate of $\dot{m}$ kg/s and dry, saturated steam is also sparged into the tank at the rate of $\dot{m}_S$ kg/s. The hot substrate solution at the sterilization temperature (T_{St}) leaves the tank at the rate ($\dot{m} + \dot{m}_S$) kg/s. It is then pumped through the long, well-insulated holding tube or coil, where it gets almost completely sterilized.

The well-sterilized substrate solution (with little microbial count) at temperature T_{St} enters the cooling chamber, which is composed of a submerged cooling coil through which a quench liquid is circulated. The substrate solution cools down almost immediately to the fermentation temperature and is then fed to the bioreactor.

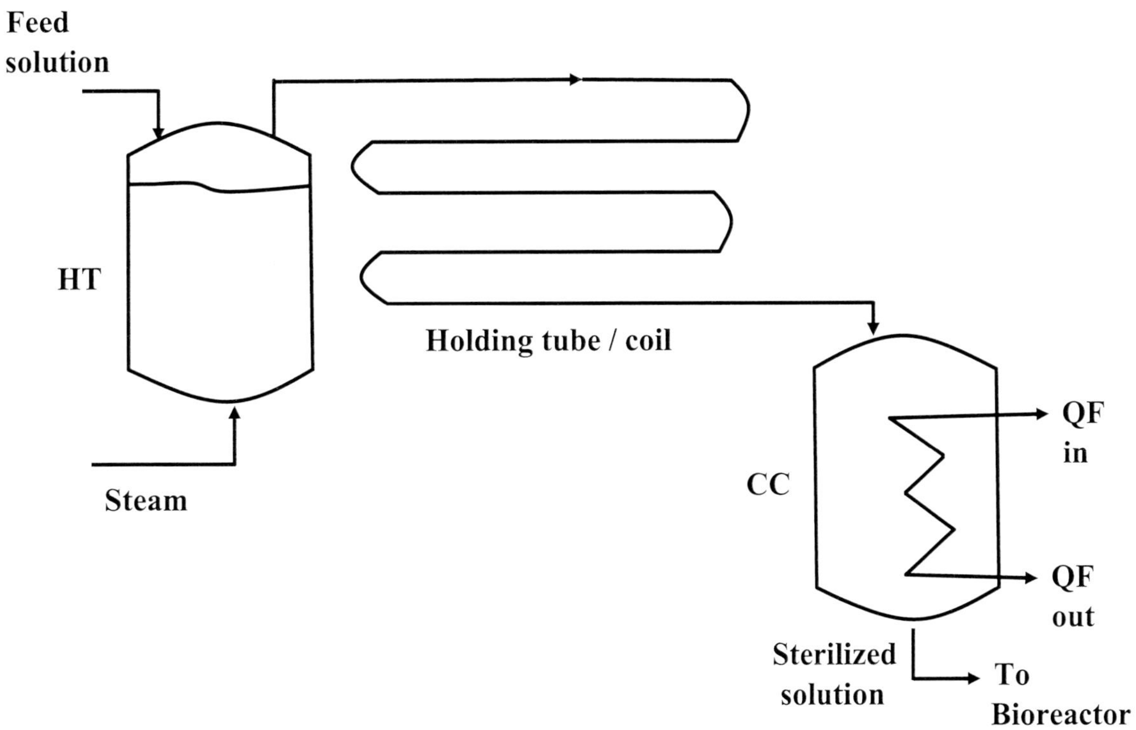

Figure 4.5: Continuous thermal sterilization (steam sparging, quench cooling). HT: Heating tank, CC: Cooling chamber, QF: Quench fluid.

As stated earlier, the required length (L) of the holding tube or coil is an important design parameter. A straightforward procedure for estimating this tube length is one that assumes true plug flow through the holding the tube or coil. Such an approach shall be, nevertheless, too idealistic and shall not necessarily predict a reliable estimate of the design parameter (namely, the tube length). A more accurate approach would be to assume that the substrate solution executes dispersed flow through the tube, the degree of axial dispersion being predicted by an appropriately defined axial dispersion coefficient (D_L). In such a case, the performance equation shall be similar to equation (3.2.18) of Chapter - 3. Thus,

$$-U_L\left(dn/dz\right)+D_L\left(d^2n/dz^2\right)=\left(k_d n\right) \quad (4.2.1)$$

where U_L = average velocity of substrate solution through the holding tube/coil, m/s.

If Q_0 is volumetric flow rate of substrate solution, then

$$U_L=Q_0/\left(\pi D_i^2/4\right) \quad (4.2.2)$$

The governing boundary conditions are,

BC 1: At z = 0 (at the inlet to the holding tube),

$$(U_L n_0)=U_L n\ (at\ z>0)-D_L\left(dn/dz\right)(at\ z>0) \quad (4.2.3)$$

BC 2: At z = L (at the outlet end of holding tube), $\left(dn/dz\right)=0$ (4.2.4)

Wehner and Wilhem [13] have reported an analytical solution to the above differential equation (based on the above prescribed boundary conditions) as

$$\left(n/n_0\right)=\left(4\beta\right)exp\left(Pe/2\right)/\left[(1+\beta)^2 exp\left(\beta Pe/2\right)-(1-\beta)^2 exp\left(-\beta Pe/2\right)\right] \quad (4.2.5)$$

where Pe = modified Peclet number = $\left(LU_L/D_L\right)$ (4.2.6)

$$\beta=\sqrt{1+\left(4k_d D_L/U_L^2\right)} \quad (4.2.7)$$

The value of axial dispersion coefficient (D_L) may be experimentally determined (using tracer experiments or RTD studies) or computed from any of the available experimental correlations such as equation (3.2.3) of Chapter - 3. For the convenience of readers, this correlation, proposed by Levenspiel [14] to predict axial dispersion in turbulent flow through straight, cylindrical pipes, is reproduced below:

For $10^4 \le Re \le 10^5$,

$$ln\left(N_d\right)=5.2886-0.8874ln\left(Re\right)+0.025\left[ln\left(Re\right)\right]^2 \quad (4.2.8)$$

For $Re>10^5$, $N_d=0.2=$ constant (4.2.9)

where N_d = dispersion number $=D_L/\left(D_i U_L\right)$ (4.2.10)

Re = Reynolds number = $\left[D_i U_L \rho_f/\mu_f\right]$ (4.2.11)

An excellent discussion on thermal sterilization using heat exchangers (double pipe, plate and shell-and-tube heat exchangers) has been presented by Narayanan [15] and also by Banerjee and Narayanan [16].

Example 4.1:

A stirred tank bioreactor (ID = 1.2 m, wall thickness = 3 mm) handles 2.0 m^3 of aqueous glucose solution at (30 °C) with an initial microbial count of 3.75 × 10^{12} per m^3 . Prior to fermentation, the glucose solution is to be sterilized such that the probability of microbes surviving sterilization shall be 1 in 100. The glucose solution is first heated to 393 K bypassing dry, saturated steam at 700 kPa (saturation temperature of steam = 438 K) through vessel jacket, the solution inside being agitated by a 120 rpm disk, flat bladed turbine impeller ($D_a = D_t / 3$). The solution is held at this temperature for a specific period of time and then transferred to a 2.5 m ID cooling tank fitted with a 12.5 mm ID cooling coil (OD = 15.5 mm, diameter of helix = 0.4 m, number of turns = 500), through which cooling water at 18 $^\circ$C is circulated at 1000 L/hr. The solution inside the tank is simultaneously agitated with a 120 rpm disk, flat bladed turbine (diameter = 0.4 m). Once the solution is cooled to 303 K, it is transferred back to the bioreactor. If the temperature dependence of specific death rate of microbes follows Arrhenius- type relationship with $k_{d0} = 5.7 \times 10^{39}$ hr^{-1}, $E_d = 2.834 \times 10^5$ kJ/kmol, determine the holding time required. Take an overall minimum drift factor of 0.0006 (m .K)/W for the cooling coil and 1.75×10^{-4}(m^2.K)/W for the jacketed reactor. The average property values of water and glucose solution such as density, viscosity, specific heat and thermal conductivity may be taken as equal to 1000 kg/m^3 , 0.001 kg/(m.s), 4.187 kJ/(kg.K) and 0.664 W/(m.K) respectively.

Solution :

Since the volume of substrate solution handled is 2.0 m ,

$$n_0 = \left(3.75\times10^{12}\right)\left(2.0\right) = 7.5\times10^{12}$$

And, $n =$ 1 in 100 = 0.01

Therefore, $\nabla(overall) = ln\left(n_0 / n\right) = 34.251$

Let us now compute the del factor attained during the heating cycle. For this, we have to first estimate the overall heat transfer coefficient (U_0), which in turn, depends on the individual heat transfer coefficients (h_i, h_0). Since steam is being circulated through vessel jacket, we can safely assume the steam-side heat transfer coefficient (h_0) as

$$h_0 = 10,000 W / \left(m^2.K\right)$$

The heat transfer coefficient (h_i) governing transfer of heat from inner surface of reactor vessel to the substrate solution depends on the type of impeller employed for agitating the substrate solution and the impeller characteristics. We may thus use equation (4.1.8), with correlation constants (C_1, a, b) retrieved from table (4.1.1). Thus, taking ϕ = viscosity correction factor = 1.0,

$$Nu = 0.54\left(Re_m\right)^{2/3}\left(Pr\right)^{1/3} \tag{i}$$

where $Re_m = \left(nD_a^{\,2}\rho_f / \mu_f\right)$ = (120/60) (0.4)^2 (1000) /(0.001) $= 3.2\times10^5$

The Reynolds number is thus within the prescribed range. Now,

$Pr = \left[C_P\mu_f / k_f\right] = 6.3057$

Therefore, from equation (i), $Nu = 4667.362$

and $h_i = (Nu)(k_f / D_t) = 2582.607$ W/$(m^2.K)$

Now, the value of overall heat transfer coefficient (U_0) can be computed as,

$$(1/U_0) = (1/h_i)\ (A_0 / A_i) + (1/h_0) + R_d \tag{ii}$$

Since volume of glucose solution taken = 2.0 m^3 and vessel ID = D_t = 1.2 m,

$$\left[\pi(1.2)^2/4\right]H = 2.0$$

or H = 1.7684 m

This is the height to which the substrate solution has been filled in the reactor tank. It is assumed that this is also the height within which the vessel has been jacketed. Therefore,

$$A_i = (\pi D_t H) + (\pi D_t^2 / 4) = 7.7976 \text{ m}^2$$

Since OD of reactor vessel = (1.2) + 2(0.003) = 1.206 m,

$$A_0 = \pi(1.206)(1.7684) + \pi(1.206)^2 / 4 = 7.8423\ m^2$$

Therefore, $(A_0 / A_i) = 1.0057$

Substituting the above value of (A_0 / A_i) and $R_d = 1.75 \times 10^{-4}(m^2.K)/W$ in equation (ii), we get

$$U_0 = 1505.068\ W/(m^2.K)$$

Now, the change of medium temperature with time during the heating cycle is predicted by equation (4.1.6):

$$T = T_S - (T_S - T_0)exp\left[-(U_0 A_0)t/(MC_{Pm})\right]$$

Substituting T_S = steam temperature = 438 K, T_0 = 303 K, M = 2000 kg and the known values of all other parameters, we get on re-arrangement,

$$T = 438 - 135\exp[-0.0014095t]$$

where t is in seconds. If t is in minutes,

$$T = 438 - 135\exp[-0.08457t] \tag{iii}$$

By substituting T = T_{St} = sterilisation temperature = 393 K in the above equation and solving for t, we get, t = 13.0 minutes

This is the required time duration of heating cycle. Now, from equation (4.5), the del factor attained during the heating cycle shall be,

$$\nabla\text{(heating)} = k_{d0}\int_0^{13} F(t)dt \tag{iv}$$

where $F(t) = exp(-E_d / RT)$.

Substituting $E_d = 2.854 \times 10^5$ kJ/kmol, R = 8.314 kJ/(kmol.K) and the expression for T from equation (iii), we get on re-arrangement,

$$F(t) = exp\left[-(77.8244)/\phi(t)\right]$$

where $$\phi(t) = 1 - 0.3082\,exp(-0.08457t). \tag{v}$$

For evaluating the integral of equation (iv) numerically, the values of $F(t)$ at different values of t are computed and are listed in Table (4.1.1), given below :

Table - 4.1.1

***t*, minutes**	***F*(*t*), dimensionless**
0.0	1.388×10^{-49}
1.0	7.01617×10^{-48}
2.0	2.04237×10^{-46}
3.0	3.7802×10^{-45}
4.0	4.80485×10^{-44}
5.0	4.45643×10^{-43}
6.0	3.166×10^{-42}
7.0	1.79217×10^{-41}
8.0	8.3462×10^{-41}
9.0	3.2839×10^{-40}
10.0	1.116×10^{-39}
11.0	3.3377×10^{-39}
12.0	8.9227×10^{-39}
13.0	2.1609×10^{-38}

Now, from trapezoidal rule,

$$I = \int_0^{13} F(t)\,dt = (h/2)\left[f_0 + 2(f_1 + f_2 + ...f_{n-1}) + f_n\right]$$

where h = step size = 1.0 min, $f_0 = 1.388 \times 10^{-49}$, $f_1 = 7.01617 \times 10^{-48}$ and so on, finally, $f_n = f_{13} = 2.1609 \times 10^{-38}$. Therefore,

$$I = 2.4614 \times 10^{-38} \text{ minutes}$$

Since $k_{d0} = 5.7 \times 10^{39}\,\text{hr}^{-1} = 0.95 \times 10^{38}\,\text{min}^{-1}$, ∇(heating) = $(0.95 \times 10^{38})\,(2.4614 \times 10^{-38}) = 2.34$

Let us now proceed to compute the del factor attained during the cooling cycle. Since a submerged cooling coil is being used, we can compute h_i (heat transfer coefficient between the inner surface of the coil and the cooling water) from equation (4.1.27) and (4.1.33). Thus, from equation (4.1.33),

$$h_i(\text{tube}) = (1057.0)C_f\left[1.352 + 0.02t_{am}\right](G_t/\rho_f)^{0.8}/(D_i)^{0.2} \tag{vi}$$

where $C_f = 0.1864\,ln(1/0.0125) + 0.22455 = 1.04136$

(G_t/ρ_f) = average velocity of cooling water = $Q_0/[\pi D_i^2/4]$ = 2.2635 m/s

To note that Q_0 = rate of circulation of cooling water = 1000L/hr or (1/3600) m^3/s. Now, batch cooling being a transient (time dependent) process, the outlet temperature of cooling water (T_{C2}) changes with time (as the medium temperature decreases with time). However, since the medium (substrate solution) is being cooled to 30 °C (303 K), the outlet temperature of cooling water must necessarily be less than 30 °C (on a safer side, less than 25 °C) to avoid temperature cross. Accordingly, we may take, t_{am} = (18+25)/2 = 21.5 °C for substitution in equation (vi). Though this does form an approximation, this shall not induce any serious error into the computation of h_i . Therefore, from equation (vi),

$$h_i \text{ (tube)} = 9057.9867 \text{ W/(m}^2\text{.K)}$$

Let us verify the magnitude of Reynolds number to ascertain the validity of the correlation selected. Thus,

$$Re_t = (0.0125)\,(2.2635)\,(1000)\,/\,(0.001) = 28{,}293.75$$

Since Re_t > 10,000, water is executing fully developed turbulent flow and therefore, the selected correlation is a valid one. Now, from equation (4.1.28),

$$C_{hc} = 1.0 + 3.5\left(D_i / D_H\right) = 1.0 + 3.5\,(0.0125/0.4) = 1.109375$$

Therefore, $h_i = h_i\left(coil\right) = \left(9057.9867\right)\left(1.109375\right) = 10{,}048.7 W/(m^2.K)$

The heat transfer coefficient (h_i) between the outer surface of the coil and the substrate solution depends on the type of impeller used and the impeller characteristics. In the present case, we may compute h_0 . from equation (4.1.18) as

$$Nu = 0.17\left(Re_m\right)^{0.67}\left(Pr\right)^{0.37}\left(D_a / D_t\right)^{0.1}\left(D_0 / D_t\right)^{0.5} \qquad \text{(vii)}$$

where Re_m = $(nD_a^{\ 2}\rho_f / \mu_f) = 3.2\times10^5$ (computed earlier)

Pr = 6.3057 (computed earlier)

(D_a / D_t) = (0.4/2.5) = 0.16

(D_0 / D_t) = (0.0155)/(2.5) = 0.0062

Therefore, from equation (vii), Nu = 107.5

And, $h_0 = \left(107.5\right)\left(0.664\right) / \left(0.0155\right) = 4605.16$ W/($m^2.K$)

The overall heat transfer coefficient (..) is now given by,

$$(1/U_0) = (1/h_i)(A_0 / A_i) + (1/h_0) + \mathrm{R}_d \qquad \text{(viii)}$$

where $(A_0 / A_i) = (D_0 / D_i)$ = (15.5/12.5) = 1.24, R_d = 0.0006 ($m^2.K$)/W.

Therefore, U_0 = 1063.378 W/($m^2.K$)

Now, how does the temperature of the medium decrease with time during the cooling cycle is predicted by equation (4.1.72):

$$ln\left[1/(1-\in)\right] = C\left[1 - exp\left(-NTU\right)\right]t \qquad \text{(ix)}$$

where $\in = (T_{St} - T)/(T_{St} - T_{C1}) = (393 - T)/(393 - 291) = (393 - T)/(102)$

$C = (\dot{m}_C C_{PC})/(MC_{Pm})$

Since $\dot{m}_C$ = mass flow rate of cooling water = (1/3600) (1000) = (5/18) kg/s, M = 2000 kg,

C = (1/7200) s^{-1} = 0.5 hr^{-1}

Now, since A_0 = , (NTU) = $(U_0 A_0)/(\dot{m}_C C_{PC})$ = 27.9736

Equation (ix), therefore, becomes on re-arrangement,

$$T = 291 + 102\ exp\ (-0.5t) \tag{x}$$

where t is in hours. If we substitute $T = 30^0 C$ = 303 K in the above equation and solve for t, we get t = 4.3 hr

This is the duration of cooling cycle. The del factor attained during the cooling cycle shall be,

$$\nabla(\text{cooling}) = k_{d0} \int_0^{4.3} F(t)\,dt$$

where $F(t) = exp\ (-E_d / RT)$.

Substituting the values of E_d, R and the expression for T (from equation - x) in the above equation, we get on re-arrangement,

$$F(t) = exp\left[-(177.1377)/\phi(t)\right]$$

where $\phi(t) = 1 + 0.3505\,exp(-0.5t)$. (xi)

The values of $F(t)$ at different values of t are now computed and the results are listed in Table (4.1.2).

Table (4.1.2)

***t*, minutes**	***F*(*t*), dimensionless**
0.0	2.144×10^{-38}
0.215	2.02897×10^{-39}
0.43	2.185×10^{-40}
0.645	2.68×10^{-41}
0.860	3.7428×10^{-42}
1.075	5.944×10^{-43}
1.290	1.0708×10^{-43}
1.505	2.18×10^{-44}
1.720	4.944×10^{-45}
1.935	1.2808×10^{-45}

t, minutes	*F*(*t*), dimensionless
2.15	3.657×10^{-46}
2.365	1.15577×10^{-46}
2.58	4.019×10^{-47}
2.795	1.5285×10^{-47}
3.01	6.319×10^{-48}
3.225	2.8233×10^{-48}
3.44	1.3554×10^{-48}
3.655	6.9533×10^{-49}
3.87	3.792×10^{-49}
4.085	2.1872×10^{-49}
4.30	1.3282×10^{-49}

Now, from Simpson's rule,

$$I = \int_0^{4.3} F(t)\,dt = (h/3)\left[f_0 + 4(f_1 + f_3 + \ldots\ldots f_{n-1}) + 2(f_2 + f_4 + \ldots . f_{n-2}) + f_n\right]$$

where h = step size = 0.215 hr, $f_0 = 2.144\times10^{-38}$, $f_1 = 2.02897\times10^{-39}$ and so on, finally, $f_n = f_{20} = 1.3282\times10^{-49}$. Therefore,

$$I = 2.1579\times10^{-39} \text{ hr}$$

Since $k_{d0} = 5.7\times10^{39}\,hr^{-1}$,

∇(heating) = $(5.7\times10^{39})(2.1579\times10^{-39})$ = 12.3

Since ∇(overall) = 34.251, ∇(holding) = (34.251) – (12.3 + 2.34) = 19.611

Since during the holding cycle, $T = T_{St} = 393K$ = constant,

∇ (holding) = 19.611 = $k_{d0}\, exp\left[-E_d/(RT_{St})\right]t$

Substituting the values of k_{d0}, E_d, R and T_{St} and solving for t, we get

t = 0.1605 hr = **9.63 min**

This is the holding time required.

Example 4.2

Corn steep liquor is being used for the manufacture of antibiotics through microbial fermentation. The liquor is to be first sterilized such that the microbial count shall not exceed 0.002 per m^3 (the original value being 8.0×10^{12} per m^3). In a batch process, 1.5 m^3 of liquor is taken in the stirred tank and high pressure steam is sparged through it until its temperature reached 386 K. It is then held at this temperature for a specific period and thereafter cooled to 308 K (operating temperature of bioreactor) by passing through an external plate heat exchanger, operating in looped flow arrangement (scheme similar to figure – 4.4). The exchanger is composed of 101 thermal plates made of stainless steel [thermal conductivity = 16.27 W/(m.K)] each of width 0.5 m, wall thickness = 1.0 mm (projected area of each plate = 0.6 m^2, gasket thickness when compressed = 3.5 mm). The liquor is admitted to the exchanger at the rate of 30 m^3/hr and cooling water (entering at 18 °C) is circulated at 100 m^3/hr. Determine the holding time required. Take a minimum overall dirt factor of 0.0004 (m^2.K)/W for the exchanger. Microbial death follows Arrhenius – type relationship with temperature, with $k_{d0} = 5.85 \times 10^{39}$ hr^{-1} and $E_d = 2.82 \times 10^5$ kJ/kmol. Property values of both diluted liquor and cooling water are more or less identical, with density = 1000 kg/m^3, viscosity = 0.001 kg/(m.s), thermal conductivity = 0.665 W/(m.K) and specific heat = 4.18 kJ/(kg.K).

Solution :

The overall del factor desired is,

$$\nabla(\text{overall}) = \ln(n_0 / n) = 35.925$$

To note that $n_0 = 8.0 \times 10^{12}$ per m^3 and $n = 0.002$ per m^3.The duration of heating cycle shall be very short, since direct steam sparging is employed (though the liquor shall get diluted by mixing with the steam condensate) and as a result, little microbial death shall occur during this period. Sterilisation of the medium shall occur during the holding period and during the cooling period. To compute the del factor attained during the cooling cycle, let us first evaluate the overall heat transfer coefficient (U), which, in turn, depends on the individual heat transfer coefficients (h_C, h_h). Since a plate heat exchanger is being employed, we can compute h_c and h_h from equation (4.1.54). Thus,

$$Nu = C_1 (Re)^m (Pr)^{1/3} \phi \quad \text{(i)}$$

The viscosity correction factor (ϕ) is taken equal to 1.0. Since the number of thermal plates is 101, the number of flow passages shall be, $(N_P + 1) = 102$. In the looped flow arrangement, therefore, both hot fluid (corn steep liquor) and cold fluid (cooling water) can be divided into equal number of streams, such that

$$N_C = N_h = (102)/2 = 51$$

Let us first consider the hot fluid (corn steep liquor). Thus,

$$\dot{m} = \dot{m}_h = (30 \times 1000)/(51) = 588.2353 \text{ kg/hr} = 0.1634 \text{ kg/s}$$

Since $D_e = (2z)$, $Re_h = (2z)(\dot{m}/Bz)/\mu_f = 2\dot{m}/(B\mu_f) = 653.6$

To note that, $B = 0.5$ m, $\mu_f = 0.001$ kg/(m.s). Now,

$$Pr = (C_P \mu_f / k_f) = 6.2857 = 6.2857$$

Since Re_h is greater than 150, $C_1 = 0.45$, $m = 0.61$. Therefore, from equation (i),

$$Nu = 43.3182$$

And $h_h = (43.3182)\,(0.665)\,/\,(0.007) = 4115.23$ W/(m².K)

In the case of cooling water, $\dot{m}_C = (100\times1000)/(51) = 1960.784$ kg/hr = 0.5446 kg/s. Therefore,

$$Re_C = 2178.649$$

$$Pr = 6.2857 \text{ (same as that for corn steep liquor)}$$

Therefore, from equation (i), $Nu = 90.286$, $h_C = 8577.2525$ W/(m².K)

Now, from equation (4.1.60a), $(1/U) = (1/h_C) + (1/h_h) + (\delta/k_P) + R_d$

where $R_d = R_d\,(\text{min}) = 0.0004$ (m².K)/W, $k_P = 16.27$ W/(m.K) and $\delta = 0.001$m. Therefore,

$$U = 1217.9323 \text{ W/(m}^2\text{.K)}$$

Now, the medium temperature (T) shall change with time (t) as per equation (4.1.76). Thus, substituting $T_{C1} = 291$ K and $T_{St} = 386$ K in the above equation, we get

$$T = (291) + (95)\,exp(K_4 t) \tag{iii}$$

Since $C_{Pm} = C_{PC} = 4.18$ kJ/(kg.K), $C = (30{,}000)\,/\,(100{,}000) = 0.3$.

Also, from equation (4.1.78), $K_3 = exp\left[FUA(1-C)/(\dot{m}C_{Pm})\right] = 2.02486$

To note that since the number of thermal plates is odd, $F = 0.942$. Also, $A = 0.6$ m², $\dot{m} = \dot{m}_h = 0.1634$ kg/s. Now, from equation (4.1.77), substituting $M = 1500$ kg,

$$K_4 = -0.233 \text{ hr}^{-1}$$

Equation (iii), therefore, reduces to

$$T = (291) + (95)\,exp(-0.233t) \tag{iv}$$

where t is in hours.

Substituting $T = 308$ K in the above equation and solving for t, we get

t (cooling) = 7.3848 hr ≈ 7.5 hr

This is the duration of cooling cycle. The del factor attained during the cooling cycle is,

$$\nabla = k_{d0}\int_0^{7.5} F(t)\,dt$$

where $F(t) = exp(-E_d/RT)$. On substituting the numerical values of E_d, R and the expression for T from equation (iv), we get

$$F(t) = \exp\left[-116.559/\varnothing(t)\right]$$

where $\varnothing(t) = 1.0 + 0.32646\,exp(-0.233t)$ (v)

The values of $F(t)$ are now computed at different values of time (t) and are listed in table (4.2.1).

Table 4.2.1

t, hr	*F*(*t*)
0.0	6.8798×10^{-39}
0.5	5.970×10^{-40}
1.0	6.028×10^{-41}
1.5	7.0856×10^{-42}
2.0	9.6807×10^{-43}
2.5	1.5324×10^{-43}
3.0	2.7975×10^{-44}
3.5	5.856×10^{-45}
4.0	1.3967×10^{-45}
4.5	3.768×10^{-46}
5.0	1.1413×10^{-46}
5.5	3.852×10^{-47}
6.0	1.4375×10^{-47}
6.5	5.8875×10^{-48}
7.0	2.6268×10^{-48}
7.5	1.2678×10^{-48}

Now, from trapezoidal rule,

$$I = \int_0^{7.5} F(t)\,dt = (h/2)\left[f_0 + 2(f_1 + f_2 + \ldots.. + f_{n-1}) + f_n\right]$$

where h = step size = 0.5 hr, $f_0 = 6.8798\times10^{-39}$,, $f_1 = 5.97\times10^{-40}$, $f_2 = 6.028\times10^{-41}$ and so on. Finally, $f_n = 1.2678\times10^{-48}$.

Therefore, $I = 2.0527\times10^{-39}$ hr

Since $k_{d0} = 5.85\times10^{39}\ hr^{-1}$, $\nabla(\text{cooling}) = (5.85\times10^{39})(2.0527\times10^{-39}) = 12.0$

Since $\nabla(\text{overall}) = 35.925$, $\nabla(\text{holding}) = (35.925 - 12.0) = 23.925$.

Since during holding period, $T = T_{St} = 386$ K = constant,

$\nabla(\text{holding}) = 23.925 = k_{d0}\, exp(-E_d / RT_{St})t$

Substituting the values of k_{d0}, E_d, T_{St} and R and solving for t, we get

t = 0.6 hr = **36.0 minutes**

This is the holding time required.

Example 4.3:

For the manufacture of industrial alcohol, a dilute, aqueous solution of molasses is to be fed continuously to the bioreactor (fermenter) at the rate of 3500 L/hr. Microbial content of fresh molasses is reported to be 1.75×10^{11} per m^3 and before feeding to the bioreactor, it is to be sterilized such that not more than one organism shall survive during one month of operation. Molasses is first heated to the sterilization temperature of 398 K by sparging dry, saturated steam at 600 kPa through it, it is then held at this temperature for a particular period by circulating through a well-insulated, 102 mm ID holding tube or coil and thereafter, cooled to the operating temperature (303 K) of the bioreactor in a flash chamber. If the temperature dependence of specific death rate of microbes follows Arrhenius – type relationship with $k_{d0} = 10 \times 10^{37}$ min^{-1} and $E_d = 2.8 \times 10^5$ kJ/kmol, estimate the total length of holding tube (coil) required. Take density and viscosity of aqueous molasses (substrate solution) to be 1000 kg/m^3 and 4.0 kg/(m.hr) respectively.

Solution:

Since heating of substrate solution (aqueous molasses) to the sterilization temperature has been done by sparging high pressure steam through it and cooling from sterilization temperature to the reactor operating temperature has been affected by flash cooling, little microbial death would have occurred during these two cycles. Microbial death has thus occurred predominantly during the holding period. Since the flow rate of molasses is 3500 L/hr or 3.5 m^3/hr, the amount of molasses handled during one month of operation shall be (3.5) (24) (30) = 2520 m^3. And, since not more than one organism is to survive during this period,

$$n = (1/2520) = 3.968 \times 10^{-4} \text{ per } m^3$$

Therefore, Del factor desired, $\nabla = \ln\left(n_0 / n\right) = 33.72$.

To note that $n_0 = 1.75 \times 10^{11}$ per m^3. Assuming dispersed flow through the tube (coil), we can use equation (4.2.5) to estimate the required length of the tube (coil). For this, let us first estimate the axial dispersion coefficient (D_L) from equation (4.2.8). Thus, the average velocity of substrate solution through the holding tube (U_L) is

$$U_L = Q_0 / \left(\pi D_i^2 / 4\right) = 0.119 \text{ m/s}$$

And $Re = \left[D_i U_L \rho_L / \mu_L\right] = 10924.2$

To note that Q_0 = (3.5/3600) m^3/s, D_i = 0.102 m, ρ_L = 1000 kg/m^3, μ_L = (4.0/3600) = (1/900) kg/(m.s). Now, from equation (4.2.8),

$$N_d = 0.44868$$

And, $\quad D_L = (0.44868)(0.102)(0.119) = 0.005446 \text{ m}^2/\text{s}$

Since the temperature dependence of specific death rate follows Arrhenius – type relationship,

$$k_d = k_{d0}\, exp\left(-E_d / RT\right) = 17.81 \text{ min}^{-1} = 0.2968 \text{ s}^{-1}$$

To note that $k_{d0} = 10 \times 10^{37}\,\text{min}^{-1}$, $E_d = 2.8 \times 10^5$ kJ/kmol, R = 8.314 kJ/(kmol.K) and T = 398 K.

Now, from equation (4.2.7), β = 1.2069

And, $(n/n_0) = (3.968 \times 10^{-4}) / (1.75 \times 10^{11}) = 2.2674 \times 10^{-15}$

Equation (4.2.5), therefore, reduces to

$$\left(2.2674\times10^{-15}\right)=\frac{(0.9912)\exp(Pe/2)}{\left[\exp(0.60345\,Pe)-0.008788\,exp(-0.60345\,Pe)\right]}$$

Solving by trial,

$$Pe = \text{Peclet number} = 326.0$$

And, $L = (326.0)\,(0.005446)\,/\,(0.119) =$ **15.0 m.**

This is the required length of the holding tube (coil).

4.3 Sterilisation Using Fibrous Filters

Gases such as air are commonly sterilized using fibrous filters made of glass wool, polyamide or polyacrylonitrile fibres. As the gas passes through the filter medium, the microbial cells are retained by the medium and the gas stream emerging from the filter shall be essentially free from microbes.

The particles in the feed gas (in this case, the microbial cells) are retained by the filter medium, either

(a) by impaction and interception,
(b) or by diffusion.

The particles (microbial cells) collide against the fibres of the filter medium and thereby get deposited on them (impaction) or are intercepted by the fibres of the filter. When the particles are very small in size (such as submicron particles), they tend to execute Brownian motion and consequently, tend to move or diffuse to the fibre surface and get deposited there. The collection efficiency (η) of the fibrous filter is defined as

$$\eta = 1 - (n/n_0) \tag{4.3.1}$$

where n_0 = microbial count (number of microbes per unit volume) in the feed gas n = that in the gas stream leaving the filter (in the sterilized gas) Pasceri and Friedlander [17] have reported an empirical correlation for the estimation of $\eta(ideal)$, as given below:

$$\eta(ideal) = 6.0\left(Re_f\right)^{1/6}(Pe)^{-2/3} + (3.0)\alpha^2\left(Re_f\right)^{0.5} \tag{4.3.2}$$

where Re_f = modified Reynolds number $= \left(dU_g\rho_g/\mu_g\right)$ (4.3.3)

Pe = modified Peclet number $= \left(dU_g/D_{Br}\right)$ (4.3.4)

D_{Br} = diffusivity due to Brownian motion $= \left(K_m kT\right)/\left(3\pi d_P\mu_g\right)$ (4.3.5)

α = interception parameter $= \left(d_P/d\right)$ (4.3.6)

d_P, d = particle diameter and fibre diameter, respectively

U_g = average superficial velocity of gas

k = Boltzmann's constant = 1.38×10^{-23} J/K (4.3.7)

T = absolute temperature, K

K_m = Stokes – Cunningham correction factor $= 1+(J\lambda / d_P)$ (4.3.8)

$J = 2.514 + 0.8\ exp(-0.55 d_p / \lambda)$ (4.3.9)

λ = mean free path of gas molecules $= \mu_g / (0.499 \rho_g U_m)$ (4.3.10)

U_m = average velocity of gas molecules $= (8RT / \pi M)^{1/2}$ (4.3.11)

M = molecular weight of gas, kg/kmol

The above expression for mean free path of gas molecules is based on Chapman – Enskog equation (which, nevertheless, assumes ideal gas behavior).

Incidentally, separate correlations for the estimation of collection efficiency due to each of the three mechanisms (impaction, interception, diffusion) have also been reported by different authors [18,19].

By $\eta(ideal)$, we mean the collection efficiency for a single fibre collector. But, we know the fibrous filters used in commercial practices are composed of a multitude of fibres and there shall be mutual interactions among them. The depth or thickness of the filter (b) and its voidage (void fraction or porosity, $\in$) are also deciding parameters. Investigations [18,20] have demonstrated that the actual collection efficiency (η) of the fibrous filter and $\eta(ideal)$ could be correlated as

$$\eta = C_d \{1 - \exp[-F\eta(ideal)]\} \tag{4.3.12}$$

where $F = (4b / \pi d)(1-\in)/\in$

$b, \in$ = depth (thickness) and voidage (porosity) of filter, respectively

C_d = correction factor to account for mutual interference among fibres

$= 1.0 + 4.5(1-\in)$ (4.3.13)

Attempts have also been made to locate the minimum gas velocity, $U_g(\min)$, below which separation of suspended particles (suspended microbial cells) by impaction shall be significantly low. For this, we define a parameter called Stokes number (N_S) such that

$$N_S = (K_m d_P^2 U_g \rho_S)/(18\mu_g d) \tag{4.3.14}$$

It has been observed [18,21] that the minimum required value of N_S, namely $N_S(\min)$, to affect effective separation of particles is around 0.0625. Substituting this value in the above equation, we get

$$U_g(\min) = (1.125\mu_g d)/(K_m d_P^2 \rho_S) \tag{4.3.15}$$

For best performance, the operating gas velocity must be maintained higher than the above value. This criterion may also be used to estimate the required diameter or cross-sectional area of the filter.

The pressure drop across the filter, $(-\Delta P)$, is another important design parameter, since it decides the operating cost of the filter. Earlier attempts to develop correlations for the estimation of $(-\Delta P)$ were based on the *capillary bundle theory* that assumes the entire void space of the filter is equivalent to a bundle of tangled capillary tubes and the gas flows only through the interspacing between fibres and not over them. Such an approach, however, accounts for resistance to gas flow due to *skin friction* only. More accurate is the *form drag theory*, which considers gas flow over the fibres as well and thereby accounts for form drag also. Based on this approach, Chen [20] has proposed the following correlation for the computation of pressure drop across a fibrous filter :

$$(-\Delta P) = \left[24.4(1-\in)/\in\right]\left(bU_g\mu_g / d_S^2\right) / \left[\pi \, ln\left(0.64/\sqrt{1-\in}\right)\right] \quad (4.3.16)$$

where d_S is the surface diameter (equivalent diameter) of each fibre. By surface diameter, we mean the diameter of a spherical particle that has the same surface area as the fibre under consideration.

Example 4.4:

Air is to be sterilized using a glass wool filter that is 0.13 m deep and is made up of fibres that are 19 microns in diameter (average voidage of filter = 0.967). Air at 25 °C and 1 atm (density = 1.187 kg/m^3, viscosity = 0.02 centipoise) enters the filter at a superficial velocity of 0.4 m/s and its microbial count is to be reduced from 5000 per m^3 to 10 per m^3. Take average size of each microbe = 1.0 micron and its density = 1020 kg/m^3. Determine whether the filter shall serve the purpose and if so, compute the required diameter of the filter to handle air at the rate of 1200 m^3/hr. What shall be the pressure drop across the bed?

Solution:

The suitability of fibre filter shall depend on its collection efficiency (η). Let us, therefore, estimate η from equations (4.3.2) and (4.3.12). To compute $\eta(ideal)$ from equation (4.3.2), we have to first evaluate the dimensionless parameters such as the Reynolds number (Re_f) and the interception parameter (α). Thus from equation (4.3.3),

$$Re_f = \left(dU_g\rho_g / \mu_g\right) = 0.451$$

To note that $d = (19 \times 10^{-16})$ m, $U_g = 0.40$ m/s, $\rho_g = 1.187$ kg/m^3, $\mu_g = (0.02 \times 10^{-3})$ kg/(m.s). Also,

$$\alpha = \left(d_P / d\right) = (1.0)/(19.0) = 0.05263$$

To compute the Peclet number (*Pe*), we have to first estimate the magnitude of diffusivity due to Brownian motion (D_{Br}). Thus, from equation (4.3.11), the average velocity of gas molecules (U_m) is,

$$U_m = \left(8RT/\pi M\right)^{1/2} = 466.4275 \text{ m/s}$$

To note that R = 8314.0 J/(kmol.K), T = 298 K, M = average molecular weight of air = 29.0 kg/kmol. Now, the mean free path of gas molecules (λ) is (from equation – 4.3.10),

$$\lambda = \mu_g / \left(0.499 \rho_g U_m\right) = 0.07239 \text{ microns}$$

Since $\left(d_P / \lambda\right) = (1.0)/(0.07239) = 13.81368$, from equation (4.3.9),

$$J = 2.5144$$

The Stokes – Cunningham correction factor (K_m) can be now estimated from equation (4.3.8) as,

$$K_m = 1.0 + (2.5144) / (13.81368) = 1.182$$

And, from equation (4.3.5), $D_{Br} = 25.7876 \times 10^{-12}$ m²/s

Therefore, $Pe = \left(dU_g / D_{Br}\right) = 2.947 \times 10^5$

Accordingly, from equation (4.3.2),

$$\eta\left(ideal\right) = 0.006767$$

Now, from equation (4.3.13), C_d = 1.0 + 4.5 (1 – 0.967) = 1.1485

Therefore, from equation (4.3.12),

$$\eta = 0.995$$

The required collection efficiency of the filter to affect the specified degree of sterilisation is,

$$\eta\left(desired\right) = 1 - (n / n_0) = 1 - (10/5000) = 0.998$$

Since the collection efficiency of the filter is very close to the desired value, the given filter can be considered suitable for the given application. The required cross – sectional area of the filter is,

$$A = (Q_g / U_g) = (1200/3600)/(0.40) = 0.833 \text{ m}^2$$

And, $D = \sqrt{4A/\pi} = \mathbf{1.03\ m}$

Nomenclature

A heat transfer area, m^2

b depth or thickness of fibrous filter, m

B width of each plate of plate exchanger, m

C_d correction factor defined in equation (4.3.13), dimensionless

C_f correction factor (see equation - 4.1.34), dimensionless

C_{hc} correction factor for helical coil (equation - 4.1.28), dimensionless

C_{Pa} specific heat of annulus fluid, $J/(kg.K)$

C_{PC} specific heat of cold fluid, $J/(kg.K)$

C_{Ph} specific heat of hot fluid, $J/(kg.K)$

C_{PL} specific heat of liquid water, $J/(kg.K)$

C_{Pm} specific heat of substrate solution, $J/(kg.K)$

d diameter of fibre, m

d_P particle diameter, m

d_S surface diameter (equivalent diameter) of fibre, m

D_a impeller diameter, m

D_{Br} diffusivity due to Brownian motion, m^2/s

D_e equivalent diameter of annulus / flow channel between parallel plates, m.

D_H diameter of helix, m

D_i ID (inside diameter) of tube/pipe, m

D_j ID (inside diameter) of jacket, m

D_L axial dispersion coefficient, m^2/s

D_0 OD (outside diameter) of tube/pipe, m

D_t ID (inside diameter) of tank, m

D_{t0} OD (outside diameter) of tank, m

E_d correlation constant (see equation- 4.4), $J/kmol$

F correction factor (see eq. 4.1.61), dimensionless

G mass velocity of fluid (between parallel plates), $kg/(m^2.s)$

G_a mass velocity of annulus fluid, $kg/(m^2.s)$

G_t mass velocity of fluid through tube/pipe, $kg/(m^2.s)$

h_C, h_h individual heat transfer coefficient (film coefficient) for cold fluid and for hot fluid respectively, $W/(m^2.K)$

h_i, h_0 individual heat transfer coefficients, $W/(m^2.K)$

h_L specific enthalpy of liquid water, J/kg

H_S scific enathalpy of saturated steam, J/kg

k Boltzmann's constant, J/K

k_d specific death rate, s^{-1}

k_{d0} correlation constant (equation- 4.4), s^{-1}

k_f thermal conductivity of fluid, $W/(m.K)$

k_{fa} thermal conductivity of annulus fluid, $W/(m.K)$

K_m Stokes-Cunningham correction factor, dimensionless

L length of holding tube, m

$\dot{m}$ mass flow rate of substrate solution, kg/s

$\dot{m}_a$ mass flow rate of annulus fluid, kg/s

$\dot{m}_C$ mass flow rate of cold fluid, kg/s

$\dot{m}_h$ mass flow rate of hot fluid, kg/s

$\dot{m}_S$ mass flow rate of steam, kg/s

M mass of substrate solution in the tank, kg

n microbial count (number of microbes per unit volume) at any time t, m^{-3}; speed (rps) of impeller, s^{-1}

n_0 microbial count at t = 0, m^{-3}

N_C number of streams of cold fluid

N_d dispersion number, dimensionless

N_h number of streams of hot fluid

N_P number of thermal plates in plate exchanger

N_S Stokes number, dimensionless

NTU number of transfer units

Nu Nusselt number, dimensionless

Nu_a annulus Nusselt number, dimensionless

Pe Peclet number, dimensionless

Pr Prandtl number of fluid, dimensionless

Pr_a Prandtl number of annulus fluid, dimensionless

Q_0 volumetric flow rate of fluid, m^3/s

R Universal gas constant, $J/(kmol.K)$

R_d overall combined dirt factor, $(m^2.K)/W$

Re Reynolds number, dimensionless

Re_a annulus Reynolds number, dimensionless

Re_f modified Reynolds number (equation- 4.3.3), dimensionless

Re_m mixing Reynolds number (impeller Reynolds number), dimensionless

Re_t tubeside Reynolds number, dimensionless

R_H hydraulic radius, m

T absolute temperature, K

T_{C1} inlet temperature of cold fluid, K

T_{h1} inlet temperature of hot fluid, K

T_0 initial temperature of substrate solution, K

T_S temperature of saturated steam/saturated liquid, K

T_{St} sterilization temperature, K

U overall heat transfer coefficient, $W/(m^2.K)$

U_g average superficial velocity of gas, m/s

U_i overall heat transfer coefficient (based on heat transfer area, A_i), $W/(m^2.K)$

U_L average fluid velocity (velocity of substrate solution) through holding tube/pipe, m/s

U_m average velocity of gas molecules, m/s

U_0 overall heat transfer coefficient (based on heat transfer area, A_0), $W/(m^2.K)$

z interspacing between plates in plate exchanger (= gasket thickness when compressed), m

Greek Letters

α interception factor, dimensionless

δ jacket clearance, m ; plate thickness, m

$(-\Delta P)$ pressure drop, N/m^2

$(-\Delta T)$(eff) effective temperature difference driving force, K

$(-\Delta T)_{ln}$ logarithmic mean temperature difference (LMTD), K

$\in$ exchanger effectiveness (see equations - 4.1.24, 4.1.73); voidage (void fraction or porosity) of fibrous filter, dimensionless

η collection efficiency, dimensionless

$\eta(ideal)$ single fibre collection efficiency, dimensionless

λ mean free path of gas molecules, m

μ_f fluid viscosity, $kg/(m.s)$

μ_{fa} viscosity of annulus fluid, $kg/(m.s)$

μ_g viscosity of gas, $kg/(m.s)$

μ_W fluid viscosity at tube wall temperature, $kg/(m.s)$

μ_{Wa} viscosity of annulus fluid at tube wall temperature, $kg/(m.s)$

ρ_f fluid density, kg/m^3

ρ_{fa} density of annulus fluid, kg/m^3

ρ_g density of gas, kg/m^3

ρ_S density of particle, kg/m^3

ϕ viscosity of correction factor, dimensionless

ϕ_a viscosity correction factor for annulus fluid, dimensionless

Exercises

4.A Briefly discuss what do you understand by,

a) Del factor,

b) Looped flow arrangement for plate exchangers,

c) Heating by direct steam sparging and flash cooling (their merits and demerits),

d) Dirt factors and fouling coefficients,

e) Hydraulic radius and equivalent diameter,

f) Ideal collection efficiency and actual collection efficiency of fibrous filters (and parameters affecting them)

4.B A 3.0 m diameter fermenter holding 40 m^3 of a glucose solution (at 298 K) is to be sterilized to bring down the bacterial count from 5.0×10^{12} per m^3 to 10^{-3} per m^3.

Sterilization is accomplished by sparging dry, saturated steam at 345 kPa at the rate of 5000 kg/hr until the temperature of the solution reaches 395 K. The substrate solution is held at this temperature adiabatically for a specific duration of time and then cooled to the operating temperature of the bioreactor (303 K) by passing cooling water at 293 K through a submerged cooling coil in the fermenter at the rate of 100 m^3/hr . The coil is made of 12.7 mm ID (OD = 16.7 mm) tube that forms a helix of 0.5 m in diameter (no. of turns = 556). During the cooling cycle, the substrate is well-agitated using a 100 rpm two-bladed paddle impeller, 0.4 m in diameter. If the temperature dependence of specific death rate of microbes follows Arrehenius-type equation with $k_{d0} = 5.85\times 10^{39} hr^{-1}$ and $E_d = 2.8 \times 10^5$ kJ/kmol, determine the holding time required. Assume an overall dirt factor of 0.0006 (m.K)/W. The average property values of glucose solution and those of cooling water within the range of temperature handled are given below:

Property	Glucose solution	Cooling water
Specific heat, kJ/(kg.K)	4.187	4.18
Density, kg/m^3	1000.0	1000.0
Viscosity, kg/(m.s)	0.0015	0.001

Thermal conductivity, W/(m.K)	0.650	0.664

4.C A bioreactor (2.0 m in diameter, wall thickness = 4 mm) containing 2.5 m^3 of starchy substrate (initially at 298 K) with a bacterial count of 3.0×10^{12} per m^3 is to be sterilized such that the probability of a microbe surviving sterilization shall be 1 in 100. The substrate is first heated to 388 K by passing dry, saturated steam at 600 kPa through the reactor jacket (jacket clearance = 25.4 mm), the medium being continuously agitated by a 120 rpm, 0.667 m diameter disk, flat bladed turbine. The substrate is then held at this temperature for a specific period of time and then cooled to 313 K by passing cooling water at 293 K through the jacket at the rate of 60 m^3/hr. If the temperature dependence of specific death rate follows

Arrhenius-type relationship with $k_{d0} = 5.7 \times 10^{39}$ hr^{-1} and $E_d = 2.834 \times 10^5$ kJ/kmol, determine the holding time required. Take an overall dirt factor of 0.00015 (m^2.K)/W. The property values of substrate solution and those of cooling water are more or less identical and their average values (within the range of temperature handled) are density = 1000 kg/m^3, specific heat = 4.187 kJ/(kg.K), viscosity = 0.001 kg/(m.s), thermal conductivity = 0.664 W/(m.K).

4.D A sample of aqueous sulfite liquor with a microbial concentration of 7.5×10^{12} cells per m^3 is to be sterilized continuously at the rate of 25 tonnes/hr to reduce its microbial content to 0.01 cells per m^3. The liquor is heated to the sterilization temperature of 398 K by sparging steam through it. It is then passed through a well-insulated, 300 mm ID holding tube/coil and thereafter cooled to 323 K by flash cooling. Determine the length of the holding tube/coil required, if temperature dependence of specific death rate of microbes follows Arrhenius-type equation with $k_{d0} = 5.7 \times 10^{39}$ hr^{-1} and $E_d = 2.834 \times 10^5$ kJ/kmol. Take property values of substrate solution (sulfite liquor) as density = 1000 kg/m^3 and viscosity = 0.001 kg/(m.s). The axial dispersion coefficient may be evaluated based on equation (4.2.8) of Section (4.2).

4.E A glass wool filter is being used to supply sterile air at 20 °C and 1 atm, at the rate of 1000 m^3/hr to an aerobic fermenter operating at the same temperature. The microbial count in the fresh air is 3000 per m^3 and microbial tolerance in sterile air is 30 per m^3. The filter is made of 15 microns glass fibres (average voidage of filter = 0.97). Estimate the required depth and diameter of the filter, taking operating air velocity to be 1.25 times the minimum. Average size of each bacterium = 1.0 micron, density of bacterium = 1080 kg/m^3. What will be the pressure drop across the bed ? Density and viscosity of air at 20 °C and 1 atm are 1.2 kg/m^3 and 0.018 cp respectively.

References

1. Brooks, A. and Su, G.J., Chem. Eng. Progr., **55** (10), 54-58, 1959.
2. Uh,V.W., CEP Symp. Series, **17**, 51, 93-98, 1955.
3. Chilton, T.H., Drew, T.B. and Jebens R.H., Heat Transfer Coefficients in Agitated Vessels, Ind. Eng. Chem., **36** (6), 510-16, 1944.

4. Brown, G.G. and others, Trans. Inst. Chem. Engrs., **25**, 181-85, 1947.
5. Oldshue, J.Y. and Gretton, A.T., Chem. Eng. Progr., **50**, 615-19, 1954.
6. McAdams, W.H., Heat Transmission, McGraw Hill, New York, 1954.
7. Sieder, E.N. and Tate, G.E., Heat Transfer and Pressure Drop of Liquids in Tubes, Ind. Eng. Chem., **28**(12), 1429-35, 1936.
8. Narayanan, C.M. and Bhattacharya, B.C., Unit Operations and Unit Processes, Vol. I, CBS Publishers, New Delhi, 2006.
9. Perry, R.H. and others, Chemical Engineer's Handbook, Sixth edition, McGraw Hill, New York, 1984.
10. Eagle A. and Ferguson, R.M., On the Coefficient of Heat Transfer from the Internal Surface of Tube Walls, **127**(806), Proc. Roy. Soc., 540-66, 1930.
11. Holland, F.A. and others, Heat Transfer, Heinemann Educational Books, London, 1970.
12. Buonopane, R.A. and others, Chem. Eng. Progr., **59**, 57-60, 1963.
13. Wehner, J.F. and Wilhelm, R.H., Boundary Conditions of Flow Reactor, Chem. Eng. Sci., **6**, 89-93, 1956.
14. Levenspiel, O., Axial Dispersion in Straight Pipes, Ind. Eng. Chem., **50**, 343-46, 1958.
15. Narayanan, C.M., Biotechnology and Bioprocess Engineering, Galgotia Publishers, New Delhi, 2011.
16. Narayanan, C.M. and Banerjee, A., Computer Aided Analysis, Simulation and Optimization of Thermal Sterilization Processes, Recent Patents on Biotechnology, **7**, 47-60, 2013.
17. Pasceri, R.E. and Friedlander, S.K., The Efficiency of Fibrous Aerosol Filters - Deposition by Diffusion of Particles of Finite Diameter, Can. J. Chem. Eng., **38**, 212- 13, 1960.
18. Strauss, W., Industrial Gas Cleaning, Pergamon Press, Oxford, 1975.
19. Friedlander, S.K., Aerosol Filtration by Fibrous Filters, in Blakebrough, N.(ed.), Biochemical and Biological Engineering Science, Vol.1, Academic Press, London, 1967.
20. Chen, C.Y., Filtration of Aerosols by Fibrous Media, Chem. Rev., **55** (3), 595-623, 1955.
21. Bosanquet, C.H., Trans. Inst. Chem. Engrs., **28**, 130-35, 1950.

CHAPTER - 5

Industrial Applications of Biotechnology and Biochemical Engineering

As stated in Chapter – 1, biotechnological processes have made their presence felt in practically all types of chemical process industries, apart from the age-old applications such as in food processing, alcohol making (distilleries) and activated sludge process (wastewater treatment). A few novel applications are outlined in this chapter, such as

a) Manufacture of biodiesels,
b) Manufacture of bioplastics (bio-compostable or *green* plastics),
c) Manufacture of phosphatic biofertilizer (PROM), concurrent with biogas manufacture,
d) Synthesis of chemical fertilizers and other valuable chemicals (hydrogen gas, methanol, synthetic liquid fuels) from biomethane,
e) Desulfurization (removal of H_2S) from air, process gases or biogas through biochemical pathway,
f) Bio-desulfurization of petroleum oils,
g) Bioleaching of ores.

The specific advantages of biochemical pathways are that they do not demand severe operating conditions (operate at ordinary temperature and pressure), provide relatively pure end products (with little formation of undesirable side products) and thereby, reduce the cost of downstream processing, are more environment-friendly. These processes are, however, slow and thereby tend to demand larger reactor volume requirement.

5.1 Manufacture of Biodiesels

The specific advantages of biodiesels, being hailed as the green fuels of the era, have been outlined in Chapter 1.

Biodiesels are manufactured through trans-esterification of vegetable oils using a low molecular weight alcohol or alkyl acetate, in presence of an acid, alkali or immobilized enzyme (lipase) catalyst. Metallic oxides and synthetic zeolites have also been proved to be promising trans-esterification catalysts [1,2]. The reactions involved are (see equation - 1.13 of Chapter 1):

$$\text{Vegetable oil} + \text{alcohol} \longrightarrow \text{Biodiesel} + \text{Glycerol} \tag{5.1.1}$$

$$\text{Vegetable oil} + \text{alkyl acetate} \longrightarrow \text{Biodiesel} + \text{Glycerol triacetate} \tag{5.1.2}$$

$$\begin{array}{l} CH_2OCOR \\ | \\ CHOCOR \\ | \\ CH_2OCOR \end{array} + 3CH_3OH \longrightarrow \underset{(Biodiesel)}{3RCOOCH_3} + \begin{array}{l} CH_2OH \\ | \\ CHOH \\ | \\ CH_2OH \end{array} \qquad (5.1.3)$$

$$\begin{array}{l} CH_2OCOR \\ | \\ CHOCOR \\ | \\ CH_2OCOR \end{array} + 3CH_3COOCH_3 \longrightarrow 3RCOOCH_3 + \begin{array}{l} CH_2OCOCH_3 \\ | \\ CHOCOCH_3 \\ | \\ CH_2OCOCH_3 \end{array} \qquad (5.1.4)$$

Glycerol (or glycerol triacetate) is the byproduct of the processes. The vegetable oil feedstock can be any of the edible or non-edible oils. Edible oils (sunflower oil, cotton seed oil, palm oil, mustard oil, coconut oil etc.) are usually avoided, since they have alternate application as cooking oils (domestic or otherwise). Among the non-edible oils, the most recommended ones are jatropha oil, neem oil and algal oil. Among these, use of algal oil as the feedstock has specific advantages, since algae grow on wasteland or in waste ponds and no skilled farming or cultivation is required. Also, the oil yield from algae is distinctly higher than that from other oil seeds (see table – 5.1.1).

Table – 5.1.1 Yield of Vegetable Oils per acre of Land

Oil	Yield, gallons / acre
Soya oil	48.0
Sunflower oil	102.0
Mustard oil	127.0
Palm oil	635.0
Jatropha oil	304.0
Algal oil	2000-5000

The alcohol used must be of low molecular weight, such as methanol, ethanol or propanol. Higher alcohols tend to increase the viscosity of the biodiesel produced. When any of the above alcohols is used for trans-esterification, the byproduct obtained shall be glycerol (see equation – 5.1.3). When an enzyme catalyst is used (for example, immobilized lipase enzyme), the glycerol produced has a tendency to form a layer around the catalyst particle and thereby deactivate it. In such cases therefore, it is preferable to use an alkyl acetate (methyl, ethyl or propyl acetate) as the second reactant. The byproduct shall then be glycerol triacetate (see reaction – 5.1.4).

The commercial biodiesels are thus the methyl, ethyl or propyl ester of a high molecular weight carboxylic acid (a carboxylic acid with a carbon number of 12 and above). They are accordingly named as JOME (jatropha oil methyl ester), NOME (neem oil methyl ester) etc.

The trans-esterification catalyst can be a homogeneous catalyst (sodium alkoxide, dilute sulfuric acid) or a heterogeneous catalyst (immobilized enzyme, solid metallic oxides, synthetic zeolites). The reaction is faster with homogeneous catalysts and since they blend with the reaction mixture, there shall be no additional diffusional resistance. But, when sodium alkoxide (sodium methoxide, ethoxide or propoxide) is being used as the catalyst, formation of soap (sodium salt of high molecular weight carboxylic acid) could occur and this contaminates the biodiesel produced. Removal of these tiny soap particles from the product biodiesel is troublesome and this increases the cost of downstream processing. Also, soap particles, being gelatinous, tend to clog the atomization nozzles.

With dilute sulfuric acid, unstable sulfonic acid esters get formed, which ultimately lead to the formation of corrosive acid sludges. This sludge is required to be separated and processed to recover its sulfur content and to affect its safe disposal. The cost of this sludge processing unit shall be additional (we have discussed on this subsequently).

When immobilized lipase enzyme is used as the catalyst, more or less uncontaminated biodiesel is obtained as the final product. The volume requirement of the bioreactor shall be, no doubt, higher, since the rate of transesterification shall be lower. The process shall be heterogeneous and column reactors such as packed bed, fluidized bed, semifluidized bed and inverse fluidized bed bioreactors are required to be employed here (design and analysis of these bioreactors have been discussed in Chapter – 3). Also, it has been observed [1,2] that excess of alcohol (say, methanol) tends to deactivate the enzyme catalyst. It is therefore, recommended to use an oil to alcohol (or oil to alkyl acetate) molar ratio of 1:3 (stoichiometric proportion) in the feed solution.

The very purpose of trans-esterification is to reduce the viscosity of the vegetable oil (keeping its combustion properties and engine performance characteristics more or less unaltered), which is the major deterrent in using it as an engine fuel (automobile fuel). Due to the enormously high viscosity of vegetable oils, atomization of these oils becomes prohibitively expensive. Abroad comparison among vegetable oil (jatropha oil), the biodiesel manufactured from it (jatropha oil methyl ester – JOME) and the petrodiesel (diesel oil manufactured from petroleum) is presented in Table (5.1.2).

Table 5.1.2 Comparison between vegetable oil, biodiesel and petrodiesel

Property	Jatropha oil	Biodiesel (JOME)	Petrodiesel
Density, kg/m^3	940	880	850
Kinematic Viscosity (centistokes)	24.5	4.8	2.6
Flash point, K	498	408	341
Pour point, K	277	275	253
Calorific value, MJ/kg	38.65	39.23	42.0
Cetane number	45.0	51.0	50.0
Boiling point, °C	210	320-350	200-350

It can be seen that the properties of biodiesel and those of petrodiesel are very much comparable. The properties of vegetable oil are also equally comparable, except for the elevated viscosity.

A condensed flowsheet of manufacture of biodiesel from vegetable oil and alkyl acetate using immobilized lipase as the catalyst is shown in figures (5.1A) and (5.1B).

The vegetable oil and alkyl acetate are first blended in the proper proportion and then admitted to the fluidized bed bioreactor from the bottom at a specified flow rate. The reactor column is composed of silica granules that accommodate immobilized lipase, and these remain fluidized in the ascending stream of substrate solution (vegetable oil - alkyl acetate mixture). As the solution moves up, the substrates diffuse into the silica granules where they meet the immobilized lipase enzyme and the trans-esterification reaction occurs. The products (biodiesel and glycerol) diffuse out into the liquid bulk. The product solution leaves from the top of the reactor column and is fed to a fractional distillation column. The unconverted alkyl acetate and vegetable oil are collected as distillate from the top and is recycled back to the blending tank (feed tank). The biodiesel and the glycerol produced boil at much higher temperatures (glycerol at around 290^{0}C and biodiesels at 310-320^{0}C) and hence are collected from the bottom of the fractionator as the bottom residue. A two-stage fractional distillation system may also be employed such that in fractionator-1, alkyl acetate is collected as the top distillate and vegetable oil, biodiesel and glycerol are collected as the bottom residue. This residue is then fed to fractionator-2, in which the unreacted vegetable oil is collected as the top distillate and biodiesel and glycerol are collected as the bottom residue.

From this biodiesel-glycerol mixture, biodiesel is separated through multistage countercurrent extraction with demineralized water. Glycerol, being an alcohol, is miscible with water and thus forms the aqueous phase (extract phase). Biodiesel, being immiscible with water, forms the oil phase (raffinate phase). Since these two phases are immiscible with each other, they separate into two different layers in the separator. Biodiesel overflows from the top of the separator, while aqueous glycerol is discharge from the bottom. The biodiesel, so obtained, may be dehydrated (almost completely) by sending to a vacuum distillation still or by passing through an adsorption column and thereafter sent to storage. The aqueous glycerol is concentrated in multiple effect evaporators and then sent to byproduct storage tank.

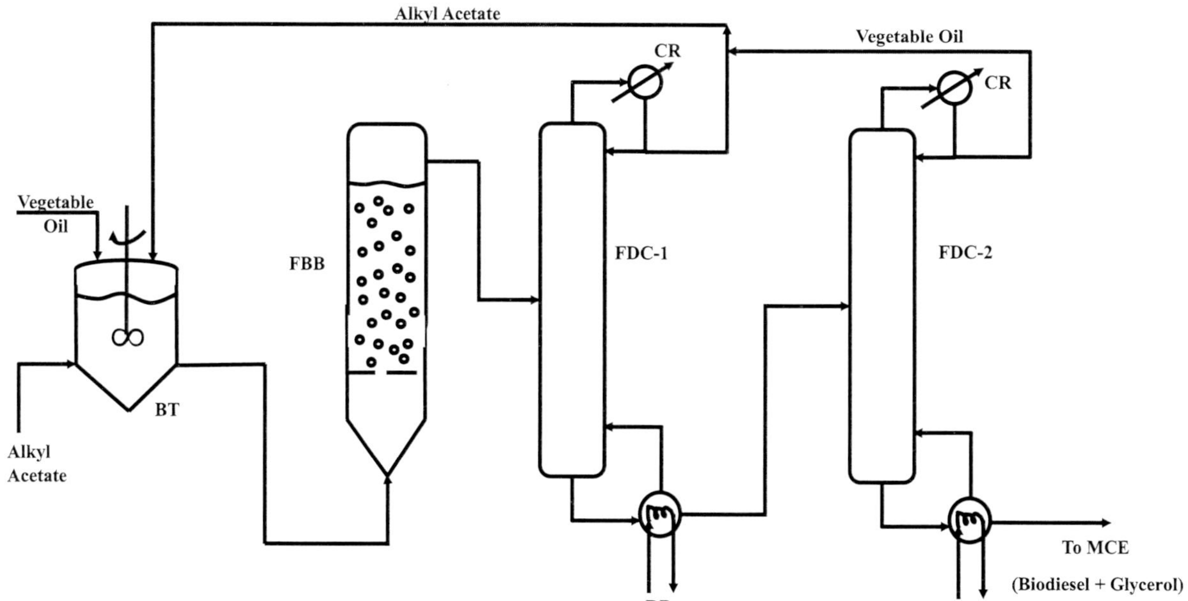

Figure 5.1A: Manufacture of Biodiesel. BT: Blending Tank, FBB: Fluidized Bed Bioreactor, FDC: Fractional Distillation Column, CR: Condenser, RB: Reboiler, MCE: Multistage Countercurrent Extraction Unit.

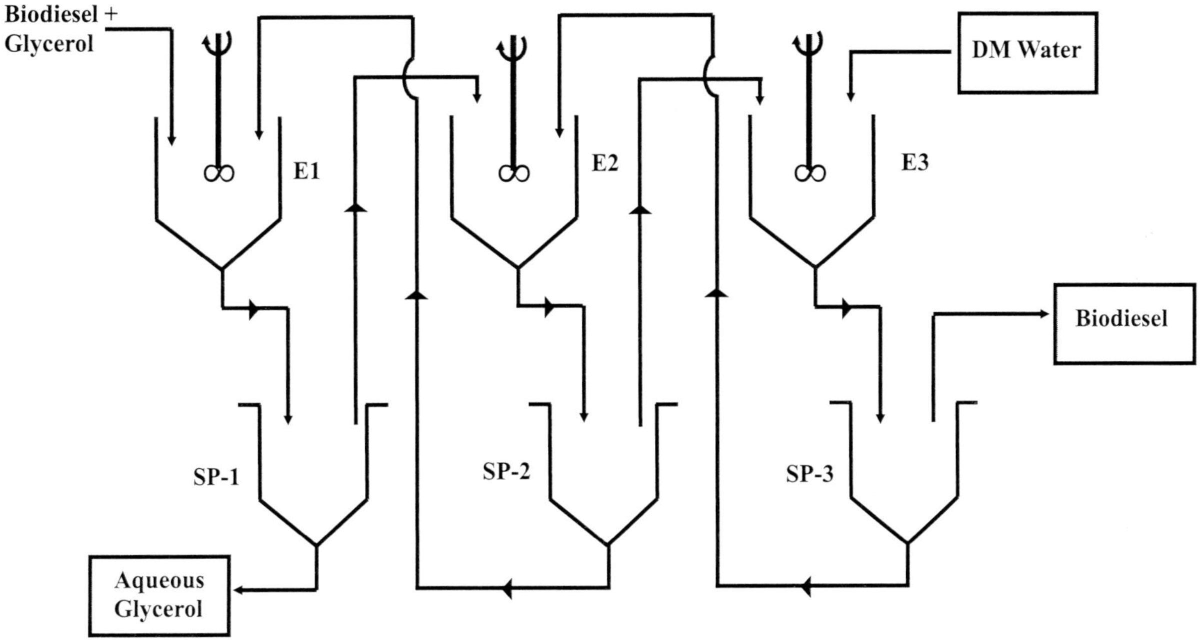

Figure 5.1B: Manufacture of Biodiesel (continued). Three stage countercurrent extraction unit. E1, E2, E3: extractors (agitated vessels), SP1, SP2, SP3: settler – separators, DM water: demineralized water.

As we have discussed in Chapter - 3, more than 80% conversion of vegetable oil into biodiesel is possible to be achieved in fluidized bed enzyme reactors of this kind at high feed flow rates (≥ 23,000 L/hr). Instead of a fluidized bed reactor, a semifluidized bed bioreactor could also be used here, which also provides high yield of biodiesel at higher feed flow rates and within a reasonably low reactor volume. For example, it has been reported [1,2] that more than 80% trans-esterification of neem oil can be achieved at feed flow rates exceeding 30,000 L/hr in a semifluidized bed bioreactor of height 1.2 m and diameter 0.5 m.

Also, as stated in Chapter 3, an inverse fluidized bed bioreactor that handles silica nanoparticles (which accommodate the immobilized lipase enzyme) and is composed of eight 25.4 mm ID vertical columns each of height 6.0 m, could reportedly provide more than 88% conversion of neem oil to biodiesel at a total feed flow rate of around 3000 L/hr [2,3]. The operating cost of this bioreactor shall be distinctly lower, since it employs downflow mode of operation.

As stated earlier, use of a homogeneous catalyst such as sodium alkoxide (alkali catalyst) or dilute sulfuric acid (acid catalyst) is also viable on industrial scale. In such cases, we may use the conventional stirred tank reactor, since the volume requirement of the reactor shall be relatively low (as the rate of trans-esterification shall be higher). However, as discussed earlier, the biodiesel produced shall be contaminated and the cost of downstream processing (product recovery) would be higher. For example, if dilute sulfuric acid is being used as the trans-esterification catalyst, then the biodiesel produced shall be contaminated with the acid sludge (which is extremely corrosive) and this sludge, after separation, would have to be processed further in a separate unit to recover its sulfur content as sodium sulfate or ammonium sulfate. A typical sludge processing plant is sketched in figure (5.2).

The cost of installation and operation of this sludge processing plant shall be then additional. On the other hand, use of conventional stirred tanks is possible and the process yields ammonium sulfate fertilizer as an additional byproduct.

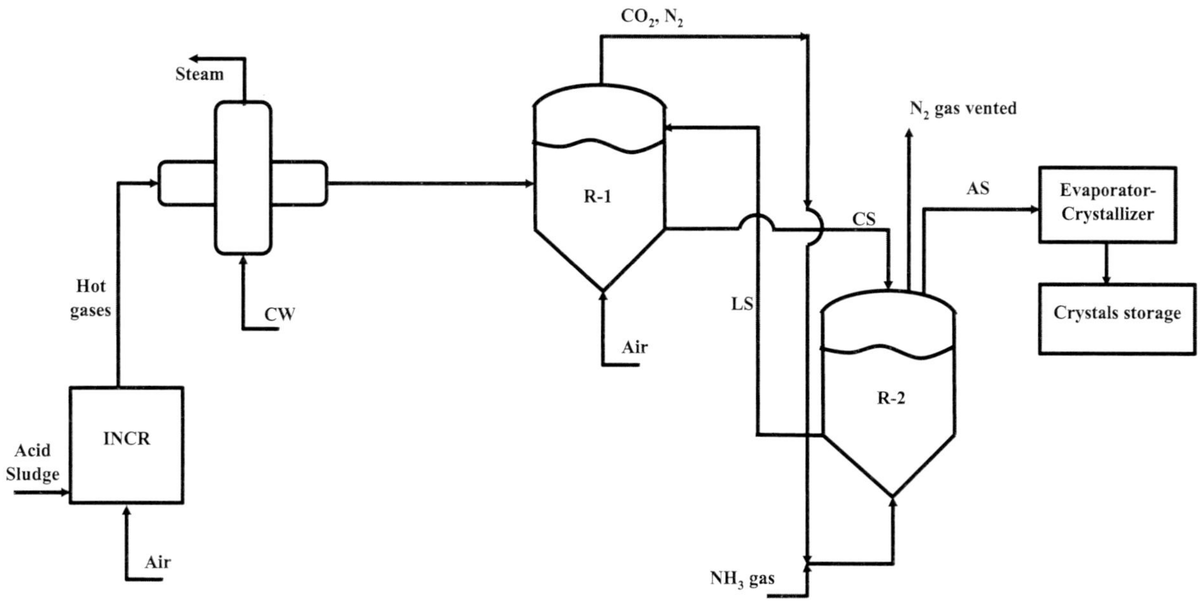

Figure 5.2: Schematic of Acid Sludge Processing Unit. INCR: Incinerator, CW: Cooling water, R – 1: Reactor – 1, LS: Limestone slurry, CS: Calcium Sulfate solution, R – 2: Reactor – 2, AS: Ammonium sulfate solution.

As shown in figure (5.2), the pre-concentrated acid sludge is burnt (combusted) completely in the incinerator, leaving behind little residue. The hot gases (product of combustion) evolved shall be composed of carbon dioxide, sulfur dioxide, nitrogen and water vapor. These are first cooled in a heat exchanger (waste heat boiler) using cooling water (which is converted into high pressure steam). The gases are now fed to reactor-1 which is also fed with recycled limestone slurry from the top. Compressed air is also admitted from the bottom. The following reaction occurs here:

$$2CaCO_3 + 2SO_2 + O_2 \longrightarrow 2CaSO_4 + 2CO_2 \qquad (5.1.5)$$

All the sulfur dioxide present in the gas mixture reacts with limestone (calcium carbonate) to produce calcium sulfate in solution. The reaction occurs quite rapidly, without the need for high temperature or high pressure. All the CO_2 gas (that were present in the feed gas mixture as well as that produced by the above reaction) along with nitrogen, leave from the top of the reactor. This gas stream is mixed with ammonia and then fed to reactor-2 from the bottom, while the calcium sulfate solution from reactor-1 is fed from the top (see figure - 5.2). The following reactions occur in this reactor:

$$2NH_3 + CO_2 + H_2O \longrightarrow (NH_4)_2CO_3 \qquad (5.1.6)$$

$$CaSO_4 + (NH_4)_2CO_3 \longrightarrow CaCO_3 + (NH_4)_2SO_4 \qquad (5.1.7)$$

These reactions also occur very fast and do not demand high pressure or elevated temperature. The calcium carbonate sludge discharged from the bottom of reactor – 2 is pumped to the top of reactor – 1. Ammonium sulfate solution that overflows from the top of the reactor-2 is concentrated to crystallize out ammonium sulfate in an evaporator-crystallizer and the crystals are sent to storage. N_2 gas is vented off from the top of the reactor. The sulfur content of the sludge is thus recovered as ammonium sulfate (chemical fertilizer). The process is more or less self – supporting (only ammonia gas is to be supplied from outside). The incinerator, however, does consume a significant amount of thermal energy. The scheme combines sludge disposal with sludge utilization.

The specific advantages of biodiesels over other liquid fuels such as petroleum oils are summarized below:

(1) Biodiesel is an energy source that is photosynthetically renewable. So long as vegetable seeds are cultivated, biodiesels shall remain available as a potential, commercial energy source. On the other hand, a degree of uncertainty is associated with the availability of fossil fuels such as petroleum oils.

(2) Being oxygenated fuels, they undergo complete combustion and there shall not be any unburnt carbon monoxide or unburnt hydrocarbons in the flue gases (combustion products). Since they are free from aromatic hydrocarbons, there shall not be any unburnt carbon particles (soot particles) also in the exhaust gases.

(3) Biodiesels do not contain sulfur compounds and accordingly, the flue gases shall not be contaminated with toxic gases such as sulfur dioxide, hydrogen sulfide or sulfuric acid mist. Biodiesels are thus much more environment-friendly.

(4) They are labelled as the green fuels of the era, since they, though contain carbon, do not contribute to carbon dioxide pollution and consequent global warming (due to the greenhouse effect exhibited by carbon dioxide gas). This is due to the fact that when biodiesels are combusted, CO_2 gas does get formed, but it gets reabsorbed by the green plants from which the vegetable oil has been derived. A carbon cycle thus gets developed in nature (see figure – 5.3) and consequently, the net amount of carbon dioxide in the atmosphere does not increase. On the other hand, during combustion of fossil fuels such as petroleum oils, the carbon that otherwise remains fixed underground, is burnt off and lot of CO_2 gas is released to the atmosphere. Carbon dioxide, being a greenhouse gas, could contribute to global warming (which is catastrophic to the life on earth).

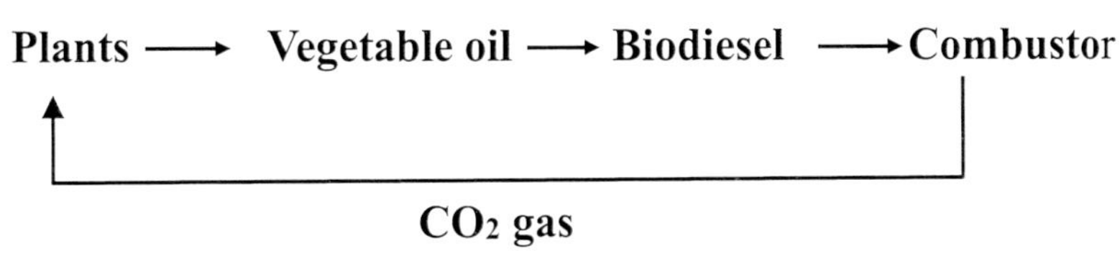

Figure 5.3: Carbon Cycle

(5) As engine fuels (automobile fuels), they provide smooth engine operation, with little noise and little tendency to knocking. They can be, therefore, conveniently used in all road vehicles (in place of petrodiesel) such as cars, buses, trucks etc. No engine modification is required and therefore, dual fuel operation is possible (the automobile can be run both on petrodiesel and on biodiesel or using a blend of the two, without any engine modifications). When run on biodiesel, the automobile exhaust shall be free from sulfur, unburnt carbon particles, unburnt hydrocarbons and shall contain little oxides of nitrogen (NO_x).

(6) Biodiesels are also being widely used in railway engines, tractors and for running irrigation pumps (used for pumping water). Test flights of air crafts powered by biodiesels have been extremely successful.

(7) Large scale use of biodiesels helps in protection of environment (as they are green fuels), eases sustainable development and assists in conservation of petroleum reserves (which is the need of the hour).

5.2 Manufacture of Bioplastics

By bioplastics, we mean plastics that are manufactured from a biological source and that are bio-compostable (biodegradable by soil bacteria). We have briefly discussed on this in Chapter – 1.

After prolonged research studies, the two biopolymers that have been identified as suitable for being used as industrial bioplastics are PHB (poly β – hydroxy butanoate) and PLLA (poly laevo lactic acid). Out of these, the former is an intracellular product and the possibility of its large-scale industrial manufacture is still being investigated. PLLA is the most promising bioplastic of the time. The major stumbling block to its manufacture had been the availability of its monomer (raw material), namely lactic acid. Synthesis of lactic acid from lactose, glucose or sucrose shall be too expensive, since neither of them is available in large tonnages and consequently, neither can act as an economical raw material for the industrial manufacture (manufacture in thousands of tonnes per day) of PLLA. Recently, successful research projects [4] have demonstrated that lactic acid can be synthesized on commercial scale (in large tonnages) from deproteinized cheese whey and clarified molasses through microbial fermentation. Both of these are waste effluents and thus form low-cost raw materials (practically, zero cost raw materials). Design of bioreactors (fluidized bed, semi fluidized bed and inverse fluidized bed biofilm reactors, DSFF bioreactors) for the synthesis of lactic acid from these two sources has already been discussed in detail in Chapter – 3.

PLLA compares excellently with PP (polypropylene) and PS (polystyrene) and can be employed as a viable substitute to these two commercial plastics (see table – 5.2.1). In addition, it is biodegradable and its wastes can be safely disposed by composting in earthen trenches.

Table – 5.2.1 Comparison between PLLA Bioplastic and Synthetic Plastics (PP, PS)

S. No.	Property	Polypropylene (PP)	Polystyrene (PS)	PLLA Bioplastic
1.	Density, kg/m^3	905 – 940	960 – 985	1210 – 1400
2.	Melting Point, ^0C	171 – 186	220 – 230	170 – 180
3.	Tensile Strength, MPa	39 – 40	46 – 60	68 – 70
4.	Glass transition temperature, ^0C	– 15.0	95 – 100	60 – 65
5.	Biodegradability	Nil	Nil	Good

As stated in earlier chapters, cheese whey is the mother liquor left behind after the separation of casein (fats or butter) from milk and after deproteinizing (by ultrafiltration), its lactose content could be converted to lactic acid by microbial fermentation using a culture of *Lactocbacillus helveticus*. Similarly, molasses is the mother liquor left behind after the crystallization of sucrose from sugar cane juice and after due clarification and filtration, its sucrose content can be biochemically converted to lactic acid using a culture of *Enterococcus faecalis*.

A condensed flowsheet illustrating manufacture of PLLA from cheese whey is shown in figures (5.4A) and (5.4B).

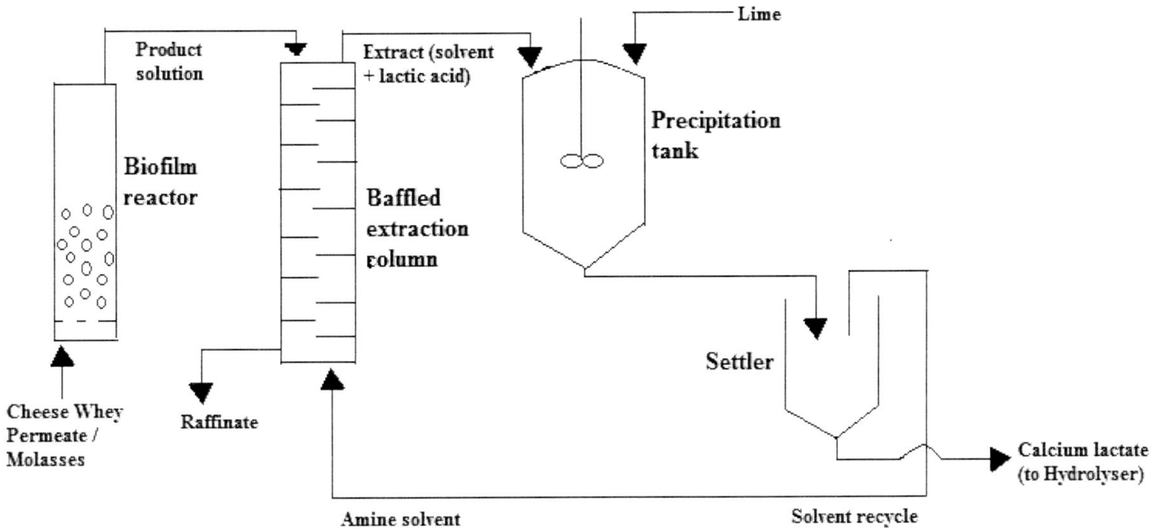

Figure 5.4A: Condensed flowsheet of PLLA bioplastic manufacture

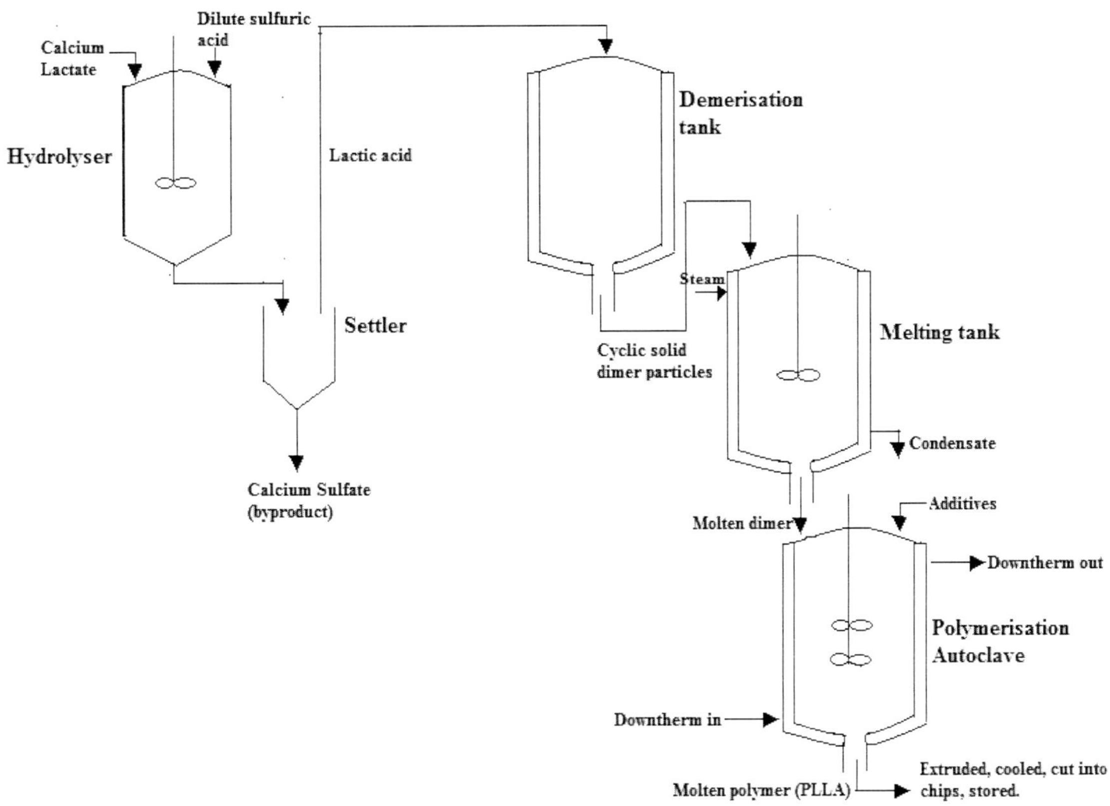

Figure 5.4B: Condensed flowsheet of PLLA bioplastic manufacture (continued)

Cheese whey, after preliminary treatments such as clarification and filtration, is pumped to the feed compartment of the UF (ultrafiltration) unit at a pressure of about 5-6 atm. The permeate compartment (that is on the downstream side of UF membrane) is maintained at atmospheric pressure. A transmembrane pressure differences of 4-5 atm is thus maintained across the membrane. All the proteins present in the whey are retained by the UF membrane (which is usually made of a polyamide composite) and the concentrated protein solution is discharged from the feed / concentrate compartment. The deproteinized whey passes through the membrane and is collected as the permeate (see figure – 5.4A)

The cheese whey permeate (deproteinized whey) is now fed to a fluidized bed biofilm reactor which is composed of particle-biofilm aggregates (1 mm silica granules each surrounded by a 0.3 mm biofilm containing microbial cells of *Lactobacillus helveticus*) and these aggregates remain suspended (fluidized) in the ascending stream of deproteinized whey. Lactose present in whey diffuses into the biofilm where it is biochemically converted to lactic acid and the acid eventually diffuses out into the liquid bulk. The product solution leaves from the top of the bioreactor and is then fed to the top of a baffled extraction column. Trioctylamine (mixed with a diluent such as octanol, decanol or isobutanol) is used as the solvent, which is fed from the bottom to affect a countercurrent contacting. The extract phase (containing lactic acid and the solvent) leaving from the top of the extraction column is fed to the precipitation tank, into which lime is added to precipitate all the lactic acid as calcium lactate. The precipitated column acetate is separated in a settling tank and then sent to the hydrolyser. The amine solution (solvent) that overflows from the top of the settling tank is recycled back to the bottom of the extraction column.

In the hydrolyzer, calcium acetate is hydrolyzed back to lactic acid by adding dilute sulfuric acid :

$$(CH_3COO)_2Ca + H_2SO_4 \longrightarrow 2CH_3COOH + CaSO_4 \qquad (5.2.1)$$

The product mixture is fed to a settling tank (sedimentor) in which all the calcium sulphate formed settles down and is discharged from the bottom of the tank as the underflow sludge (this calcium sulphate forms the byproduct of the process). Lactic acid solution overflows from the top of the sedimentor, it is concentrated and fed to the dimerization tank. Lactic acid dimerizes easily at ordinary temperature and pressure to form a cyclic, solid dimer.

The dimer, so obtained, is first melted in a jacketed melting tank (using steam as the heating fluid) and the molten dimer is fed into the polymerization autoclave, which operates at 200^0C and 1.5 atm pressure. The autoclave is jacketed and dowtherm (at around $250\text{-}300^0C$) is circulated through the jacket (as the heating fluid).

In the polymerization autoclave, ring-opening polymerization occurs (which is exactly similar to the polymerization of caprolactam to Nylon-6) and the molten polymer is extruded from the bottom of the polymerization tank, cooled in a blast of nitrogen gas and further under water. The polymer ribbon, so obtained, is then dried, cut into chips and stored.

The attractive characteristics of PLLA bioplastics are summarized below:

1. As described above, PLLA can be economically manufactured on commercial scale, using cheese whey or molasses as the raw material. Both of them are basically waste effluents discharged from the milk processing industries or sugar manufacturing units.

The cost of raw material can be thus treated as more or less nil. This brings down the manufacturing cost of PLLA significantly.

2. The process of polymerization of lactic acid dimer to PLLA is exactly similar to the process of polymerization of caprolactum to Nylon-6, the latter being a very well-established process (to note that Nylon-6 is a polyamide and is the oldest synthetic fibre manufactured in the world). Accordingly, once the availability of lactic acid monomer in large tonnages (from an economical feedstock) has been ascertained, manufacture of PLLA through its polymerization shall not pose any technical difficulties.
3. As shown in table (5.2.1), PLLA resembles popular synthetic plastics such as PP and PS in most of its properties. It thus forms an excellent substitute to PP and PS in all industrial / commercial applications. It can also be blended with other plastics / polymers to form different polymer composites.
4. Unlike other synthetic plastics, PLLA is biodegradable (it is degraded by soil bacteria). As a result, PLLA waste can be effectively composted in earthen trenches (in the same way as paper wastes and vegetable wastes are being composted). The lament over plastic-made pollution can be thus put to arrest to a large content.
5. The prospects of using PLLA as a commercial material of construction are thus extremely bright.

5.3 Manufacture of Phosphatic Biofertilizer (PROM) concurrent with Biogas

The technology of biogas production is around a century old. However, it had been for long restricted to small scale production for use as cooking gas in households and its large scale manufacture (on commercial scale) for use as an industrial fuel (or as a commercial raw material for chemical synthesis) came into practice and became popular only during the recent years.

Today, biogas can be manufactured from a variety of substrates (see tables 5.3.1A to 5.3.1C), apart from the time old, conventional raw material such as the cow dung [5,6].

No doubt, cow dung has the exclusive characteristic that it can develop the anaerobic microbes by itself and can undergo anaerobic digestion when confined out of contact with air. No inoculum of microbes is required to be seeded from outside.

Due to this exclusive characteristic, cow dung can act as the inoculum for all other alternate substrates. In other words, a particular amount of cow dung is mixed with the substrate-water slurry before it is fed to the bioreactor (anaerobic digester) and this shall initiate growth and multiplication of all the required anaerobic microbes. Addition of cow dung as the inoculum shall be, nevertheless, required only at the startup, since all commercial anaerobic digesters (stirred tank bioreactors) in this connection operate with a partial recycle of microbial sludge.

It may also be noted that addition of inoculum is required only when tank reactors are being employed (in which suspended growth of microbes occurs). When column reactors employing attached growth of microbes are being used (such as fluidized bed, semifluidized bed, inverse fluidized bed biofilm reactors, DSFF and UASB bioreactors), the question of selection and addition of inoculum shall not arise, neither shall there be any need for recycle of microbial sludge (since microbial cells do not leave the reactor).

Table – 5.3.1A Biogas yield from different substrates

Substrate(s)	Gas yield, ml CH_4/(gm.VS)	Reference
Algae (*Scendesmus* & *Chlorella*)	168.0	Golueke et.al [7]
Sewage sludge (88%) + algae (12%)	309.0	Olsson et.al [8]
Sewage sludge (85%) + algae (15%)	294.0	Yuan et.al [9]
Food waste (88%) + algae (12%)	420.0	Krustok et.al [10]
Corn straw + algae	325.0	Weizhang [11]
Wastepaper (40%) + algae (60%)	321.0	Yen and Brune [12]

Table – 5.3.1B Biogas yield from alternate substrates (contd.)

Substrate(s)	Gas yield, ml CH_4/(gm.VS)	Reference
Wheat straw	120.0	
Wheat straw + dairy waste + poultry manure	400.0	Wang et.al [13]
Rice straw + Kitchen waste + pig manure	300.0	Ye et.al [14]
Water hyacinth (40%) + Poultry droppings (45%) + Pig dung (15%)	307.0	Okewale & Adesina [15]
Bagasse (35%) + food waste (65%)	80% higher than with bagasse alone	Vats et.al [16]
Bagasse + poultry droppings (1:1)	3-fold increase in gas yield	Vats et.al [17]

Table – 5.3.1C Comparison of biogas yield from mono-substrate, two substrate and three substrate digestion

Substrate(s)	Yield of biogas, ml/d	Methane content, %
Wastepaper	338.0	69.5
Sewage sludge	190.0	70.0
Wastepaper + sewage sludge (1:1)	520.0	69.3
Wastepaper + sewage sludge + waste grown algae (1:1:1)	700.0	70.0

Source: References (18), (19).

Use of multiple substrates is always advantageous (see tables 5.3.1A to 5.3.1C), since this helps in achieving higher biogas yield and better biogas quality (enhanced methane content of biogas). The yield and quality of biogas generated depends heavily on the C:N ratio of the substrate(s) selected. Too high C/N ratio tends facilitate the activity of acidogens (acidogenic microbes) leading to excessive acid formation and this, in turn, tends to lower the effective pH of the substrate slurry (solution). This adversely affects the growth and multiplication of methanogens (which are obligate microbes) and the production of biogas (and its methane content) gets diminished. Too low C/N ratio (large N-content of substrate used) leads to formation of sizeable amount ammonia during the process and this contaminates the biogas produced. Ammonia is not only poisonous, but it also incombustible. The optimum C/ N ratio has been observed to be around 20 to 30. Typical values of C/N ratio for different substrates are listed in the table (5.3.2). As shown in table (5.3.2), cellulosic wastes (wheat straw, rice straw) possess high C/N ratio (67 to 87). Waste paper has a C/N ratio as high as 165 to 170, while the C/N ratio of bagasse (the cellulosic residue left behind after the extraction of juice from sugarcane stalks) is around 95 to 105. On the other hand, poultry droppings (poultry manure) possess very low C/N ratio (around 5 to 5.5), followed by pig manure (12.5 to 13). Food wastes and cattle dung have C/N ratios close to optimum (18 to 25).

As a result, if we blend substrates of different C/N ratios together, we could get a feed mixture of C/N ratio very close to optimum (20 to 30).

Table – 5.3.2 Typical values of C/N ratio of different substrates

Substrate	C/N ratio
Wheat straw	85-87
Rice Straw	65-67
Wastepaper	165-170
Bagasse	95-105
Water hyacinth	20-40
Food waste	15-25
Poultry manure	5.0-6.0
Cattle dung	18-25
Pig manure	12-13

Effluents from different process industries could also be subjected to anaerobic digestion to produce biogas. Biogas manufacture thus combines waste disposal with waste utilization.

It is important to keep in mind that only dissolved organic matter gets subjected to microbial attack and thereby to anaerobic digestion (though solid state fermentation is not impossible). As a result, while dealing with cellulosic raw materials (wheat straw, rice straw, bagasse, waste, paper, water hyacinth, algae), it should be crushed and shredded into very small fragments (less than 1mm in size) and then kept soaked under water for sufficient time (called the incubation period) so that they blend with water and most of the organic matter goes into solution. The incubation period could vary from 3-4 hr to 12-14 hr, depending on the type of feedstock handled.

Acid or alkali pretreatment of the feedstock is also worthwhile in many cases. For example, bagasse is usually given thermal acid pretreatment (boiling with dilute sulfuric acid at 140 to 150 °C) so that it becomes softer and blends more easily with water. Similarly, wheat straw is mostly subjected to pre-digestion in boiling water or with caustic soda solution. Digestion with caustic soda tends to delignify straw and makes it more amenable to microbial attack. Even biological pretreatment of cellulosic matter using selected fungi has been reported to facilitate anaerobic digestion.

As discussed in Chapter 3, apart from the conventional tank reactor (anaerobic digester) which employs suspended growth of microbes, many column reactors that employ attached growth of microbes can also be efficiently used for large scale production of biogas. Examples are fluidized bed and semi-fluidized bed biofilm reactors, DSFF bioreactors and UASB bioreactors. Performance features of these bioreactors have been discussed in Chapter - 3.

Once biogas is manufactured on a large scale from different multiple substrates, it could be utilized for diversified industrial applications, after subjecting it to desulphuriation (removal of even traces of H_2S present) through a biochemical pathway (discussed subsequently in the next section, see figure 5.6) and thereafter to enrichment (removal of all CO_2 present) in an absorption-desorption unit (also described subsequently through figure 5.7) so as to convert it to more or less pure methane (biomethane). Biomethane has excellent commercial applications such as

a) as an industrial fuel (in boilers, furnaces),
b) as an engine fuel (in automobiles), as a viable substitute to CNG (compressed natural gas) and LNG (liquefied natural gas),
c) as a raw material for the manufacture of a host of chemicals (methanol, hydrogen gas, chemical fertilizers) through syngas route (discussed in subsequent sections, namely sections 5.5 and 5.6),
d) as a raw material for the synthesis of liquid fuels (liquid hydrocarbons, mainly motor gasoline or petrol) through Fischer-Tropsch process (see section 5.7).

When a tank reactor is being employed, the sludge discharged from the reactor (digester), called anaerobic digester sludge (ADS), by itself could be used as a low-cost, nitrogenous fertilizer. After dewatering, drying and grinding into fine particles, it may be fed to the plants. However, it is a low-grade fertilizer, since its N-content is only 2.5 to 3.0 percent. In contrast, chemical fertilizers such as urea has around 42% N, ammonia nitrate 33% N, and ammonium sulfate has 25 to 27 % N.

Another excellent means of utilizing ADS is as one of the starting materials for the manufacture of a phosphatic biofertilizer called PROM - Phosphorus Rich Organic Manure [20,21].

A condensed flow-sheet illustrating manufacture of PROM, concurrent with biogas manufacture is shown in figures (5.5A) and (5.5B). The raw materials for the PROM manufacture are,

(i) Refined rock phosphate ore,
(ii) An organic manure (ADS, vermicompost, cow dung).

Rock phosphate ore (also called chlorapatite or fluorapatite ore) is a complex phosphate of calcium and is the *universal* raw material for the manufacture of phosphoric acid, elemental phosphorus and phosphatic fertilizers. The same raw material is used for PROM manufacture as well. However, in this case, the ore is biochemically converted to water soluble phosphate. The process takes place at ordinary temperature and pressure and no chemicals are required to be consumed during the process.

To note that ADS shall be available only from those biogas plants that employ conventional tank reactor for anaerobic digestion. Column reactors such as fluidized bed and semi-fluidized biofilm reactors, DSFF bioreactors and UASB bioreactors, do not discharge any such microbial sludge, since microbial cells do not leave the reactor chamber but remain attached to the support particles or to the inner surfaces of the flow channels or remain as sludge granules within the reactor. A good substitute to ADS is vermicompost, which is the residue (composted material) obtained after composting vegetable wastes or cellulosic wastes using earthworms.

The well-refined rock phosphate ore is first pulverized in a closed circuit grinding unit (see Figure 5.5A) to obtain very fine particles of uniform size (less than 1 mm). It is then mixed with well-ground (finely powdered) ADS or vermicompost in the ratio 1:2 and thereafter blended with water in the ratio 7:3 (water : solids). The resultant slurry forms the feed slurry to bioreactor-1. The slurry is added to the reactor tank and continuously agitated. Since the consistency of the slurry is maintained at 7:3, the pH of the medium remains close to 7 (optimum value). The temperature is kept under control at around 30-35^{0}C and is not permitted to raise above 60^{0}C under any circumstances. The process is aerobic, but since the reactor is kept open, atmospheric oxygen diffuses into the reaction mixture and that meets the oxygen requirement of the process. Sparging of atmospheric air into the reactor from outside using heavy air compressors is not required. Agitation of the slurry also facilitates dissolution of oxygen. The microbes developed in the solution act on the particles of rock phosphate and convert them into water-soluble phosphate (PROM granules). The process is invariably slow and shall demand a residence time of 5 to 7 days (120 to 170 hours). After this initial start-up period, the reactor could operate continuously, the feed slurry being admitted at a specified flow rate and the product slurry leaving the reactor at the same rate.

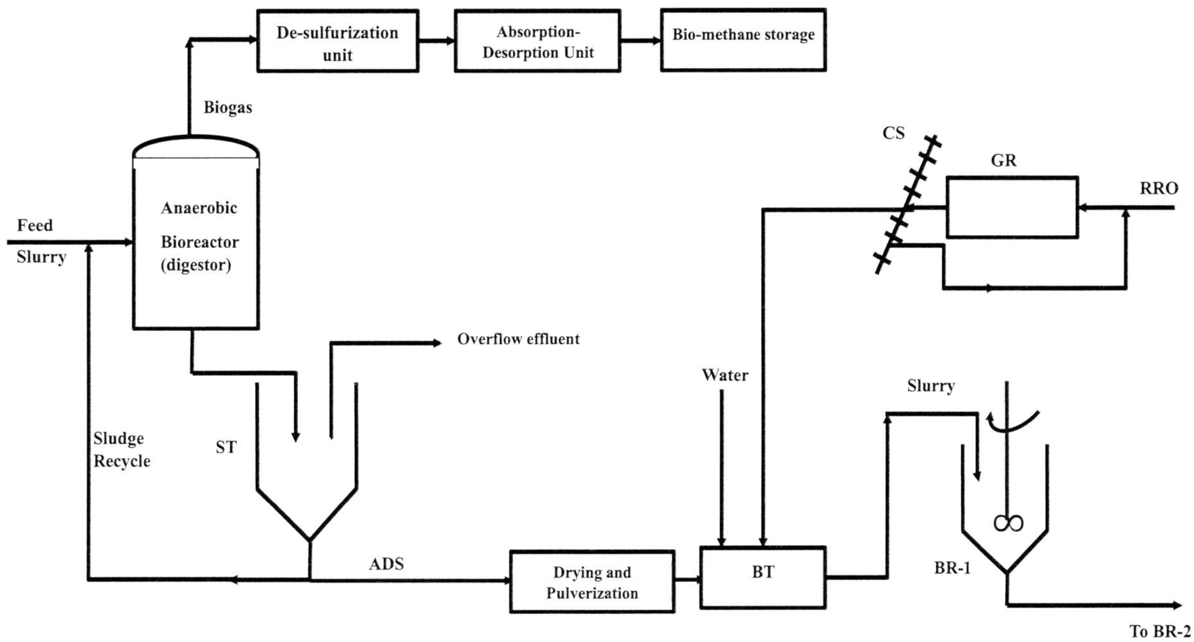

Figure 5.5A: PROM Manufacturing Unit (integrated with Biogas Manufacturing Unit). ST: Sedimentation tank, BT: Blending tank, GR: Grinding Unit, CS: Classifying Screen, RRO: Refined Rock Phosphate Ore, BR1, BR2, BR3: Bioreactors.

PROM formation is usually more than 70-75% complete in this bioreactor (bioreactor-1). To increase the yield of PROM and to improve the product quality, the product slurry from this reactor is sent to bioreactor-2, which is also an open, agitated tank and is seeded with an inoculum of a phosphate solubilizing bacteria such as *Bacillus Megatherium var Phosphaticum* and the reaction mixture is adequately agitated (size of inoculum = 3 to 5 %). Here also, a residence time of 5 to 6 days (120 to 150 hours) is to be provided by adjusting the inflow rate of feed slurry and the outflow rate of the product slurry. The biochemical conversion of rock phosphate particles into PROM particles gets more or less complete in this reactor. The bioreactor-2 is thus used to supplement bioreactor-1.

The use of bioreactor-3 (see figure 5.5B) is, in fact, optional. This step is used to improve the overall quality of the fertilizer produced. This reactor is also an open, agitated tank, which is seeded with an inoculum of a nitrogen-fixing bacteria such as *Azotobacter* or *Rhizobia.* A residence time of 3 to 4 days (70 to 100 hours) is provided in this reactor also and adequate agitation of slurry is practiced. This would improve the N-content of PROM and makes it more nutrient-rich and an improved quality fertilizer.

The final product slurry from bioreactor-3 is dewatered, dried, granulated and sent to storage. PROM, so produced, shall typically contain 16.5% P (expressed as soluble P_2O_5) and shall have a C/N ratio of around 19.0. It is thus fit for direct use in agriculture.

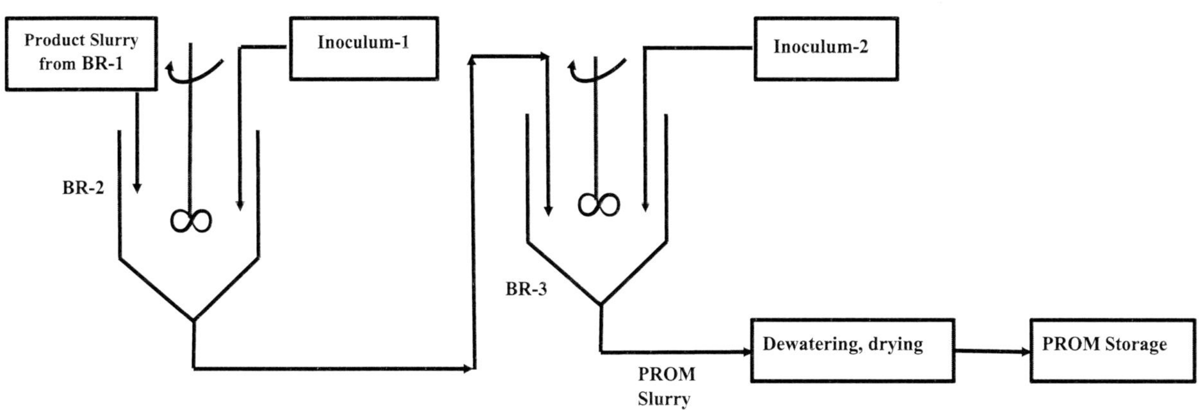

Figure 5.5B: PROM Manufacturing unit, integrated with biogas plant (continued).

The distinct advantages of PROM over other synthetic chemical fertilizers (such as single superphosphate, triple superphosphate and ammonium phosphates) are,

(i) The manufacturing cost of PROM is significantly lower than that of synthetic chemical fertilizers. The process does not demand high temperature or high pressure nor is any chemical catalyst required. Also, there is no consumption of any valuable chemicals such as sulfuric acid or phosphoric acid. The entire cost of production of phosphoric acid is dispensed with.

(ii) PROM competes excellently with synthetic phosphatic fertilizers such as single superphosphate (SSP), triple superphosphate (TSP) and diammonium phosphate (DAP) in agriculture (see table – 5.3.3). PROM provides comparable, if not higher, yield of paddy, grams, fruits and vegetables, when compared with DAP or SSP. It can be thus successfully employed as an excellent, economical substitute to synthetic fertilizers in agriculture. The data presented in table (5.3.3) are those compiled by different investigators through successful field trials [20,21].

(iii) Integration of biogas manufacturing unit with PROM production unit is an attractive proposition, since the former produces a very versatile biofuel for domestic as well as industrial applications, while the latter manufactures an efficient, but low-cost, biofertilizer from the sludge left.

Table – 5.3.3 Comparison of PROM with synthetic fertilizers in agriculture

Crop yield, quintal / hectare

	SSP	DAP	PROM
Wheat	28.9	----	29.0
Soyabeans	10.37	----	11.65
Groundnut	----	30.29	31.20
Rice (kg/hectare)	3000	3910	4375
Barley	----	56.23	70.0
Cabbage (quintal/acre)	----	46.66	54.44
Onion	----	165.19	166.66

Source : References (20), (21).

Production of PROM through *solid state fermentation* of rock phosphate particles using a culture of *Bacillus Megatherium var Phosphaticum* (or any other available phosphate solubilizing bacteria) has been technically successful on laboratory scale. Its industrial adaptation is being awaited. This helps in reducing the cost of downstream processing significantly. During filtration or dewatering of PROM slurry, a part of the fertilizer could get lost in solution and this could be avoided when solid state fermentation is employed.

5.4 Manufacture of Biomethane and its Commercial Utilization as Industrial or Automobile Fuel

As stated in the earlier section, once biogas is manufactured on a large commercial scale, it can be used for the manufacture of biomethane by absorbing off all the carbon dioxide present. And biomethane forms a promising biofuel (boiler fuel, furnace fuel, engine fuel) and a promising raw material for chemical synthesis.

5.4.1 Biochemical Desulfurization of Biogas

For the manufacture of biomethane, biogas is to be first desulphurized. Sulphur is present in biogas mainly as hydrogen sulfide and even if present in small concentrations, H_2S can create severe corrosion problems in power plants and process industries. Sulfur is also poisonous to many catalysts used in chemical synthesis. When combusted, it could release sulfur dioxide and often sulfuric acid mist into the atmosphere, both of which are highly injurious to public health. As a result, it is essential to ensure that the biomethane produced is free from even traces of hydrogen sulfide.

Since hydrogen sulfide is present at low concentrations (though still critically harmful), conventional Claus process becomes unsuitable as well as uneconomical. Claus process is best suitable for handling pure hydrogen sulfide stream (for converting H_2S to elemental sulfur) or gases containing relatively large percentage of hydrogen sulfide.

Abiochemical process [22] that is not only cost-effective but is also capable of affecting almost complete desulphurization even when the H_2S content of the feed gas is very low, is recommended here. This process was originally developed at National Environmental Engineering Research Institute, Nagpur, India and could be employed for the desulphurization of process gases, exhaust gases and even atmospheric air.

The process is sketched in figure (5.6). The scheme uses two reactors. Reactor-1 is a chemical reactor and reactor-2 is an aerobic bioreactor. Reactor-1 could be a packed bed reactor (an acid- resistant, inert packing material is to be used) that operates in the countercurrent mode (a well agitated, stirred tank reactor may also be employed here). Biogas containing H_2S is admitted from the bottom of the packed column and recycled ferric sulphate solution is fed from the top so as to affect countercurrent contacting between the two. The packing is also employed to ensure effective and intimate contacting between the two fluids. As the biogas bubbles through the descending stream of ferric sulphate solution, H_2S present in the biogas reduces ferric sulphate to ferrous sulphate, getting by itself oxidized to elemental Sulfur. The reaction may be represented as

$$Fe_2(SO_4)_3 + H_2S \longrightarrow 2FeSO_4 + 2H_2SO_4 + S\downarrow \qquad (5.4.1)$$

The reaction is quite fast and occurs at ordinary temperature and pressure and without the help of any catalyst. The precipitated sulfur is in the colloidal form (pale yellow or dull white in color) and is separated in a pressure filter and sent to storage (could be treated as the byproduct of the process). The clear, acidic ferrous sulphate solution (filtrate) is pumped to reactor-2 (bioreactor). The desulphurized biogas leaves from the top of the packed bed reactor.

Reactor-2 is a stirred tank reactor (agitated vessel) into which an inoculum of recombinant *Thiobacillus ferrooxidant* culture (sulfur-eating bacteria) is seeded (no doubt, seeding of inoculum is required only at the startup). These microbes multiply in mild acidic environment. Compressed air is sparged from the bottom. The microbial culture catalyses oxidation of ferrous sulphate to ferric sulphate in solution :

$$4FeSO_4 + 2H_2SO_4 + O_2 \longrightarrow 2Fe_2(SO_4)_3 + 2H_2O \qquad (5.4.2)$$

The product solution from reactor-2 is sent to a sedimentation tank, in which the entire microbial sludge settles down and is discharged from the bottom of the tank. A part of the sludge is recycled back to reactor-2 (to maintain adequate concentration of microbial cells in the reactor) and the rest is wasted. The ferric sulfate solution that overflows from the top of the sedimentation tank, is recycled back to reactor-1.

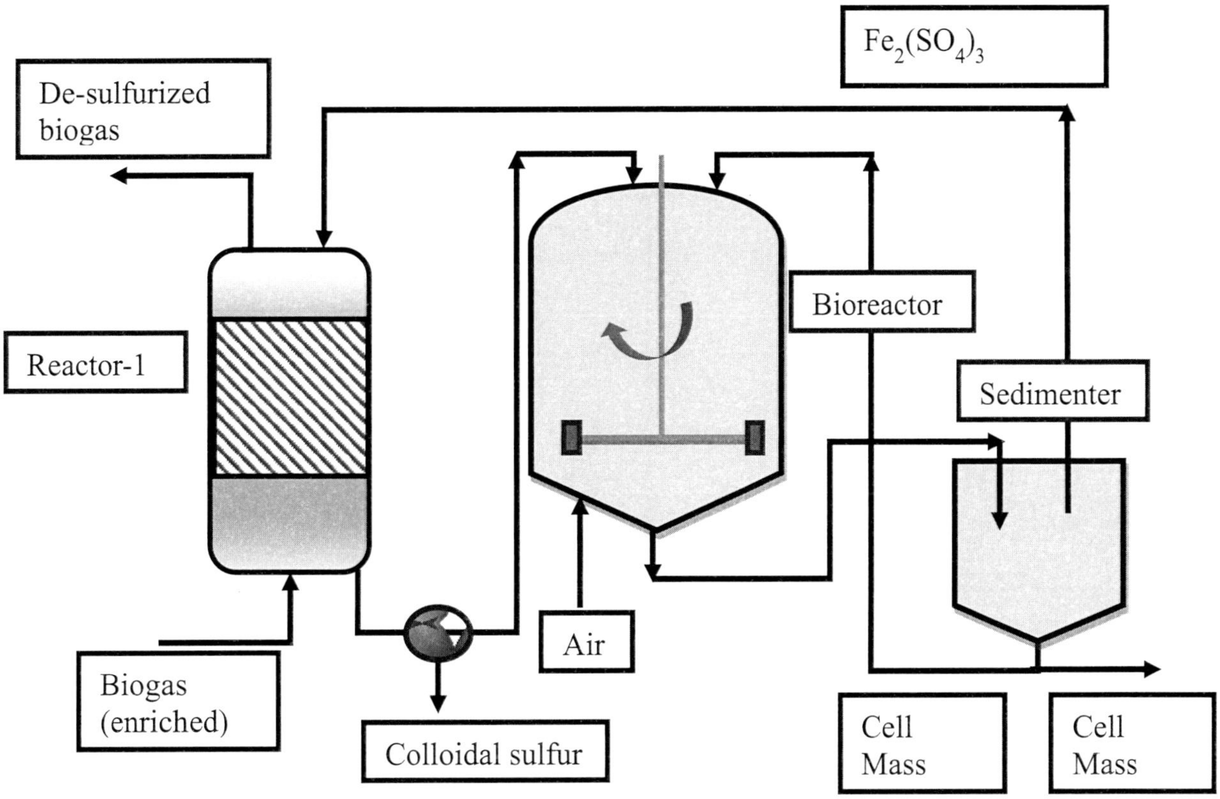

Figure 5.6: Schematic of process of biochemical desulfurization of biogas

The process, as can be seen, is more or less self – supporting. Both reactors operate at ordinary temperature and pressure and there is no consumption of any valuable chemicals. The ferric sulfate solution is continuously regenerated and recycled back for reuse. No chemical catalyst is required and more or less complete desulfurization (more than 98%) of feed gas (biogas) is affected, even when the H_2S concentration in the feed gas is quite low. Colloidal sulfur is obtained as the byproduct.

No toxic gases are evolved during the process, nor is there any solid residue or corrosive sludge discharged. The process thus utilizes fully green technology.

As stated earlier, this process can be conveniently used for H_2S removal from any other process gases, exhaust gases (flue gases) or even atmospheric air.

5.4.2 Production of Biomethane

Once desulfurized, the biogas can be enriched (its CO_2 content absorbed off) to produce more or less pure methane gas (biomethane) in an absorption – desorption system. A typical absorption – desorption unit is shown schematically in figure (5.7).

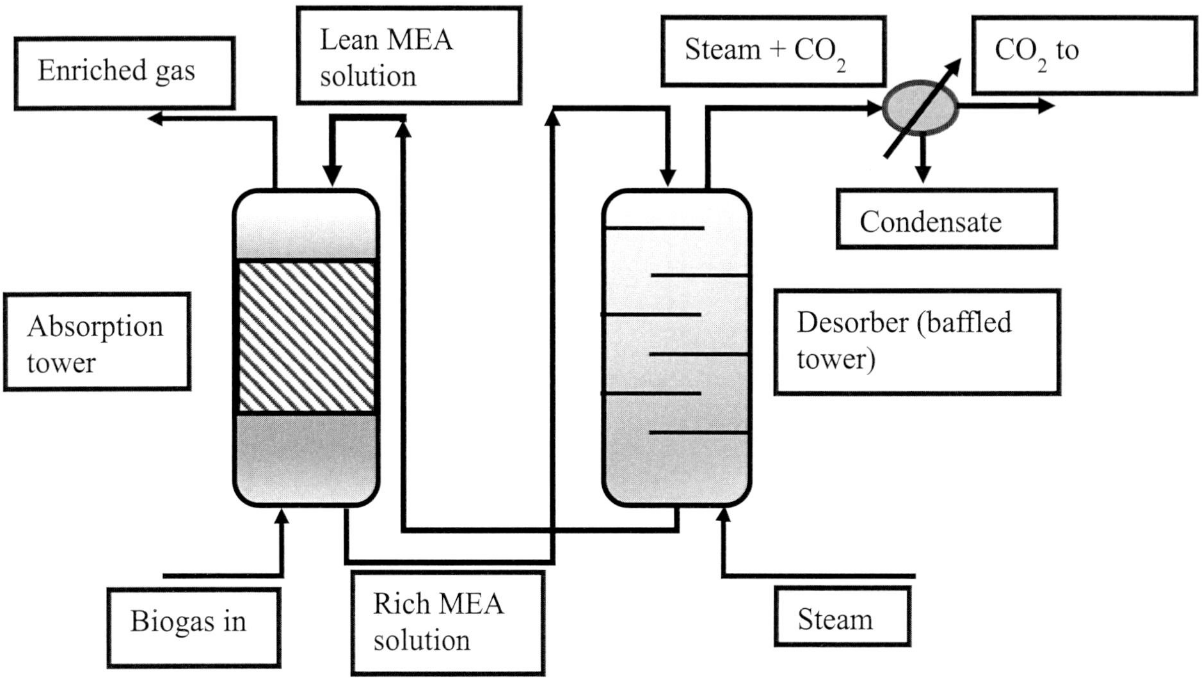

Figure 5.7: Schematic of Absorption – desorption Unit.

Aqueous MEA (mono ethanolamine) solution is an excellent, selective solvent for carbon dioxide absorption. Carbon dioxide has high solubility in aqueous MEA even at atmospheric pressure. Further, carbon dioxide not only dissolves in aqueous MEA solution, but also reacts with it (chemisorption) and due to the chemical reaction, the rate of absorption gets enhanced. From the rich MEA solution that contains all the dissolved carbon dioxide, the solvent (absorbent), namely aqueous MEA, can be effectively and economically recovered by steam stripping. Continuous recovery and recycle of solvent (absorbent) is thus possible.

As shown in figure (5.7), the desulfurized biogas is admitted to the absorption tower from the bottom and recycled, lean MEA solution is fed from the top of the column, so as to affect a continuous countercurrent contacting between the gas stream and the liquid solution. The tower is packed with an inert, acid – resistant packing material, such that both the fluid streams take a tortuous path through it and thereby affect an intimate contacting between the two. Biogas containing little CO_2 and essentially methane, leaves from the top of the tower and is sent to biomethane storage.

The rich MEA solution containing the dissolved carbon dioxide is discharged from the bottom of the absorption tower and is fed to the top of the desorption tower (steam stripper). It is a baffled tower into which high pressure steam is sparged from the bottom. As the steam bubbles through the descending stream of rich MEA solution, most of the dissolved CO_2 is stripped off and the steam – CO_2 mixture leaves the top of the tower. The lean MEA solution containing very little dissolved carbon dioxide is discharged from the bottom of the desorption tower and is recycled back to the top of the absorption tower. The steam – CO_2 mixture is condensed, all the condensate drained off and the carbon dioxide gas left is sent to storage.

More or less complete carbon dioxide removal can be affected by this process [23]. The absorption tower operates at ordinary temperature and pressure. The solvent (absorbent) is continuously recovered and recycled and there is no net consumption of solvent, except for marginal plant losses.

5.4.3 Biomethane as Fuel

Biomethane has a high calorific value of around 55000 kJ/kg and could very well be used as a boiler fuel or furnace fuel.

Biomethane is also a promising engine fuel and it can function as a viable substitute to CNG (compressed natural gas) and LNG (liquefied natural gas) in all road vehicles. Three wheeler and four wheeler passenger vehicles powered by CNG are already on the road all over the world. To note that natural gas is also almost pure methane (more than 97% methane). But, unlike natural gas that is obtained from petroleum reserves, biomethane is manufactured from biological wastes (the cost of raw material is almost nil). Also, as stated in the earlier sections, manufacture of biogas (or biomethane) combines waste disposal with waste utilization.

Use of biomethane as an automobile fuel has distinct advantages. Being a gaseous fuel, it does not demand atomization. As we know, atomization is invariable for the combustion of all liquid fuels such as petroleum oils (motor gasoline, diesel oil) and it, by itself, is an expensive process, since high pressure is required to force the oil through the fine atomization nozzles. The cost of atomization is thus completely dispensed with, when compressed biomethane is used as the automobile engine fuel,

Biomethane undergoes almost complete combustion, leaving behind no solid residues. As a result, the exhaust gases from the automobile shall not contain any unburnt carbon particles or soot particles (presence of soot particles in the exhaust gases is due to the aromatic content of the fuel burnt and biomethane is free from aromatic hydrocarbons), neither there shall be any unburnt hydrocarbons in the exhaust. The exhaust gases shall also be free from toxic sulphur compounds (sulfur dioxide, sulfuric acid mist), once the biomethane has been effectively desulfurized by the biochemical process described in the earlier section (Section 5.4.1). There shall also be little oxides of nitrogen in the exhaust. Biomethane is thus extremely environment- friendly and is a green fuel.

Being highly inflammable, biomethane catches fire almost immediately, thereby reducing the delay period (or ignition period) to a minimum and this provides smoother and more steady engine performance.

Being a gaseous fuel however, biomethane demands high degree of pressurization and large, thick-walled fuel tanks. Safety precautions would have to be made more intricate, since the gas is stored at high pressure and even a tiny leakage could cause serious accidents.

Biomethane could also be used as an automobile fuel in the liquefied form, as a substitute to LNG (liquefied natural gas). However, liquefaction of methane is an expensive process, since critical temperature of methane is very low (–82 °C or 191 K) and its critical pressure is as high as 42 atm. As a result, the gas would have to be cooled to very low temperature

(191 K) and compressed to very high pressure (42 atm) for converting it to the liquefied state. No doubt, once liquefied, it would occupy very low volume and consequently, a large amount of fuel could be stored within a low volume fuel tank. The fuel tank has to be made thick-walled in this case also.

Being a liquefied gas, liquefied biomethane also does not demand atomization. It is also a clean, environment-friendly fuel that undergoes complete combustion, discharges exhaust gases that are practically free from unburnt carbon particles, unburnt hydrocarbons and sulfur compounds. As in the case of CNG, safety precautions would have to be tightened in this case also, since even a tiny leakage could turn out to be highly disastrous.

5.5 Production of Synthesis Gas from Biomethane and its Utilization for Chemical Synthesis

Biomethane, once manufactured in large quantities as described above, can be conveniently used as a raw material for the production of synthesis gas (Syngas) and this, in turn, becomes the primary raw material for the manufacture of a host of valuable industrial chemicals [5,6].

Syngas is a mixture of carbon monoxide and hydrogen (in the molar ratio, 4 : 5). It can be produced by steam – carbon reaction (from coal) or by steam – hydrocarbon reaction (from naphtha or natural gas or biomethane). The reaction between biomethane and steam to produce syngas may be represented as

$$CH_4 + H_2O \longrightarrow CO + 3H_2 \tag{5.5.1}$$

The reaction is termed as *steam reforming* or *steam cracking* of methane. Accordingly, the reactor is named as the *steam cracker* or *primary reformer*. The process takes place at high temperature (750-950^{0}C) and at high pressure (10-15 atm) and in presence of nickel catalyst. As shown in figure (5.8), the primary reformer is composed of a packed bed of nickel catalyst pellets (a fluidized bed reactor could also be used here) that has already been preheated to the operating temperature of 750 – 950^{0}C. Both biomethane and high pressure steam are admitted from the bottom of the reactor column (since both the reactants are in the gaseous or vapor state, the reactor is operated in the co-current mode). As the steam – biomethane mixture moves up the preheated catalyst bed, the steam – hydrocarbon reaction occurs and syngas gets produced. Syngas with some excess steam leaves the top of the reactor column (see figure 5.8).

The steam – hydrocarbon reaction is endothermic in nature and as a result, the temperature of the catalyst bed tends to drop and this, in turn, brings down the rate of reaction. It then becomes necessary to preheat the catalyst bed to the required operating temperature (750 – 950°C). It is, therefore, advisable to operate two packed bed reactors in parallel, such that when one of them is under operation, the other shall be under regeneration (shall be getting preheated). For example, in one cycle, biomethane and steam enter reactor – 1 (where the reaction occurs and syngas is produced), while preheated air is admitted to reactor – 2 so as to heat up the catalyst bed. In the next cycle, the feed mixture (biomethane and steam) is admitted to reactor – 2, whose catalyst bed has already been preheated to the operating temperature

(750 – 950°C), while preheated air is passed through the catalyst bed of reactor – 1. In cycle – 2, syngas is discharged from reactor – 2. Production of syngas thus occurs continuously.

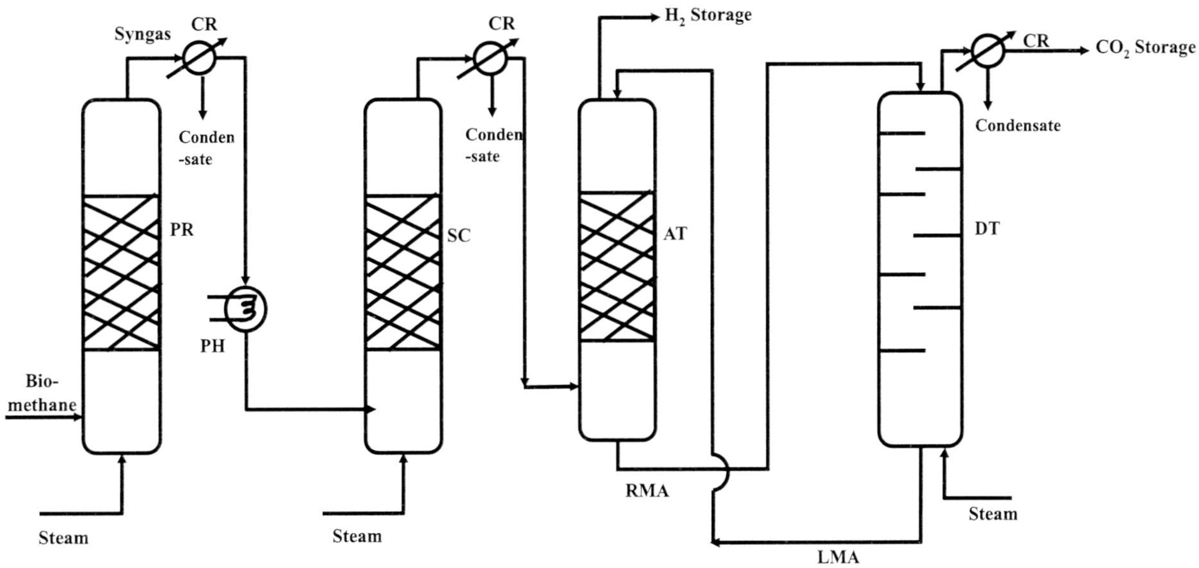

Figure 5.8: Manufacture of Hydrogen from Biomethane. PR: Primary reformer (Steam cracker), CR: Condenser, PH: Preheater, SC: Shift Converter, AT: Absorption tower, DT: Desorption tower (Steam stripper), RMA: Rich MEA solution, LMA: Lean MEA solution.

Catalyst de-activation due to deposition of carbon particles (formed during the process of steam cracking) on the catalyst surface is a possibility. In such cases, when preheated air is passed through the catalyst bed, these carbon particles are burnt off and the heat released during their combustion helps in heating up the catalyst bed. The catalyst also gets regenerated (re- activated). With biomethane however, such carbon deposition on catalyst surface is not serious. This is of concern when coal or naphtha is used as the feedstock for the manufacture of syngas.

Syngas must be desulfurized prior to being used as a feedstock for the manufacture of chemicals, since, as discussed earlier, sulphur is extremely poisonous to the catalysts used in subsequent processes of chemical synthesis. In the present case however, since biomethane has already been made free from sulphur, no further desulfurization of syngas shall be required.

Once syngas has been manufactured, a host of valuable chemicals can be synthesized using it as the starting material. For example, manufacture of hydrogen gas from biomethane through the syngas route is sketched in figure (5.8). Hydrogen is an important raw material in many process industries such as for hydrodesulfurization of petroleum oils, for the hydrogenation of vegetable oils, for the hydrogenation of coal (for the manufacture of synthetic liquid fuels) and so on. The syngas route has been accepted as an efficient and economical pathway for the large scale production of hydrogen gas.

For hydrogen production, the clean, sulfur - free syngas from the primary reformer (steam cracker) is fed to reactor – 2, which is called the *shift converter* and is also a column reactor containing a packed bed of iron oxide catalyst (with chromium oxide as the promoter). The reactor operates at 425C. Syngas and high pressure steam are admitted from the bottom of the reactor column (here also a co-current operation is employed) and as the syngas – steam mixture moves up the catalyst bed, all the carbon monoxide present in the syngas reacts with excess steam, getting converted to carbon dioxide and producing more hydrogen gas. The reaction may be represented as

$$CO + H_2O\ (g) \longrightarrow CO_2 + H_2 \tag{5.5.2}$$

The gas mixture that leaves the shift converter shall be thus composed of carbon dioxide and hydrogen and it is sent to an absorption – desorption system (described in the earlier section, see figure – 5.7), where all the carbon dioxide present is absorbed off in aqueous MEA solution. Pure hydrogen gas that leaves from the top of the absorption tower is sent to storage. The steam – CO_2 mixture that leaves the top of the desorption tower (steam stripper) is cooled in a condenser, all the steam thus drained off as condensate and the carbon dioxide gas left is sent to storage.

Methanol is another valuable industrial chemical that can be manufactured on large scale from biomethane through the syngas route. Here also, we use a packed bed catalytic reactor that is composed of a bed of CuO – ZnO catalyst (supported on alumina). The reactor operates (methanol synthesis occurs) at high temperature (250C) and high pressure (50 to 60 atm). The preheated, compressed syngas is admitted from the bottom of the reactor column and as it moves up the catalyst bed (at the specified temperature and pressure), carbon monoxide and hydrogen present in the syngas react with each other producing methanol. The reaction maybe represented as

$$CO + 2H_2 \longrightarrow CH_3OH \tag{5.5.3}$$

Methanol vapors leave the top of the reactor column and are condensed and sent to storage.

Methanol is widely used as an industrial solvent. It is also a popular solvent in medicine and in drugs and pharmaceutical industries. As a feedstock, it is used in the manufacture of phenol- formaldehyde resins (bakelite) and polyester films and fibres (PET). One of its recent, large scale industrial applications is in the manufacture of biodiesels (discussed in Section 5.1).

5.6 Manufacture of Chemical Fertilizers with Biomethane as Feedstock

A large number of chemical fertilizers (synthetic fertilizers) can be manufactured using biomethane as the starting material [6]. The syngas route is followed her also, but with a difference. Since the objective is to produce *ammonia synthesis gas* (a 1:3 mixture of nitrogen and hydrogen), two reactors are employed here, such as the *primary reformer and the secondary reformer* (see figures 5.9A and 5.9B).

Apart of the biomethane (sulphur – free) is fed to the primary reformer, in which, as discussed in the previous section, steam – methane reaction occurs at high temperature (750-950^{0}C) and high pressure (10-15 atm) and in presence of Ni – catalyst to produce syngas (a mixture of carbon monoxide and hydrogen).

The rest of biomethane is fed to the secondary reformer, into which calculated amount of preheated, compressed air is also admitted co-currently. This is also a catalytic reactor, packed with a bed of Ni – catalyst particles and is operated at 500-600C and 5-6 atm pressure. At this specific temperature and pressure and in presence of the chosen catalyst (Ni – catalyst), methane gets catalytically oxidised to carbon monoxide and hydrogen :

$$2\ CH_4 + (\ N_2 + O_2\ ) \rightarrow 2\ CO + 4\ H_2 + N_2 \qquad (5.6.1)$$

All the nitrogen present in the air supplied shall remain behind in the product gases. Since air has been supplied in calculated amount, combustion of methane hardly occurs and little carbon dioxide is formed. The gas mixture leaving the secondary reformer shall be, therefore, essentially composed of carbon monoxide, nitrogen and hydrogen. This is mixed with the syngas from the primary reformer and is then fed to the shift converter (reactor – 3).

Into the shift converter, as discussed in the previous section, high pressure steam is also admitted and it operates at 425C and in presence of iron oxide catalyst (with chromium oxide promoter). Here, all the carbon monoxide present in the feed gas mixture is oxidized to carbon dioxide by reacting with the excess steam supplied (see reaction - 5.5.2). Accordingly, the product gas mixture leaving the shift converter shall be composed of nitrogen, hydrogen and carbon dioxide.

This gas mixture is sent to an absorption – desorption unit (same as that described in section- 5.4.2, see figure 5.7), in which all the carbon dioxide present in the gas mixture is absorbed off in aqueous MEA solution (absorbent).

The gas stream leaving the top of the absorption tower shall be, therefore, a mixture of nitrogen and hydrogen. By adjusting the amount of steam admitted to the primary reformer and the amount of air fed to the secondary reformer, it is possible to obtain a product gas mixture (N_2 - H_2 mixture) that contains nitrogen and hydrogen in the molar ratio, 1:3. It is then called the *ammonia synthesis gas*. The steam – CO_2 mixture leaving the top of the desorption tower (steam stripper) is cooled to remove all the steam as condensate and the resulting carbon dioxide gas is sent to storage.

Once ammonia synthesis gas has been produced, the rest of the process is straightforward. This gas mixture is compressed to high pressure (10 to 15 atm) and preheated to 375C (optimum temperature for the Haber catalyst) and then fed to the Haber reactor. It is once again a column reactor packed with a bed of iron catalyst (Haber catalyst) and operating at high pressure and high temperature (mentioned above). As the ammonia synthesis gas ascends through the bed of Haber catalyst, nitrogen and hydrogen react with each other to produce ammonia :

$$N_2 + 3H_2 \longrightarrow 2NH_3 \qquad (5.6.2)$$

The product gas mixture consisting of ammonia vapour and all the unreacted nitrogen and hydrogen gases, leaving the Haber reactor at high temperature, is cooled in a heat exchanger (could be a waste heat boiler), when ammonia condenses and the resultant liquid ammonia is sent to storage. The nitrogen – hydrogen mixture is recycled back to the Haber reactor (after compression and preheating).

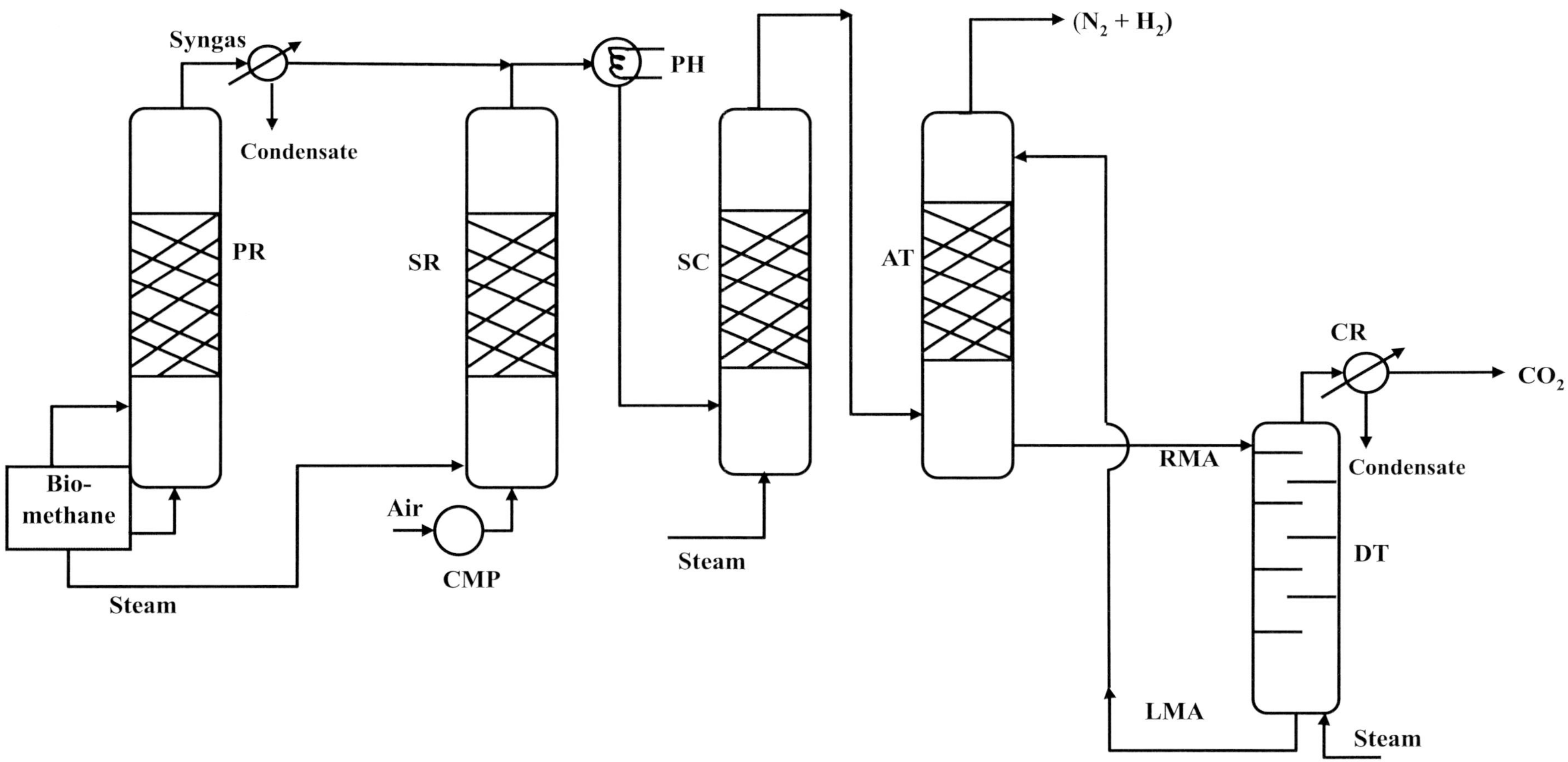

Figure 5.9A: Manufacture of Ammonia from Biomethane. PR: Primary reformer (Steam cracker) SR: Secondary reformer, CMP: Compressor, PH: Preheater, SC: Shift converter, AT: Absorption tower, DT: Desorption tower (Steam stripper), LMA: Lean MEA solution, RMA: Rich MEA solution, CR: Condenser.

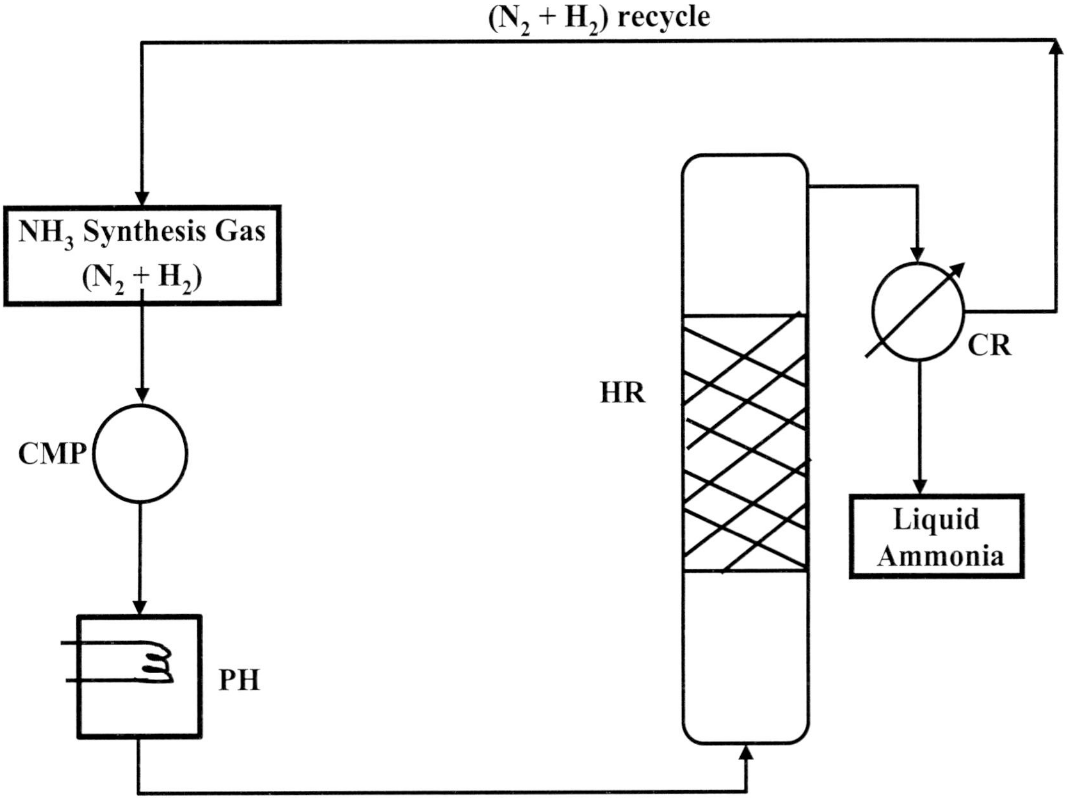

Figure 5.9B: Manufacture of ammonia from biomethane (continued). HR: Haber reactor, CMP: Compressor, PH: Preheater, CR: Condenser.

Ammonia is the universal starting material for the manufacture of a large number of nitrogenous fertilizers and nitrophosphates. Examples are ammonium sulphate, ammonium nitrate (sold as calcium ammonium nitrate or nitrolime), urea and ammonium phosphates (nitrophosphates). Most of them are synthesized by the chemical reaction between ammonia and a selected acid (sulphuric acid, phosphoric acid, nitric acid) :

$$2NH_3 + H_2SO_4 \longrightarrow (NH_4)_2SO_4 \qquad (5.6.3)$$

Ammonium sulphate is the only synthetic nitrogenous fertiliser that contains sulphur and is thus specifically suitable for use in the cultivation of citrus plants (oranges, lemons), which need a specific N : S ratio in the soil for efficient growth.

Ammonia can be catalytically oxidised by air in presence of Pt / Rh catalyst at high temperature (600-800C) and at high pressure (5-8 atm) to nitric oxide (often called the Ostwald process) :

$$4\,NH_3 + 5\,O_2 \longrightarrow 4\,NO + 6\,H_2O\,(g) \qquad (5.6.4)$$

Nitric oxide can be easily oxidized to NO_2 by air at around 50-60C and the NO_2 gas on absorption in water yields nitric acid :

$$2\,NO + O_2 \longrightarrow 2\,NO_2 \qquad (5.6.5)$$

$$3\,NO_2 + H_2O \longrightarrow 2\,HNO_3 + NO \qquad (5.6.6)$$

The nitric oxide liberated in the absorption column (see reaction – 5.6.6) could be recycled back to the oxidation tower. Nitric acid so produced reacts with the rest of ammonia producing ammonium nitrate fertilizer :

$$NH_3 + HNO_3 \longrightarrow NH_4NO_3 \qquad (5.6.7)$$

Since ammonium nitrate is explosive in nature, it is blended with lime and is marketed as calcium ammonium nitrate (CAN) or nitrolime.

Urea (NH_2CONH_2) is another chemical fertiliser widely used in agriculture and is manufactured from ammonia and carbon dioxide :

$$2NH_3 + CO_2 \rightarrow NH_2CONH_2 + H_2O \qquad (5.6.8)$$

Urea is a nutrient – rich fertilizer as it contains 46% N, as compared to ammonium nitrate which contains 33-34% N and ammonium sulphate with 21% N.

Ammonium phosphates such as MAP (mono-ammonium phosphate), DAP (di-ammonium phosphate) and TAP (tri-ammonium phosphate) are nitrophosphates and they contain two vital nutrients needed for plant growth, such as nitrogen and phosphorus. They are manufactured by the reaction between ammonia and phosphoric acid :

$$NH_3 + H_3PO_4 \longrightarrow (NH_4)H_2PO_4 \qquad (5.6.9)$$

$$2\ NH_3 + H_3PO_4 \longrightarrow (NH_4)_2HPO_4 \qquad (5.6.10)$$

$$3\ NH_3 + H_3PO_4 \longrightarrow (NH_4)_3PO_4 \qquad (5.6.11)$$

DAP is the fertiliser that is popularly used in the cultivation of paddy (rice, wheat, barley).

5.7 Manufacture of Synthetic Liquid Fuels from Biomethane (Fischer – Tropsch Synthesis)

Fischer – Tropsch synthesis (F – T synthesis) has a long past. It was first developed on the occasion of the second world war to synthesise liquid fuels (chiefly, high octane motor gasoline) from coal (production of oil from coal), mainly for military purposes. The first pilot plant ofF-T process was set up in Kharagpur, India, in 1950 but it was eventually shut down based on the misguided notion that it is not economically acceptable. The first commercial plant producing oil from coal using F – T process was started in Johannesburg, South Africa, during 1950 – 51 and it remains a commendable fact that this plant is still in operation at almost full capacity.

Rejecting F-T process by labelling it as uneconomical or non-cost-effective is one among the largest technological blunders committed by the world community. What happened was that the world went madly after oil with the misconception that being a better quality fuel, it alone shall meet the entire energy demands of the world. It is when the oil wells started drying up and the oil crisis (energy crisis) started gripping the world, we started going back to coal, and hectic research was initiated on F-T process all over the world to synthesise oil from coal.

Today, a large number of industrial units on F-T process are operating almost in full capacity in different parts of the world.

The process is illustrated schematically in figure (5.10). The F-T synthesis gas is qualitatively the same as syngas (discussed in earlier sections) and is a mixture of carbon monoxide and hydrogen, but in the molar ratio 1:2 (it is around 4:5 in syngas). It can be, therefore, produced through steam-carbon reaction (if produced from coal) or through steam – hydrocarbon reaction (if produced from natural gas or biomethane). In the present case therefore, it is produced by the reaction between biomethane and high pressure steam in presence of Ni –catalyst, the reactor (reactor – 1) operating at 750-950C and 10-15 atm pressure. The reaction is same as that shown in equation (5.5.1). The amount of steam injected into the reactor is adjusted such that the molar ratio of CO and H_2 in the product gas mixture shall be around 1 : 2.

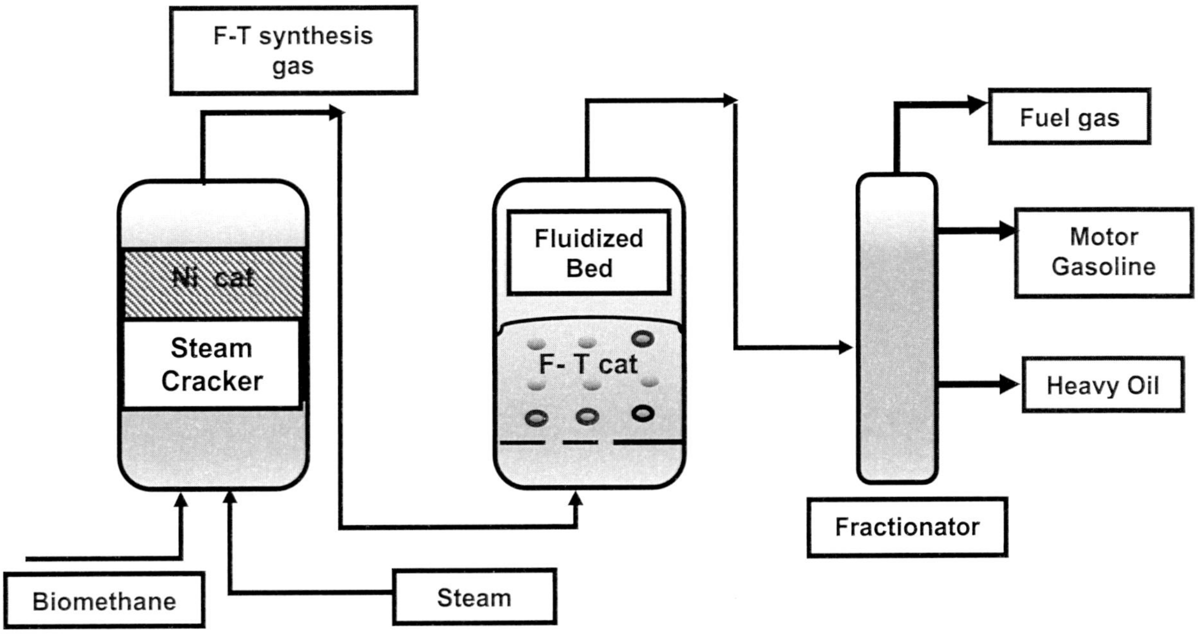

Figure 5.10: Schematic of Fischer – Tropsch Synthesis

Since biomethane has already been made free from sulphur, no further desulfurisation ofF-T synthesis gas shall be required in the present case. In cases otherwise, it is essential to ensure that the gas mixture is completely free from sulphur, since sulfur is highly injurious to the F-T catalyst (used subsequently in the F-T reactor).

The F-T reactor could be a packed bed reactor composed of a fixed bed of F-T catalyst (it is an alkali promoted iron catalyst) or a fluidized bed reactor. The latter is shown in figure (5.10). The reactor operates at 300 – 350C and the process does not demand high pressure ($P \approx$ 1atm). In fact, at elevated values of operating pressure, the product quality has been observed to deteriorate, as the wax content of the oil produced has been found to increase substantially.

The F-T catalyst particles remain suspended in the ascending stream ofF-T synthesis gas and at the specified operating temperature of 300-350C, carbon monoxide and hydrogen present in the gas mixture react with each other producing liquid hydrocarbons (mainly those boiling in the motor gasoline boiling range). The reaction may be represented as

$$n\,CO + (n + m/2)\,H_2 \longrightarrow C_nH_m + n\,H_2O \qquad (5.7.1)$$

Vapors of liquid hydrocarbons (mainly composed of high octane gasoline) leave from the top of the fluidized bed reactor and are fed to a fractional distillation column. The vapors may be condensed to separate any unreacted CO present, preheated and then fed to the fractionator. Motor gasoline of high octane number (principal product) is collected from the top of the fractionating tower and is sent to storage. Small amounts of middle distillates and heavy oil could also be produced and they are withdrawn as separate fractions.

High wax content of the product had been the major concern associated with this process. However, with the development of improved catalysts (such as cobalt-based catalysts), this problem has been more or less rectified. It has been reported [24, 25] that instead of the traditional iron catalyst, cobalt-based catalysts could be used in the F-T reactor and these reportedly provide better product quality and higher yield of motor gasoline, within a reasonably low reactor volume. More investigations in this direction are undoubtedly worthwhile.

5.8 Biodesulfurization of Petroleum Oils

Sulfur compounds are one among the most objectionable constituents of petroleum oils. Apart from leading to the formation of toxic sulfur dioxide gas and sulphuric acid mist (which cause severe corrosion problems and injury to public health when exhausted to atmosphere), sulfur compounds are extremely poisonous to the different chemical catalysts used during the different stages of petroleum refining (processing). Desulfurization of petroleum oils is, therefore, a must prior to being used as a fuel (boiler fuel, furnace fuel, automobile fuel) or as a feedstock for the manufacture of industrial chemicals (petrochemicals).

Sulfur is present in petroleum oils in the form of mercaptanes or thioalcohols (RSH, which cause foul odor as well) or as thiophene derivatives. Hydro-desulfurization is the presently employed process in petroleum refineries for sulfur removal. Technologically, it is an efficient process, but it is also an expensive process as it demands severe operating conditions (T = 250-300C, P = 25-30 atm), needs a chemical catalyst such as cobalt molybdate and it consumes large volumes of an expensive raw material such as hydrogen gas. Oil vapors are to be mixed with hydrogen gas, compressed to high pressure and then passed through a bed of cobalt molybdate catalyst, maintained at high temperature. Sulfur compounds react with hydrogen gas and are removed as hydrogen sulfide (H_2S). But, hydrogen sulfide can not be exhausted directly to atmosphere (as it, by itself, acts as a severe pollutant) and is required to be processed further in a separate Claus process plant to convert it to elemental sulfur. The cost of installation and operation of this plant shall be additional.

Bio-desulfurization process [26, 27], which is an aerobic microbial process that uses a culture of a sulfur-eating microbe such as *Rhodococcus sp.* or *Rhodococcus erythropolis*, on the other hand, operates at ordinary temperature and pressure, does not demand a chemical catalyst, does not consume any expensive chemical such as hydrogen gas. Also, the entire cost of installation and operation of Claus process unit is dispensed with.

Sulfur is removed from the oil-stock by converting to sulphate in solution (see the reactions sketched below) and can be recovered as sodium sulphate by adding caustic soda or as ammonium sulphate by adding ammonium hydroxide. Each of them forms a valuable byproduct of the process.

$$RSH + H_2O + 2\ O_2 \longrightarrow ROH + H_2SO_4 \qquad (5.8.1)$$

Dibenzothiophene ($C_{12}H_8S$) + H_2O + 2 O_2 $\longrightarrow$ p-hydroxy biphenyl ($C_{12}H_{10}O$) + SO_4^{2-}

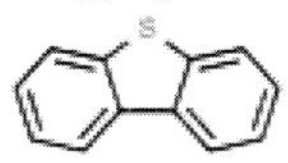 $+ H_2O + 2\ O_2 \rightarrow$ 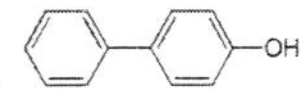 $+ SO_4^{2-}$ (5.8.2)

More or less complete desulfurization is affected. No doubt, being a biochemical process, it is slow and would demand large volume bioreactors. The process is illustrated schematically in figure (5.11).

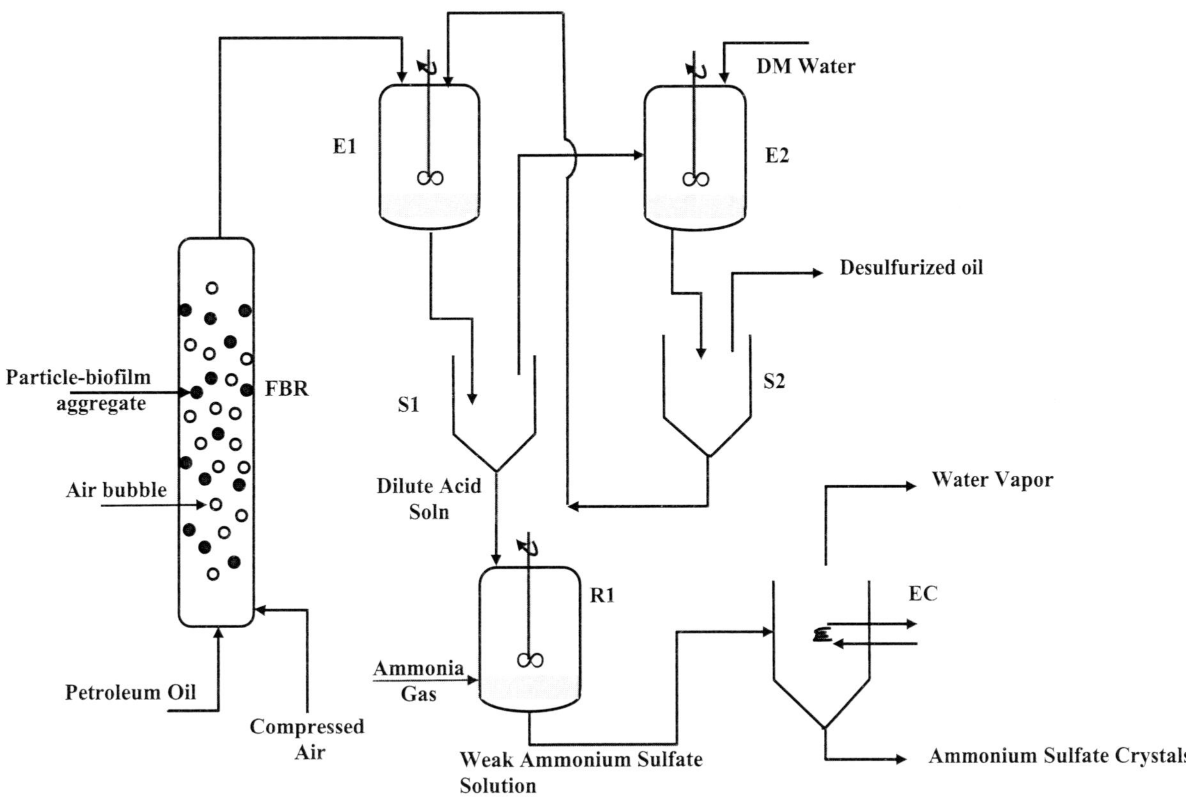

Figure 5.11: Schematic of biodesulfurization of petroleum oils. FBR: Fluidized bed biofilm reactor, E1, E2: Extraction tanks, S1, S2: Separators (settling tanks), R1: Reactor (neutraliser), EC: Evaporator – crystallizer.

As shown in figure (5.11), the petroleum oil (gas oil, diesel oil) is fed to the fluidized bed biofilm reactor (FBR) from the bottom. The bioreactor is composed of support particles (usually made of silica) each surrounded by a thin biofilm containing *Rhodococcus sp.* culture. These particle-biofilm aggregates remain suspended in the ascending stream of petroleum oil. Compressed air is also admitted from the bottom and it moves up the reactor column in the form of tiny, discrete bubbles. The bioconversion of di-benzothiophene to hydroxy biphenyl occurs within the biofilm.

The product liquor leaving the top of the reactor column (air is vented off) is sent to a two stage extraction unit (mixer – settler cascade) that operates in the countercurrent mode (a three stage or four stage operation may also be employed depending on the capacity of the plant). Demineralised water (DM water) is fed to the second (in fact, last) extraction vessel

(E2), while the product liquor from the bioreactor enters the first extractor (E1). Desulfurized oil (oil phase) overflows from the second settler or separator (S2). The aqueous phase, which is a dilute acid solution, leaves from the bottom of settler-1 (S1) and is fed to a stirred tank reactor (R1), into which ammonia gas is also blown in. Ammonia reacts with dilute sulphuric acid to form ammonium sulphate.

The weak aqueous solution of ammonium sulphate discharged from the bottom of the reactor is sent to an evaporator-crystalliser unit (EC), where the solution is concentrated and ammonium sulphate crystallises out. The crystals are dried and sent to storage.

As stated earlier, ammonium sulphate fertilizer forms the important byproduct of this process.

5.9 Bioleaching of Ores

Biotechnology has made its presence felt in mineral beneficiation and hydrometallurgical applications as well. Bioleaching of ores is an excellent example in this connection.

The ore or mineral is agitated with water or with a dilute acid solution in the leaching tank, into which an inoculum of the selected microbial culture is seeded. The metal selectively goes into solution due to the microbial action. The solution is filtered to separate all the undissolved, suspended gangue particles and from the clear metallic solution (metallic salt solution), the metal is recovered by precipitation or electrolysis.

For example, iron and copper are being successfully extracted from their corresponding sulphide ores such as pyrite (FeS_2) and chalcopyrite ($CuFeS_2$) by bioleaching using a culture of *Thiobacillus ferrooxidant* (aerobic, sulfur-eating microbes). The reactions involved are given below :

$$2\,FeS_2 + 2\,H_2O + 7\,O_2 \longrightarrow 2\,FeSO_4 + 2\,H_2SO_4 \tag{5.9.1}$$

$$4\,CuFeS_2 + 2\,H_2SO_4 + 17\,O_2 \longrightarrow 4\,CuSO_4 + 2\,Fe_2(SO_4)_3 \tag{5.9.2}$$

Cultures of *Aspergillus sp.* and *Psuedomonas sp.* have also been successfully employed to leach out iron from clay, sand and kaolin and also to extract metals such as copper, nickel and zinc from their respective oxide ores [28].

The specific advantages of bioleaching (microbial hydrometallurgy) are,

(i) the process does not demand high temperature or high pressure (occurs at ambient temperature and atmospheric pressure),

(ii) there is no emission of toxic gases or particulate matter into the atmosphere during the process (is environment-friendly),

(iii) no valuable chemicals are required to be consumed during the extraction process,

(iv) pure metal can be economically extracted from simple as well as complex ores, from both low grade ores as well as mineral concentrates,

(v) less energy-intensive, high cost-effectiveness.

Biodesulfurization of coal has also been equally successful [29]. Sulfur is present in coal mainly as iron pyrite (FeS_2). Pyrite is one among the most objectionable constituents of coal. It is a mineral matter associated with coal and is an incombustible constituent. During combustion of coal, it gets oxidized to iron oxide (which becomes a constituent of coal ash) and releases harmful gases such as sulfur dioxide (this can also form sulphuric acid mist on interaction with moisture or steam) :

$$4\,FeS_2 + 11\,O_2 \longrightarrow 2\,Fe_2O_3 + 8\,SO_2 \quad (5.9.3)$$

$$2\,SO_2 + 2\,H_2O + O_2 \longrightarrow 2\,H_2SO_4 \quad (5.9.4)$$

Iron oxide is the fusible constituent of coal ash. Ash, being fine powder, could be carried off by the flue gases and when rejected to atmosphere could cause serious harm to public health. Fusible ash is still more objectionable since it can melt and form clinkers, that are more troublesome to dispose off and they also absorb significant amount of heat and cause wastage of heat. Fusible ash could also melt on the surface of coal particles and thereby render them incombustible.

Heavy medium separation (commonly called *coal washing*) is the conventional technology used for reducing the ash content of coal. Biodesulfurisation competes efficiently with coal washing technology.

Using a culture of *Thiobacillus ferrooxidant*, the pyrite present in coal can be oxidized to soluble ferrous sulphate (see reaction shown in equation 5.9.1). The pyrite content of coal thus gets leached off into solution.

The process is sketched in figure (5.12). Pulverized coal particles are agitated with water or with a dilute acidic solution in a leaching tank, into which compressed air is also sparged from the bottom. The tank is seeded with an inoculum of *Thiobacillus ferrooxidant* at the start-up. As stated earlier, in presence of the microbial culture (biocatalyst), pyrite present in coal is catalytically oxidized to ferrous sulphate, which goes into solution. The product slurry from the leaching tank is fed to a sedimentation tank (sedimenter-1), in which all the desulfurized coal particles settle down and are discharged from the bottom of the tank and sent to storage.

The acidic solution that overflows from sedimenter-1 is fed to sedimenter-2, in which all the microbial cells settle down as sludge. This microbial sludge (discharged from the bottom of the sedimentor) is partially recycled back to the leaching tank and the rest is wasted. The clear, acidic ferrous sulphate solution that overflows from the top of sedimenter-2, is sent to storage.

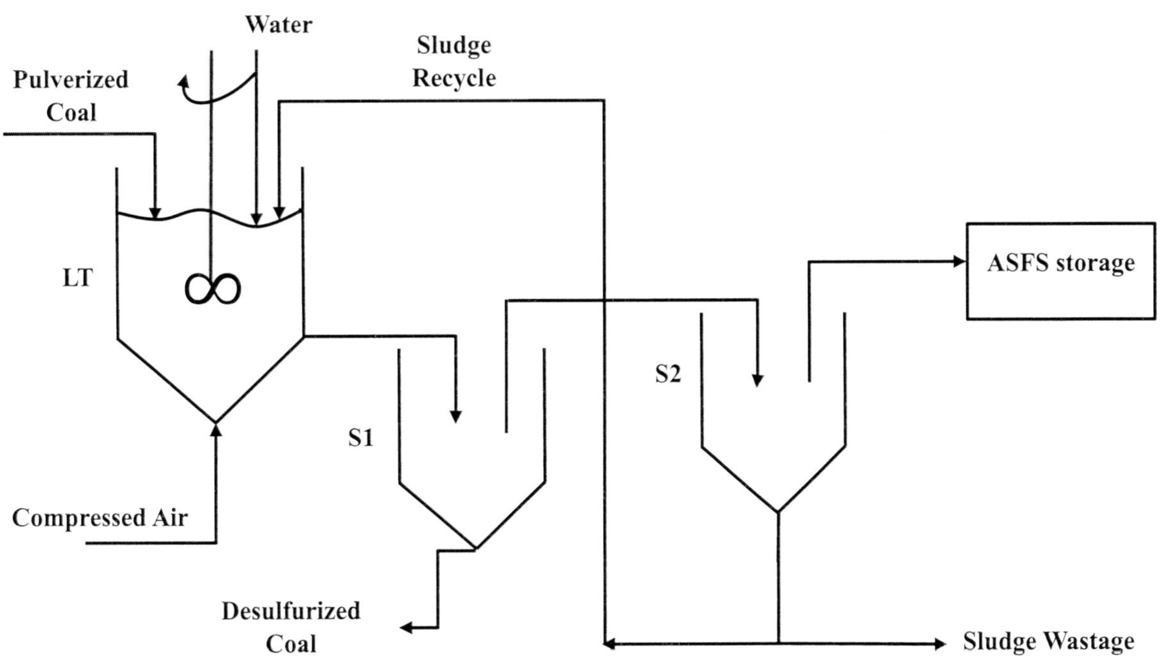

Figure 5.12: Schematic of Bio-desulfurization of Coal: LT: Leaching Tank, S1, S2: Sedimentation tanks, ASFS: Acidic solution with dissolved ferrous sulfate.

This dilute, acid solution containing dissolved ferrous sulphate is the mother liquor (effluent) left behind. Its economical re-utilization should also be a matter of interest. Direct disposal to the soil shall cause serious contamination to soil and to ground water. Disposal after neutralization with an alkali (say, lime) shall not also be safe enough, since pollution due to solid waste shall still remain behind. Also, lot of sulfur gets lost to the environment. Sulfur is a precious material to countries like India, which does not have any sulfur mines and where elemental sulfur is cent per cent an imported material. A cost-effective process for the recovery of free acid, along with ferrous sulphate, from this acid solution based on the principle of differential solubility has been developed by Narayanan et.al [30]. This process is briefly sketched in figure (5.13).

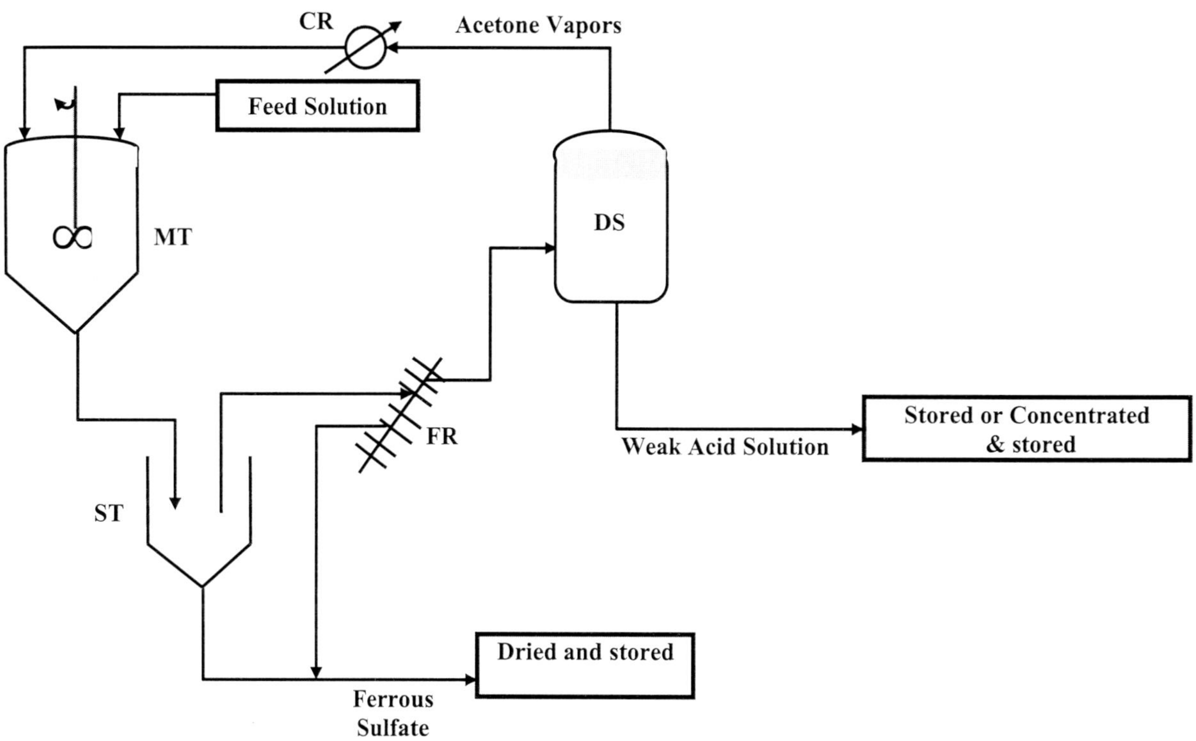

Figure 5.13: Condensed flowsheet of recovery of free acid and ferrous sulphate from acidic effluent. MT: Mixing tank (precipitation tank), ST: Sedimentation tank, FR: filter, DS: distillation still, CR: Condenser.

Ferrous sulphate has very low solubility in organic solvents like acetone. As a result, when the acid solution containing dissolve ferrous sulphate is mixed with acetone, the solubility of ferrous sulphate decreases and it gets precipitated. By using an adequate (in fact, optimum) ratio of acetone to feed solution (1.5 : 1.0), most of the ferrous sulphate (more than 98%) can be made to precipitate out [30]. The precipitated ferrous sulphate is then separated either by filtration or in a settling tank, leaving behind a blend of dilute acid (free acid) and acetone as the filtrate or overflow liquor. From this liquor, acetone can be easily separated in a conventional distillation still. Since acetone is highly volatile (BP $\approx$ 60C), it vaporises easily (and almost completely) and these vapors are condensed and then recycled back to the precipitation tank. Consequently, no acetone is consumed in the process, it is continuously recovered and recycled. The dilute free acid (dilute sulphuric acid solution) discharged from the distillation still is sent to storage, either as such or after concentration.

The above process can also be utilized for the recovery of free acid and ferrous sulphate from alternate effluents such as the spent pickle liquor discharged from iron and steel industries. The surface of hot steel is washed with a dilute sulphuric acid solution to improve its passivity or oxidation resistance (resistance to atmospheric oxidation) and this process is termed as *pickling*. A layer of iron oxide gets formed on the surface of the steel block or sheet and this resists further oxidation of the metal (provides passivity to the metal surface). The wash liquor (spent pickle liquor) discharged shall be, therefore, a very dilute solution of sulphuric acid containing some dissolved ferrous sulphate (formed during the pickling operation). The above- described process can be, therefore, effectively used for the recovery of free acid and ferrous sulphate from this spent pickle liquor as well.

Ferrous sulphate is used as a coagulant in wastewater treatment. It is also used in medicine for treating iron deficiency and as an additive in fabric dyes. Ferrous sulphate is often mixed with chemical fertilizers to provide an optimum sulfur content in soil. This is particularly important for the growth of citrus plants (lemons, oranges).

Another interesting application of ferrous sulphate is in the construction and operation of salt- gradient solar ponds. Conventionally, these ponds utilize sodium chloride solutions (brine) of different concentrations, but brine is acutely corrosive particularly at high concentrations and at high temperatures. In tropical climate, the temperature of the bottom-most layer of solar ponds (made of around 10-12% brine) often reaches 80-85C and at this temperature and concentration, brine becomes extremely corrosive and cracks get developed on the concrete wall of the pond. Though a PTFE (poly-tetrafluoroethene) lining is often provided, its life has been observed to be not too large. Use of ferrous sulphate solutions, that are much less corrosive than brine, as the media in solar ponds could alleviate these problems. Successful investigations [31] have shown that the performance characteristics of salt-gradient solar ponds employing ferrous sulphate solutions are comparable with, if not superior to, those employing sodium chloride solutions (within the same range of solar insolation and wind velocities).

Exercises

1. Distinguish between :
   (a) Vegetable oil and biodiesel,
   (b) Biodiesel and petrodiesel,
   (c) PLLA bioplastic and PHB bioplastic,
   (d) PROM and diammonium phosphate,
   (e) F-T synthesis gas and ammonia synthetic gas,
   (f) Hydrodesulfurization and biodesulfurisation of petroleum oils.
2. (a) Discuss on the different types of raw materials and different types of catalysts that could be used for the manufacture of biodiesels. How do you select among them?
   (b) Why are biodiesels called green fuels ?

3. What are the advantages of bioplastics over synthetic plastics ? Why are they called green plastics ?
4. Discuss on the recommended raw materials (including biocatalysts) for the manufacture of PLLA bioplastic. Highlight on their availability and characteristics.
5. What is PROM ? What are the raw materials required for its manufacture ? How does PROM compare with synthetic fertilizers such as DAP and SSP ?
6. Briefly outline the technique of biochemical desulfurisation of biogas. What are its specific advantages ?
7. How is syngas synthesized from biogas ? Briefly sketch the chemical reactions / conversions involved.
8. Briefly discuss on how is syngas utilized,
   (a) as a source for hydrogen production,
   (b) for the manufacture of chemical fertilizers.
9. What is the basic principle behind Fischer-Tropsch synthesis ? Any technical problems associated with it ?
10. Write short notes on :
    (a) Biodesulfurisation of petroleum oils,
    (b) Biodesulfurisation of coal,
    (c) Biomethane as automobile fuel,
    (d) Multiple substrates for biogas production.

References

1. De, T., Sikder, J. and Narayanan, C. M., Biodiesel Synthesis and Bioreactor Design – An Overview, Ind. J. Chem. Technol., **24**, 575 – 92, 2017.
2. Narayanan, C. M., Sikder, J. and De, T., Characteristics and Applications of Biodiesels and Design of Bioreactors for their Industrial Manufacture, Chapter – 13, 285 – 320, in N. K. Rathinam and R. K. Sani (eds.), Next Generation Biomanufacturing Technologies, ACS Symposium Series, Washington, 2019. **DOI: 10.1021/bk-2019- 1329-ch013.**
3. Narayanan, C. M. and Pandey, A., Studies on Biodiesel Synthesis Using Nanosilica Immobilized Lipase in Inverse Fluidized Bed Bioreactors, J. Adv. Chem., **15**(1), 6072 – 86, 2018.
4. Narayanan, C. M., Das, S. and Pandey, A., Green Technologies for Manufacture of Valuable Products from Food Wastes and Agricultural Wastes, Volume – 2, Chapter – 1, 1 – 54, Handbook of Food Bioengineering, Elsevier, 2017.
5. Narayanan, C. M. and Vikas, N., Recent Developments in Biogas Manufacture and Biogas Utilization: A Review. European J. Sustainable Dev. Res., **4**(4), em0135, 2020. https://doi.org/10.29333/ejosdr/8366.
6. Narayanan, C. M. and Vikas N., Advances in Biogas Manufacture, Bioreactor Design and Biogas Utilization, Lambert Academic Publishing, Mauritius, 2020. **ISBN : 978 – 620 – 3 – 19653 – 5.**

7. Golueke, C.G., Oswald,W.J. and Gotaas, H.B., Anaerobic Digestion of Algae, Applied Microbiol., **5**(1), 47-55, 1957.
8. Olsson, J., Feng, X.M., Ascue, J., Gentili, F.G., et al., Co-digestion of Cultivated Microalgae and Sewage Sludge from Municipal Waste Water Treatment, Bioresour. Technol., **171**, 203–10, 2014.https://doi.org/10.1016/j.biortech.2014.08.069.
9. Yuan, X., Wang, M., Park, C., Sahu, A.K. and Ergas, S.J., Microalgae Growth using High Strength Wastewater followed by Anaerobic Co-digestion, Water Environ Res, **84**, 396–404, 2012.https://doi.org/10.2175/106143011X13233670703242.
10. Krustok, I., Nehrenheim, E. and Odlare, M., Cultivation of Microalgae for Potential Heavy Metal Reduction in a Wastewater Treatment Plant, in Proc. Int. Conf. Applied Energy, Suzhou, July 5-8, 2012.
11. Weizhang, Z., Zhongzhi, Z., Yijing, L., Wei, Q., Meng, X. and Zhang, M., Biogas Productivity by Co-digesting Taihu Blue Algae with Corn Straw, Bioresour. Technol., **114**, 281-86, 2012.https://doi.org/10.1016/j.biortech.2012.02.111.
12. Yen, H. W. and Brune, D. E., Anaerobic Co-digestion of Algal Sludge and Waste Paper to Produce Methane, Bioresour. Technol., 98, 130–34, 2007. https://doi.org/10.1016/j.biortech.2005.11.010.
13. Wang, X., Yang, G., Feng, Y., Ren, G. and Han, X., Optimizing Feeding Composition and Carbon to Nitrogen Ratios for Improved Methane Yield during Anaerobic Co- digestion of Dairy, Chicken Manure and Wheat Straw, Bioresour. Technol., **120**, 78- 83, 2012. https://doi.org/10.1016/j.biortech.2012.06.058.
14. Ye, J., Li, D., Sun, Y., Wang, G., Yuan, Z., Zhen, F. and Wang, Y., Improved biogas production from rice straw by co-digestion with kitchen waste and pig manure, Waste Manag., **33**(12), 2653-58, 2013.https://doi.org/10.1016/j.wasman.2013.05.014.
15. Okewale, A.O. and Adesina, O.A. (2019). Evaluation of Biogas Production from Co-digestion of Pig Dung, Water Hyacinth and Poultry Droppings, Waste Disposal & Sustainable Energy, **1**, 271-77, 2019.https://doi.org/10.1007/s42768-019-00018-8**.**
16. Vats, N., Khan, A.A. and Ahmad, K., Observation of Biogas Production by Sugarcane Bagasse and Food Waste in Different Composition Combinations, Energy, 2019. https://doi.org/10.1016/j.energy.2019.07.080.
17. Vats, N., Khan, A.A., and Ahmad, K., Anaerobic Co-digestion of Thermal Pretreated Sugarcane Bagasse using Poultry Waste, J. Environ. Chem. Eng., 2019. https://doi.org/10.1016/j.jece.2019.103323.
18. Ajeej, A., Thanikal, J.V., Narayanan, C.M. and Kumar, R.S., An Overview of Bio-augmentation of Methane by Anaerobic Co-digestion of Municipal Sludge along with Microalgae and Waste Paper, Renew. Sust. Energy Rev., **50**, 270-76, 2015. https://doi.org/10.1016/j.rser.2015.04.121.
19. Ajeej, A., Thanikal, J.V. and Narayanan, C.M., Studies on Production of Biogas by Co-digestion of Sewage Sludge, Wastepaper and Waste Grown Algae, J. Mod. Chem. Chem. Technol., **7**(1), 74–81, 2016.

20. Narayanan, C.M., Production of Phosphate Rich Biofertiliser Using Vermicompost and Anaerobic Digester Sludge – A Case Study, Adv. Chem. Eng. Sci., 2(2), 187-191, 2012. **DOI:10.4236/aces.2012.22022.**
21. Sekhar, D.M.R. and Aery, N.C., PROM Manual, Himanshu Publishers, Udaipur, 2005.
22. Rajvaidya, A.S., Pre-combustion Desulfurisation of Gaseous Fuels, Proc. All India Seminar on Recent Trends in Automotive Fuels, August 23-24, Nagpur, 2002.
23. Narayanan, C.M. and Bhattacharya, B.C., Computer Aided Analysis and Optimization of Biogas Enrichment Process, J. Energy Heat Mass Transfer, 12 (1), 17 – 24, 1990.
24. Asalieva, E.Y., Kul'chakovskaya, E.V., Sineva, L.V. and Mordkovich, V.Z., Effect of Zeolite on Fischer–Tropsch Synthesis in the Presence of a Catalyst based on Skeletal Cobalt, Petroleum Chem , **60**(1), 69-74, 2020. https://doi.org/10.1134/S0965544120010028
25. Mahmoudi, H., Mahmoudi, M., Doustdar, O., Jahangiri, H., Tsolakis, A., Gu, S. and Wyszynski, M.L., A Review of Fischer Tropsch Synthesis Process, Mechanism, Surface Chemistry and Catalyst Formulation, Biofuels Eng., **2**(1), 11-31, 2017. https://doi.org/10.1515/bfuel-2017-0002.
26. Narayanan, C.M., Design of Fluidised Bed Bioreactors for Bio-desulfurization of Petroleum Oils – Development of CAD (Software) Packages, Int. J. Chem. Eng. Res., 2(2), 97-108, 2010.
27. Ohshiro, T. and Izumi, Y., Microbial Desulfurization of Organic Sulfur Compounds in Petroleum, Biosci. Biotechnol. Biochem., 63(1), 1-9, 1999. https://doi.org/10.1271/bbb.63.1.
28. Torma, A.E. and others (eds.), Bio-hydrometallurgical Technologies, TMS Publications, Pennsylvania, 1993.
29. Prayuenyong, P., Coal Bio-desulfurization Processes, Songklanakarin J. Sci. Tchnol., 24(3), 2002.
30. Narayanan. C.M., Kurmaiah,N., Lal, K., Shinde, S.B., Rath, L. and Kumar, A., Studies on Recovery and Reuse of Free Acid and Copperas as Byproduct from Spent Pickle Liquor, Ind. Chem. Engr., **34** (2&3), 163-69, 1992.
31. Narayanan, C.M., Studies on Performance Characteristics of Salt-gradient Solar Ponds using Aqueous Solutions of Copperas, Int. J. Green Herbal Chem., Section – A, **5**(1), 67-74, Dec.2015 – February, 2016.

INDEX